Lecture Notes in Computer Science 2235

Edited by G. Goos, J. Hartmanis, and J. van Leeuwen

Springer

Berlin
Heidelberg
New York
Barcelona
Hong Kong
London
Milan
Paris
Tokyo

Cristian S. Calude   Gheorghe Păun
Grzegorz Rozenberg   Arto Salomaa (Eds.)

Multiset Processing

Mathematical, Computer Science,
and Molecular Computing Points of View

 Springer

Series Editors

Gerhard Goos, Karlsruhe University, Germany
Juris Hartmanis, Cornell University, NY, USA
Jan van Leeuwen, Utrecht University, The Netherlands

Volume Editors

Cristian S. Calude
University of Auckland, Department of Computer Science
Center for Discrete Mathematics and Theoretical Computer Science
E-mail: cristian@cs.auckland.ac.nz, cristian@attglobal.net

Gheorghe Păun
Institute of Mathematics of the Romanian Academy
E-mail: gpaun@imar.ro, gp@astor.urv.es

Grzegorz Rozenberg
Leiden University, Leiden Institute for Advanced Computer Science
E-mail: rozenber@liacs.nl

Arto Salomaa
Turku Center for Computer Science, TUCS
E-mail: asalomaa@utu.fi

Cataloging-in-Publication Data applied for

Die Deutsche Bibliothek - CIP-Einheitsaufnahme

Multiset processing : mathematical, computer science, and molecular
computing points of view / Cristian S. Calude ... (ed.). - Berlin ;
Heidelberg ; New York ; Barcelona ; Hong Kong ; London ; Milan ; Paris ;
Tokyo : Springer, 2001
  (Lecture notes in computer science ; Vol. 2235)
  ISBN 3-540-43063-6

CR Subject Classification (1998): E.1, F.1, F.4.2-3, J.3, J.2

ISSN 0302-9743
ISBN 3-540-43063-6 Springer-Verlag Berlin Heidelberg New York

Springer-Verlag Berlin Heidelberg New York
a member of BertelsmannSpringer Science+Business Media GmbH

http://www.springer.de

© Springer-Verlag Berlin Heidelberg 2001
Printed in Germany

Typesetting: Camera-ready by author, data conversion by Steingräber Satztechnik GmbH, Heidelberg
Printed on acid-free paper   SPIN: 10845800   06/3142   5 4 3 2 1 0

Preface

The *multiset* (a set with multiplicities associated with its elements, in the form of natural numbers) is a notion which has appeared again and again in many areas of mathematics and computer science, sometimes called a *bag* (some historical information appears in the enclosed paper by A. Syropoulos). As a data structure, this notion stands "in-between" *strings/lists*, where a linear ordering of symbols/items is present, and *sets*, where no ordering and no multiplicity is considered; in a multiset, only the multiplicity of elements matters, not their ordering. Actually, in between lists and multisets we also have *pomsets*, partially ordered multisets.

Confining ourselves to computer science, we may mention many areas where multisets are used: formal power series, Petri nets, data bases, logics, formal language theory (in relation with Parikh mapping, commutative grammars, etc), concurrency, and so on. In the last few years, the notion has occurred in a rather natural way in the molecular computing area. An aqueous solution of chemical compounds, swimming together in a given space, without any given spatial relation between individual elements, is just a multiset. Actually, this chemical metaphor was used several years before the occurrence of what is now called molecular computing, as the basic ingredient of the Gamma language and the Chemical Abstract Machine (a comprehensive survey of these ideas is provided by J.-P. Banâtre, P. Fradet, D. Le Metayer). Then, multisets were used in relation with DNA computing, especially in the context of computing by splicing (H systems): taking into account the number of DNA molecules proved to be a very powerful feature of H systems, leading to computational completeness.

In the prolongation of the chemical metaphor, the membrane computing area has recently emerged, as an abstraction of the living cell structure and biochemistry: in the compartments defined by a membrane structure, one processes multisets of chemical compounds, denoted by symbols or by strings over a given alphabet (up-to-date details about this area can be found on the web page http://bioinformatics.bio.disco.unimib.it/psystems, mastered by C. Zandron, in Milan).

In spite of all these "applications" and of many other related topics, the notion of a multiset has not received any systematic, monographic attention. In August 2000, a workshop was organized in Curtea de Argeş, Romania, explicitly with the aim of discussing the mathematical backgrounds, the computer science and molecular computing relevance of the notion of a multiset, with emphasis on multiset processing in the distributed framework of membrane computing. The editors of the present volume constituted the program committee of the workshop, while the organizing institutions were the "Simion Stoilow" Institute of Mathematics of the Romanian Academy, Bucharest, the Artificial Intelligence Department of the Polytechnical University of Madrid, the Centre for Discrete Mathematics and Theoretical Computer Science of Auckland University, New Zealand, and the "Vlaicu-Vodă" High School of Curtea de Argeş, Romania. A technical report (no 140/2000) of CDMTCS, Auckland University, was published as a pre-proceedings working volume, in anticipation of obtaining improved versions of the papers during the workshop, through the interaction of the authors. Actually, several new papers were started during the meeting as a result of cooperation among participants. Some of these papers were collected in a special double issue of the *Romanian Journal of Information Science and Technology* (ROMJIST), published as nr. 1-2 of volume 4, from 2001. Other papers, as well as several additional invited papers, are contained in the present volume.

While the ROMJIST issue is mainly devoted to DNA and membrane computing, the present volume is focused on fundamentals, on mathematical formalizations, and computer science (including molecular computing) applications of the notion of a multiset. Besides the two papers mentioned above, there are papers about the mathematical status of the notion of a multiset (S. Marcus, S. Miyamoto, T.Y. Nishida), about a theoretical, formal language-like study of multisets (by E. Csuhaj-Varju, M. Kudlek, C. Martin-Vide, V. Mitrana, Gh. Păun), with the ambition of founding a consistent rewriting Chomsky-like theory of multisets, as well as applicative papers: multisets and data bases (a rather comprehensive paper by G. Lamberti, M. Melchiori, M. Zanella), multisets in visual languages (P. Bottoni, B. Meyer, F. Parisi Presice), multisets and dynamic systems (W. Reising), multisets and formal languages (J. Dassow). Several papers are related to membrane computing, discussing the relation with constraint programming (A. Dovier, C. Piazza, G. Rossi), the power (M. Madhu, K. Krithivasan) and efficiency of membrane systems (A. Obtulowicz), a possible implementation on the usual electronic computer (F. Arroyo, A.V. Baranda, J. Castellanos, C. Luengo, L.F. Mingo), or applications in Artificial Life, mainly via extensive simulations on the computer (Y. Suzuki, Y. Fujiwara, H. Tanaka).

These papers, besides a large quantity of background information (basic notions and results, bibliography, further developments not presented in an extensive manner), also contain research topics and problems, so that the volume is intended to be useful both to the "users" of the notion of a multiset and to those who want to contribute to the development of its theory, with special attention being paid to the computer science approaches (including the membrane computing area).

Of course, this is only a step towards "FMT" (Formal Macroset/Multiset Theory), to borrow one of the syntagmas used as the title of a paper included in the volume. Hopefully, the next step will be done by the reader.

* * *

We warmly thank all contributors to this volume for their timely and high-quality work, as well as A. Hofmann, from Springer-Verlag, Heidelberg, for the efficient and pleasant cooperation.

September 2001

Cristian S. Calude
Gheorghe Păun
Grzegorz Rozenberg
Arto Salomaa

Table of Contents

Structures and Bio-language to Simulate Transition P Systems on Digital Computers

Fernando Arroyo[1], Angel V. Baranda[2], Juan Castellanos[2],
Carmen Luengo[1], and Luis F. Mingo[3]

[1] Dpto. de Lenguajes, Proyectos y Sistemas Informáticos
Escuela Unversitaria de Informática, Universidad Politécnica de Madrid
Crta. de Valencia Km. 7, 28031 Madrid, Spain
{farroyo, cluengo}@eui.upm.es
[2] Dpto. de Inteligencia Artificial, Facultad de Informática
Universidad Politécnica de Madrid
Campus de Montegancedo, 28660 Boadilla del Monte, Madrid, Spain
jcastellanos@fi.upm.es
[3] Dpto. de Organización y Estructura de la Información
Escuela Unversitaria de Informática, Universidad Politécnica de Madrid
Crta. de Valencia Km. 7, 28031 Madrid, Spain
lfmingo@eui.upm.es

Abstract. The aim of this paper is to show that some computational models inspired from biological membranes, such as P systems, can be simulated on digital computers in an efficient manner. To this aim, it is necessary to characterize non-determinism and parallel execution of evolution rules inside regions. Both these issues are formally described here in order to obtain a feasible description in terms of data structures and operations able to be implemented in a functional programming language. Static and dynamic structures of transition P systems are formalised in order to define a bio-language to represent them. Finally, a draft of a language for describing a transition P systems is presented. It will facilitate the description of transition P systems in terms of sentences in a high level programming language; such sentences will define a program. A process of compilation will parse the program to appropriate data structures and will launch the execution of the simulation process.

1 Introduction

Membrane computing is a new research field, of a biological inspiration, introduced by Gh. Păun in [4]. This new computational paradigm tries to imitate the way nature computes at the cellular level. Any biological system can be viewed as a hierarchical construct where the interchange of materials takes place. This process of material interchange can be interpreted as a computing process [4]. Taking into account this point of view, we can say that membrane computing is a general architecture of "living organisms", in the sense of Artificial Life [5]. Computing with membranes uses a hierarchical membrane structure in the same sense that it is used in the chemical abstract machine of Berry and Boudol [3],

C.S. Calude et al. (Eds.): Multiset Processing, LNCS 2235, pp. 1–15, 2001.

and the basic ideas of evolution rules as multisets transformation are inspired in Γ systems introduced by Banâtre [1]. Here we present the classical general description of transition P systems for determining their main components. The transition P systems structure is recursively defined, as composed by several membranes enclosed by a unique membrane named "skin". This structure defines a hierarchical compartmental structure that can be easily represented as a Venn diagram. The membrane structure is the main component of a transition P system. Membranes are uniquely labelled and they delimit regions or vesicles where there exist objects, represented by a multiset, and a set of evolution rules. The objects are represented by symbols from a given alphabet. Usually, an evolution rule from a determined region has the form $r : ca \rightarrow cb_{in_j} da^2_{out}$ and it says that a copy of a together with a copy of the catalyst c (catalysts are objects which are never modified but they must be present to apply the rule) is replaced by a copy of the object d which remains in the membrane, while two copies of the object a are sent out through the membrane of the region, and a copy of the object b is sent to the region labelled by j. Note that the rule r can be applied if and only if the region includes directly the membrane j. In general, the use of the expression "a copy of the object" is reduced to "the object" in order to make descriptions of computations in membrane systems simpler. However, it is important to bear in mind that copies of objects are used when evolution rules are applied in regions.

P systems evolution is performed in a non-deterministic and massively parallel manner. That is, inside every region of the P system, the objects to evolve and the evolution rules that can be applied are chosen in an exhaustive and non-determinist way. Evolution in membranes is synchronous, i.e., evolution inside regions is governed by an external clock. In each time unit, every region evolves and changes its multiset of objects by application of evolution rules of the region. This means that in each time unit the P system changes its state or configuration.

Computations in P systems are achieved by sequences of transitions between configurations. A configuration is *halting* if no rule is applicable in any region. We say that a computation is successful if we reach a *halting* configuration, and the result of the computation is the number of objects sent to the environment through the skin or to a given elementary membrane during the computation.

To this basic model many modifications can be added. Here we consider only one more feature: a priority relation among rules in regions. This new feature means that only rules with the higher priority can be used in a given evolution step. This feature permits to establish an order relationship among rules and in some sense to project a kind of execution program in the evolution rules set.

2   Static Structure of Transition P Systems

In this section we present the simplest variant of P systems, transition P systems. It is the basic model with the added feature of a priority relation on the set of evolution rules.

We introduce the main components of transition P systems in order to define them in a constructive manner and then to define the appropriate data structures for representing transition P systems on a functional programming language.

2.1 Membrane Structures

As we have said above, the membrane structure is the main component of a P system and it can be pictorially represented by a *Venn* diagram; it can also be mathematically represented by a tree or by a string of matching parentheses. In order to define appropriately strings that represent membrane structurse let us consider the language MS over the alphabet $\{[,]\}$, whose strings are recurrently defined as follows:

- $[\,] \in MS$,
- if $\mu_1, \cdots, \mu_n \in MS$, $n \geq 1$, then $[\mu_1 \cdots \mu_n] \in MS$,
- nothing else is in MS.

Over MS we define the equivalence relation $\sim$ as follows.

Let x, $y \in MS$; then $x \sim y$ if and only if $x = \mu_1\mu_2\mu_3\mu_4$ and $y = \mu_1\mu_3\mu_2\mu_4$, where $\mu_1, \mu_2, \mu_3, \mu_4 \in MS$. Let us denote $MS/\sim = \overline{MS}$.

The elements of $\overline{MS}$ are called membrane structures. Each matching pair of parentheses $[,]$ of a membrane structure is called a membrane. The external membrane of a membrane structure is called *skin*. A membrane without internal membranes (i.e., having the form $[\,]$) is called an elementary membrane.

The number of membranes in a membrane structure $\mu \in \overline{MS}$ is called the *degree* of the membrane structure, it is represented by $deg(\mu)$, and it is recursively defined by:

- $deg([\,]) = 1$,
- $deg([\mu_1 \cdots \mu_n]) = 1 + \sum_{i=1}^{n} deg(\mu_i)$.

The number of "levels" of a membrane structure $\mu \in \overline{MS}$, the *depth* of μ, can be recursively defined as follows:

- $dep([\,]) = 1$,
- $dep([\mu_1 \cdots \mu_n]) = 1 + max\{dep(\mu_1), \cdots, dep(\mu_n)\}$.

As it was stated above, membranes delimit regions. In a membrane structure μ, the number of regions is equal to $deg(\mu)$. The *Venn* diagram representation of a membrane structure helps to clarify the notion of a region. A region is any closed space delimited by membranes in a membrane structure. It is also clear that a membrane structure of degree n contains n regions, one associated with each membrane. Moreover, we will say that two regions are adjacent if and only if there is only one membrane between them. This fact is important because communication between two regions is possible only if they are adjacent.

2.2   Multisets

Regions defined by membranes are vesicles able to contain copies of objects from a given alphabet. Therefore, a natural way to represent objects inside regions is by using multisets. A compact representation of a multiset is by means of words over a given alphabet.

A multiset over a given object set U can be formally defined as follows:

Let U be an arbitrary set and let $\mathbf{N}$ be the natural number set: a multiset M over U is a mapping from U to $\mathbf{N}$, $M : U \rightarrow N$, which we will use to write in the form $a \rightarrow Ma$.

Every mapping can be also represented in form of a set of coordinated pairs:

$$M = \{(a, Ma) \mid a \in U\}.$$

Multisets can be also represented by a polynomial with exponents in the natural number set; this third representation is more compact and natural, and it is used in most of the papers about $\mathbf{P}$ systems; the polynomial representation is more intuitive and it expresses better than the others the sense of number of copies of objects from the alphabet in the region:

$$a^{n_a} b^{n_b} \cdots z^{n_z} = \{(a, n_a), (b, n_b), \cdots, (z, n_z)\}.$$

It is important to know which objects are really present in the region with at least one copy. That subset of a multiset is called the *support* of the multiset: $Supp\, M = \{a \in U \mid Ma > 0\}$.

Moreover, we distinguish the **Empty Multiset** represented by $\mathbf{0}$, and defined as $\mathbf{0} : U \rightarrow \mathbf{N}$ with $a \rightarrow \mathbf{0}a$.

Several *operations* on multisets can be defined. Let M, M_1, and M_2 be multisets over U. Then:

Multiset Inclusion: $M_1 \subseteq M_2$ if and only if $\forall a \in U$, $(M_1\, a) \leq (M_2\, a)$.
Multiset Union: $\forall a \in U$, $(M_1 \cup M_2)\, a = (M_1\, a) + (M_2\, a)$.
Multiset Difference: $\forall a \in U$, $(M_1 - M_2)\, a = (M_1\, a) - (M_2\, a)$, if $M_1 \subseteq M_2$.

2.3   Super-cell

The super-cell concept provides the possibility of adding objects to a membrane structure. This fact transforms the membrane structure in a computing device [4].

Let U be a denumerable set of objects. Let μ be a membrane structure with its membranes injectively labelled with natural numbers from 1 to $deg(\mu)$. The regions associated with membranes are labelled with the same label as their external membranes.

A partial order relation $<_\mu$ can be defined over the label set of μ $LABELS = \{1, \cdots, deg(\mu)\}$ as follows: Let $i, j \in LABELS$, then we can say that $i <_\mu j$ iff region i contains region j directly or i contains a region k which contains j.

We define now the concept of adjacent regions. Let $i, j \in \{1, \cdots, deg(\mu)\}$ be labels for two regions of μ. We say that region i is adjacent to region j iff:

$$(i <_\mu j \land \not\exists k \in \{1, \cdots, deg(\mu)\} \mid i <_\mu k \land k <_\mu j) \lor$$
$$(j <_\mu i \land \not\exists k \in \{1, \cdots, deg(\mu)\} \mid j <_\mu k \land k <_\mu i).$$

A super-cell is obtained from a membrane structure μ by associating with each region of the membrane structure μ a multiset over U in such a manner that for i varying from 1 to $n = deg(\mu)$, the region i has associated the multiset $M_i : U \to \mathbf{N}$. The multiset M_i associated with the region i is said to be the contents of region i.

2.4   Evolution Rules

Evolution rules make evolve the region with which they are associated, changing the region contents and sending objects to other regions; they can also make disappear the external membrane of the region. An evolution rule is formed by one antecedent and by one consequent. Both of them can be represented by multisets over different sets. An evolution rule can be defined as follows.

Let L be a label set, let U be an object set, and let $D = \{out, here\} \cup \{in \ j \mid j \in L\}$. An evolution rule is a triple (u, v, δ), where u is a multiset over U, v is a multiset over $U \times D$ and $\delta \in \{dissolve, not \ dissolve\}$.

We can define several operations on evolution rules.

Definitions: Let $r = (u, v, \delta), r_1 = (u_1, v_1, \delta_1)$, and $r_2 = (u_2, v_2, \delta_2)$ be evolution rules with labels in L and objects in U. Let $a \in U$ and n be natural number.

Addition of Evolution rules: $r_1 + r_2 = (u_1 \cup u_2, v_1 \cup v_2, d_1 \lor d_2)$.

Product of an evolution rule by a natural number: $n \ r = \sum_{i=1}^n r$.

Inclusion of evolution rules: $r_1 \subseteq r_2$ if and only if $u_1 \subseteq u_2$.

Input of an evolution rule: $Input \ r = u$.

Dissolve: $Dissolve \ r = \delta$.

Also, it is necessary to define some functions over evolution rules related to the rule consequent. These functions will provide important information about what happens when the rule is applied.

The set of **membranes where the rule is sending objects** is

$$Ins \ r = \{j \in L \mid (\_, in \ j) \in Supp \ v\}.$$

Evolution rule's outputs:

$$(OutputToOut \ r)a = v(a, out),$$
$$(OutputToHere \ r)a = v(a, here),$$
$$((OutputToIn \ j)r)a = v(a, in \ j).$$

Functions $OutputToOut, OutputToHere, OutputToIn$, return the multiset that the rule is sending to his father ($ToOut$), to itself ($ToHere$) and to a determined region ($ToIn \cdots$). These functions will be very useful in order to define the changes produced during the evolution of a transition P system.

2.5   Regions

A region is an inter-membranes space; the regions form compartments which contain objects and a set of evolution rules able to make evolve the region. Therefore, a region is the area delimited by a membrane r and any membrane directly inside r. Regions in the membrane structure of a transition P system are uniquely labelled by elements from some set L, contains objects and a partially ordered set of evolution rules. In order to define more precisely a region, let L be a label set and U be an object set. A region with a label in L and objects in U is a triple $(l, \omega, (R, \rho))$ where $l \in L$, ω is a multiset over U, R is a set of evolution rules with labels in L and objects in U and ρ is a partial order relation over R.

2.6   Transition P Systems

In order to provide a formal definition of transition P systems [4], it will be necessary to take into account the previous given definitions: super-cell and evolution rules.

Let U be a set (alphabet) whose elements are named objects. A transition P system over U is a construct

$$\Pi = (\mu, \omega_1, \cdots, \omega_{deg(\mu)}, (R_1, \rho_1), \cdots, (R_{deg(\mu)}, \rho_{deg(\mu)}), i_0),$$

where:

i. μ is a membrane structure with membranes labeled in a one-to-one manner with natural numbers from 1 to $deg(\mu)$.
ii. ω_i, $1 \leq i \leq deg(\mu)$, is a multiset (word) over U associated with region i.
iii. R_i, $1 \leq i \leq deg(\mu)$, is a finite set of evolution rules associated with region i.
iv. ρ_i, $1 \leq i \leq deg(\mu)$, is a partial order over R_i.
v. $1 \leq i_0 \leq deg(\mu)$, is the label that identifies the output membrane of the system.

It is possible to distinguish two different parts in a transition P system: $(\mu, \omega_1, \cdots, \omega_{deg(\mu)})$ defines a super-cell, it is the static part of the system, while the second part, $((R_1, \rho_1), \cdots, (R_{deg(\mu)}, \rho_{deg(\mu)}), i_0)$, is the dynamic component of the P system, and provides the evolving capacity to it.

This definition corresponds to a transition P system with a membrane output, but it is possible to define different variants, for instance, without an output membrane, or with multiples output membranes.

3   Data Structures for Representing Transition P Systems

Until now, we have presented all the components of transition P systems described in terms of algebraic structures, with their corresponding operations and functions [2]. Now it is necessary to translate such algebraic structures into data structures. In order to do this, we have several possibilities, because:

i. The same data structure with different operations on it can represent different algebraic structures.
ii. The same algebraic structure can be represented by different combinations of data structures with operations.

Therefore, the main difference between two data structures and their associated operations representing the same algebraic structure will be efficiency in the computing time and allocated memory. To represent an algebraic structure into a data structure the best choice will be the one that allows the highest efficiency maintaining the algebraic structure original properties.

From a computational point of view, the algebraic structure used for defining transition P systems has one essential characteristic: regions and their contents are independents, but they are linked by labels. So, these facts make easy to design a structure of references (tree, graph, etc.) that automatically build a structure for referred items.

These kinds of structures are called *independent references*. In what follows, we present a data structure which does not need to refer to regions in order to define its structure, but it builds directly a region structure. It is called a *regions tree*.

A tree, algebraically speaking, can be represented by multiple data structures. Among them:

i. An array or vector of pairs of labels in the form (father's label, son's label).
ii. A list of pairs of labels in the form (father's label, son's label).
iii. A square matrix whose rows and columns are marked will be the labels.

However, the variants which permit a more efficient treatment due to a minimal unnecessary search are the following:

i. An array or vector of pairs in the form (father's label, list of sons' labels) with labels as indices. The main advantage of this structure is that it permits a direct access from every region to its father's region, and this is due to the fact that it represents a bi-directional tree. The structure has a certain redundancy degree, the same label can appear as a father index of some node and as the son index of another node.
ii. A general non-empty tree whose elements are labels. In this case, labels appear only once in the structure. This structure is more similar in form to the original definition of transition P systems.

Definitions: A tree over an arbitrary set S is a pair where the first component is an element of S and the second one is a set of trees over S.

In this definition, the second element of the pair can be the empty set, a case when the element is a leaf.

Over trees it can be defined the multiplicity function which provides the number of occurrences of a given element $a \in U$ in the tree.

Multiplicity: Let $T = (u, S)$ be a tree, where $u \in U$ and S is a set of trees over U. Then

$$(Mult\ T)a = \begin{cases} 1 + \sum_{s \in S}(Mult\ s)a & \text{if } a = u, \\ \sum_{s \in S}(Mult\ s)a & \text{if } a \neq u. \end{cases}$$

The membrane structure of a transition P system defines a tree whose root is the "skin" and elementary membranes are the leaves.

Finally, for defining a transition P system, it is necessary to provide a content to membranes. With all given definitions, a transition P system can be defined as a tree of regions uniquely labelled. In a formal way, we have:

Let L be a label set, let U be a set of objects. A transition P system with labels in L and objects in U is a tree Π of regions with labels in L and objects in U provided that:

$$\forall l \in L, (Mult\, \Pi)(l, \_, \_) < 2.$$

An alternative definition can be given: A transition P system with labels in L and objects in U is a pair whose first element is a region with labels in L and objects in U and the second one is a set of transition P systems with labels in L and objects in U, whose regions are uniquely labelled:

$$\Pi = ((l, \omega, (R, \rho)), \Pi\Pi) \text{ and } \forall l \in L, (Mult\, \Pi)(l, \_, \_) < 2,$$

where ω is a multiset over U, R is a set of evolution rules over L and U, ρ is a partial order relationship over R and $\Pi\Pi$ is a set of transition P systems with labels in L and objects in U.

This algebraic representation define precisely the static structure of transition P systems. We have named it static structure because this representation do not take into account the evolution in transition P systems, it only represents transition P systems in a static manner.

4   Transition P Systems: The Dynamic Structure

We call dynamic structure of transition P systems everything which is related to evolution. During the evolution of a transition P system there are produced many changes. For instance, the static structure may be changed because of the execution of some evolution rules that dissolve membranes; moreover, the contents of the regions are changed by the parallel application of evolution rules. If a rule dissolves the external membrane of a region, then the region disappears from the static structure of the P system, the region sends its objects to its father region, and its evolution rules disappear from the system.

The dynamics of a transition P system has been described in terms of parallel and non-deterministic events: "evolution is done in parallel for all objects able to evolve", and "all these operations are done in parallel, for all possible applicable rule $u \to v$, for all occurrences of multiset u in the region associated with the rules, for all regions at the same time" [4].

During evolution in P systems, it is possible to distinguish between at least two different types of parallelism. The first one is the regional parallelism and it is related to what is happening inside a region. The second one is the system global parallelism and it is related to what is happening in every region of the P system at the same time. Both of them are involved in the system evolution. They are interrelated because regions may disappear from the static structure of the P system and evolution rules send objects to other regions; but in some way, they can be independently described.

4.1   Regional Parallelism

This point is related to the need to characterize the parallel application of evolution rules. We will describe the dynamic evolution in a given region of a P system in a determined step of its evolution. The aim is to determine a set of evolution rules able to make evolve the region by the application of one of these rules.

A region has been defined as a triple:

$$Region : (label, multiset, (Rules, Partial\ Order\ Relation)) = (l, \omega, (R, \rho)).$$

The evolution inside the region involves objects consumption by rules, sending objects to other regions and, possibly, membrane dissolution, which consequently implies that the region vanishes. All these operations are done in parallel with only one constraint, that a rule with a higher precedence inhibits rules of a lower precedence. Therefore, only rules with a higher precedence can be applied in order to make evolve regions.

If we want to define exactly the rules able to make evolve a region, then we need the following definitions:

Maximal Set: Let $(U, <)$ be a partial order over U; then

$$((Maximal\ U)\ <) = \{u \in U \mid \not\exists v \in U, u < v\} = \{u \mid (u \in U) \wedge (\neg(u < \_))\}.$$

Useful Rules: An evolution rule in a rules set is useful over a given labels set L if and only if the set $Ins\ r$ is included in L. In a more formal way, let R be an evolution rule set with labels in L and objects in U. Let $L' \subset L$. Then

$$(Useful\ R)L' = \{r \in R \mid Ins\ r \subseteq L'\} = \{r \mid (r \in R) \wedge (Ins\ r \subseteq L')\}.$$

By the definition of a transition P system, the objects can pass through only one membrane. Therefore, a rule can send objects to regions separated by only one membrane. In order to calculate the rules useful in a region, let L' be the set of labels corresponding to adjacent regions to the given region.

Applicable Rules: A rule is applicable to a multiset ω if and only if its antecedent is included in ω. Let R be a set of evolution rules with labels in L and objects in U. Let ω be a multiset over U. Then

$$(Applicable\ R)\omega = \{r \in R \mid Input\ r \subseteq_u \omega\} = \{r \mid (r \in R) \wedge (Input\ r \subseteq_u w)\}.$$

Therefore, a rule is applicable in a region if and only if the rule antecedent is included in the multiset contained by that region. In addition, in a determined evolution step of a transition P system, applicable rules are the only ones that can be used.

Active Rules: A rule $r \in R$ is active with respect to a partial order relation ρ defined on R if and only if there is no rule in R higher than r according ρ. Therefore, actives rules are characterized as follows.

Let ρ be a partial order relation defined over a set R of evolution rules with labels in L and objects in U; then:

$$(Active\ R)\rho = (Maximal\ R)\rho.$$

Adjacent regions to a region: Let reg be a region of a transition P system Π with labels in L and objects in U. Let $(reg, \Pi\Pi)$ be the sub-tree of Π with the root reg. We define the adjacent region in Π as:

$$ADJACENT_\Pi\ reg = \{r \mid (r, \Pi\Pi') \in \Pi\Pi\} = \{r \mid (r, \_) \in \Pi\Pi\}.$$

Rules able to make evolve the region: With these definitions, it is possible to determine exactly the set of evolution rules that can be applied in order to make evolve the region reg. By definition, they are those rules satisfying the following equations:

$$L' = \{l \in L \mid (l, \_, \_) \in ADJACENT_\Pi\ reg\}$$
$$Active(Applicable((Useful\ R)L')\omega)\rho. \tag{1}$$

The evolution of a transition P system is described inside a region as follows: every rule satisfying (1) can be applied in parallel for all occurrences of rule antecedent multiset, for all regions at the same time. Therefore, we must search for a linear combination of rules satisfying (1) being complete; that is, no further rule satisfying (1) can be added to the linear combination. Let us explain this last sentence.

The application of two rules r_1, $r_2 \in R$ at the same evolution step in a region, in parallel, is equivalent to applying the rule $r_1 + r_2$. In this sense, we can say that to apply n times the same rule in parallel has the same effect as applying the rule defined by nr. Therefore, the parallel application of evolution rules inside a region can be substituted by the application of one linear combination of evolution rules of the region. What is necessary is that such a linear combination is maximal in covering the multiset of objects in the region, i.e., it is a complete linear combination of evolution rules in the region. Finally, the region evolves in one step by the application of only one rule formed by a complete linear combination of evolution rules in R satisfying (1).

Any linear combination of evolution rules can be represented as a multiset over R, the set of evolution rules of the region, as follows. Let $R = \{r_1, \cdots, r_n\}$ be the set of evolution rules of region reg. Then any linear combination of evolution rules in R and scalars in $\mathbf{N}$ can be represented by a multiset over R as follows:

$$\sum_{r_i \in R} n_i\, r_i \Rightarrow MR : R \to \mathbf{N},$$

$$r_i \to n_i.$$

4.2   Characterization of Complete Multisets of Evolution Rules in a Region

Once it is known how can be possible to make evolve a region in one step, we need to characterize multisets of complete evolution rules. Until now, we have provided an exact definition of rules that can be applied to evolve a region of a transition P system in one evolution step. Moreover, the parallel execution of rules inside the region can be replaced by the execution of only one rule. Those rules are a linear combination of active rules of the region expressed in terms of rules from R – the set of evolution rules of the region. What follows are several definitions over the multiset of evolution rules aiming to obtain an expression for identifying complete multisets of evolution rules in the given region.

Definitions: Let R be a set of evolution rules with labels in L and objects in U. Let MR, MR_1, and MR_2 be multisets over R and let w be a multiset over U.

MultiAdd: This operation performs the transformation of a multiset of evolution rules into a linear combination of evolution rules. From now on, we will keep in mind this equivalence between linear combinations of evolution rules and multisets of evolution rules.

$$\oplus MR = \sum_{r \in R}(MRr)r.$$

MultiInclusion: This binary relation defines a partial order relation over the multiset of evolution rules over R. It will be very useful in order to determine which multiset over R will be complete.

$$MR_1 \sqsubseteq_u MR_2 \equiv (\oplus MR_1) \subset_u (\oplus MR_2).$$

MultiApplicable: This function defines the set of multisets over R (the evolution rules set) that can be applied to w.

$$(MultiApplicable\ R)w = \{MR \mid (MR : R \rightarrow N) \wedge ((Input\ (\oplus MR)) \subseteq w)\}.$$

MultiComplete:

$$(Multicomplete\ R)w = (Maximal((MultiApplicable\ R)w)) \sqsubseteq_u .$$

The above expression describes the set of complete multisets over $(R, \sqsubseteq_u)$ and w.

Finally, let Π be a transition P system with labels in L and objects in U, let $reg = (l, w, (R, \rho))$ be one region of Π and let $(reg, \Pi\Pi)$ be the sub-tree of Π whose root is reg. The set of complete multisets of evolution rules is defined by the following equations:

$$LABELS = \{l \in L \mid ((l, \_, \_), \_) \in \Pi\Pi\},$$
$$ACTIVES = Active(Applicable((Useful\ R)LABELS)w)\rho,$$
$$COMPLETES = (Multicomplete\ ACTIVES)w.$$

The $LABELS$ set defines the set of labels of regions adjacent to region reg. The $ACTIVES$ set defines the set of evolution rules that can be applied in order to make the region evolve. The $COMPLETES$ set defines the set of complete multisets over R (the set of evolution rules of the region) capable of making evolve the region in one step. Therefore, there are as many different possibilities for the region evolution as many elements $COMPLETES$ has.

4.3   Regional Non-determinism and Global Non-determinism

The above section has given a procedure for characterizing the set of evolution rules multiset over R (the set of evolution rules of a given region) such that their transformation into linear combinations by application of the $MultiAdd$ function is able to make evolve the region in one step. It has been also stated that there are many possibilities for making evolve the region. The number of different possibilities is equal to the cardinal of $COMPLETES$ set. Therefore, a non-deterministic selection of a multiset in $COMPLETES$ is enough for having a non-deterministic evolution in the region.

The global non-determinism in the system is achieved through the regional local non-determinism. P systems are devices with a high degree of parallelism. Evolution is done in parallel in every region of the system. When regions execute the selected linear combination of evolution rules, according to a non-deterministic selection of a complete multiset of rules belonging to the region and evolve to a new region configuration, then a new configuration for the P system is achieved and the step of evolution is finished. We have to be aware that P systems are synchronous devices, i.e., in order to get a new configuration for the system, all the regions must finish their evolution and only when the new system configuration is reached a new evolution step may start.

5   The Software System

Based on this formalism, we are developing a set of software modules in order to implement transition P systems on digital computers. The software system has two different kinds of modules, the first one related to algebraic structures, and the second one related to the transition P systems architecture.

Modules included in the algebraic structures are tuples, sets, multisets, and relations. In the P system architecture we can find modules of directions, rules, regions, and transition P system. The functional dependencies among modules are shown in Figure 1.

6   Bio-language for Representing Transition P Systems

Programming a transition P system is not writing a code in a determined programming language, but developing a transition P system architecture (membrane structure) and projecting on it data (an object multiset) and operations

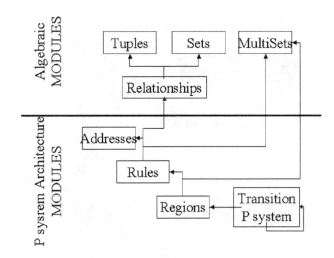

Fig. 1. Functional dependencies among modules of the software system

– evolution rules with a priority relation among them. Then, the code execution is obtained making evolve the P system, and results are obtained by examination of a given region in the system. Of course, the code execution is made in a non-deterministic and massively parallel way. However, it will be very useful if P systems designers could have a special language for developing P systems architectures. Such a language must express the special characteristics of P systems and, through a process of compilation, should translate the architecture to the needed data structures for simulating the execution on a digital computer and launch the execution of the code. This is the usual process of compilation in programming languages (Figure 2).

The process of compilation assumes the task of main function or launcher of the distributed processed embedded in evolution rules of P systems regions.

Now we will describe the syntax of our bio-language by rewriting rules:

- V a set of objects,
- L a set of labels,
- Transition P system (TPS): $\Pi \Rightarrow [_l\ Region; \{\Pi\}]_l$,
- Region $\Rightarrow$ Objects, Rules, Priorities,
- Objects $\Rightarrow \{o^n\}$, where $o \in V, n \in N$,
- Rules $\Rightarrow \lambda | rule\{, rule\}$,
- Priorities $\Rightarrow \lambda | r_i < r_j \{, r_i' < r_j'\}$, where $i, i', j, j' \in \mathbf{N}$,
- Addresses $\Rightarrow here | out | l$ where $l \in L$,
- ObjTarg $\Rightarrow \{(o, Addresses)^n\}$, where $o \in V$ and $n \Rightarrow 1|2|\cdots$,
- Rule $\Rightarrow r_m : Objects \rightarrow ObjTarg\ \delta$, where $\delta \in \{dissolve,\ not\ dissolve\}$.

These rewriting rules define the syntax of the TPS bio-language for determining the static structure of a transition P system at any step of the evolution. The internal code execution is embedded in the transition P system architecture

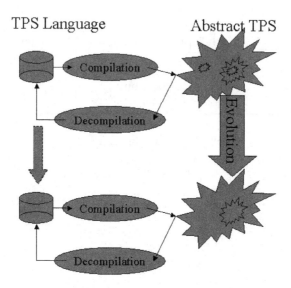

Fig. 2. Compilation and decompilation processes during transition P system evolution

through the evolution rules of the transition P system rules, which define the internal distributed functions of the P system. We have named bio-language this language because sentences are much closed to the particular components of P systems (membranes, regions, objects and evolution rules), and such systems have a biological inspiration.

7   Conclusions

This work deals with the formalisation problem of the static and dynamic structures of transition P systems in terms of computational data structures. It has been also needed to solve the non-deterministic and parallel execution of evolution rules in regions. A first approach to a bio-language for transition P systems representation has been presented. Sucha bio-language will facilitate the task of developing programs, representing P systems and to execute them on digital computers.

We believe that simulations of some models of membrane computing may help to spread these new computational paradigms (membrane computing) outside the actual group of researchers and to present them to the computer science community as feasible models for solving NP-complete problems even in linear time.

References

1. J.P. Banâtre, A. Coutant, D. Le Metayer, A parallel machine for multiset transformation and its programming style, *Future Generation Computer Systems*, **4** (1988), 133–144.

2. V. Baranda, J. Castellanos, R. Molina, F. Arroyo, L.F. Mingo, Data structures for implementing transition P systems in silico, *Pre-proceedings of Workshop on Multiset Processing*, Curtea de Argeş, Romania, TR140, CDMTCS, Univ. Auckland, 2000, 21–34.
3. G. Berry, G. Boudol, The chemical abstract machine, *Theoretical Computer Science*, **96** (1992), 217–248.
4. Gh. Păun, Computing with membranes, *Journal of Computer and Systems Sciences*, **61**, 1 (2000), 108–143.
5. M. Sipper, Studying artificial life using a simple, general cellular model, *Artificial Life Journal*, **2**, 1 (1995), 1–35.

Gamma and the Chemical Reaction Model: Fifteen Years After*

Jean-Pierre Banâtre[1], Pascal Fradet[2], and Daniel Le Métayer[3]

[1] Université de Rennes I, Campus de Beaulieu, 35042 Rennes, France
and INRIA, Domaine de Voluceau-Rocquencourt, 78153 Le Chesnay Cedex, France
jpbanatre@inria.fr
[2] INRIA / IRISA, Campus de Beaulieu, 35042 Rennes, France
fradet@irisa.fr
[3] Trusted Logic S.A. 5, rue du Bailliage, 78000 Versailles, France
Daniel.Le_Metayer@trusted-logic.fr

Abstract. Gamma was originally proposed in 1986 as a formalism for the definition of programs without artificial sequentiality. The basic idea underlying the formalism is to describe computation as a form of *chemical reaction* on a collection of individual pieces of data. Due to the very minimal nature of the language, and its absence of sequential bias, it has been possible to exploit this initial paradigm in various directions. This paper reviews most of the work around Gamma considered as a programming or as a specification language. A special emphasis is placed on unexpected applications of the chemical reaction model, showing that this paradigm has been a source of inspiration in various research areas.

1   The Basic Chemical Reaction Model

The notion of sequential computation has played a central rôle in the design of most programming languages in the past. This state of affairs was justified by at least two reasons:

- Sequential models of execution provide a good form of abstraction of algorithms, matching the intuitive perception of a program defined as a "recipe" for preparing the desired result.
- Actual implementations of programs were made on single processor architectures, reflecting this abstract sequential view.

However the computer science landscape has evolved considerably since then. Sequentiality should no longer be seen as the prime programming paradigm but just as one of the possible forms of cooperation between individual entities.

---

* This paper is a revised version of *Gamma and the chemical reaction model: ten years after* [10]. It has been reorganized and includes additional sections on applications of the chemical reaction model. Sections presenting large examples, extensions of the formalism and implementations issues have been seriously shortened. The reader is referred to [10] and the original papers for further details on these topics.

C.S. Calude et al. (Eds.): Multiset Processing, LNCS 2235, pp. 17–44, 2001.
© Springer-Verlag Berlin Heidelberg 2001

The Gamma formalism was proposed fifteen years ago to capture the intuition of computation as the global evolution of a collection of atomic values interacting freely. Gamma is a kernel language which can be introduced intuitively through the chemical reaction metaphor. The unique data structure in Gamma is the multiset which can be seen as a chemical solution. A simple program is a pair (*Reaction condition, Action*). Execution proceeds by replacing in the multiset elements satisfying the reaction condition by the products of the action. The result is obtained when a stable state is reached, that is to say when no more reactions can take place. The following is an example of a Gamma program computing the maximum element of a non-empty set.

$$max \ : \ x, \ y \rightarrow \ y \ \Leftarrow x \leq y$$

The side condition $x \leq y$ specifies a property to be satisfied by the selected elements x and y. These elements are replaced in the set by the value y. Nothing is said in this definition about the order of evaluation of the comparisons. If several disjoint pairs of elements satisfy the condition, the reactions can be performed in parallel. Let us consider as another introductory example a sorting program. We represent a sequence as a set of pairs (*index, value*) and the program exchanges ill-ordered values until a stable state is reached and all values are well-ordered.

$$sort \ : \ (i, x), \ (j, y) \rightarrow \ (i, y), \ (j, x) \ \Leftarrow (i > j) \ and \ (x < y)$$

The possibility of getting rid of artificial sequentiality in Gamma confers a very high level nature to the language and allows the programmer to describe programs in a very abstract way. In some sense, one can say that it is possible in Gamma to express the very "idea" of an algorithm without any unnecessary linguistic idiosyncrasy (like "exchange any ill-ordered values until all values are well ordered" for the sorting algorithm). This also makes Gamma suitable as an intermediate language in the program derivation process: Gamma programs are easier to prove correct with respect to a specification and they can be refined for the sake of efficiency in a second stage. This refinement may involve the introduction of extra sequentiality but the crucial methodological advantage of the approach is that logical issues can be decoupled from implementation issues.

To conclude this introduction, let us quote E.W.Dijkstra thirty years ago [27]: *"Another lesson we should have learned from the recent past is that the development of "richer" or "more powerful" programming languages was a mistake in the sense that these baroque monstrosities, these conglomerations of idiosyncrasies, are really unmanageable, both mechanically and mentally. I see a great future for very systematic and very modest programming languages"*. We believe that this statement is more relevant than ever and the very minimal nature of the original Gamma formalism is one factor which has made possible the various developments that are sketched in this paper.

We start by providing in section 2 the basic intuitions about the programming style entailed by Gamma as well as various programming examples and extensions.

Gamma was originally proposed in the context of a work on systematic program derivation [8]. Section 3 describes how Gamma can be used as an intermediate language in the derivation of efficient implementations from specifications.

Since the initial developments, Gamma has been a source of inspiration in unexpected research areas. Section 4 shows several applications of the chemical reaction model in various domains such as the semantics of process calculi, imperative programming and software architectures. Section 5 concludes by a sketch of related work.

2   Gamma as a Programming Language

Using the chemical reaction model as a basic paradigm can have a deep effect on our way of thinking about algorithms. We first try to convey the programming style favored by Gamma through some very simple examples. Then we introduce five basic programming schemes, called "tropes" which have emerged from our experience in writing Gamma programs. We proceed by presenting large applications written in Gamma and reviewing various linguistic extensions.

2.1   A New Programming Style

We first come back on the straightforward *max* program defined in the introduction to illustrate some distinguishing features of Gamma:

$$max \; : \; x, \, y \to \; y \; \Leftarrow x \leq y$$

In order to write a program computing the maximum of a set of values in a "traditional" language, we would first have to choose a representation for the set. This representation could typically be an array for an imperative language or a list for a declarative language. The program would be defined as an iteration through the array, or a recursive walk through the list. The important point is that the data structure would impose constraints on the order in which elements are accessed. Of course, parallel versions of imperative or functional programs can be defined (solutions based on the "divide and conquer" paradigm for example), but none of them can really model the total absence of ordering between elements that is achieved by the Gamma program. The essential feature of the Gamma programming style is that a data structure is no longer seen as a hierarchy that has to be walked through or decomposed by the program in order to extract atomic values. Atomic values are gathered into one single bag and the computation is the result of their individual interactions. A related notion is the "locality principle" in Gamma: individual values may react together and produce new values in a completely independent way. As a consequence, a reaction condition cannot include any global condition on the multiset such as ∀-properties or properties on the cardinality of the multiset. The locality principle is crucial because it makes it easier to reason about programs and it encapsulates the intuition that there is no hidden control constraints in Gamma programs.

Let us now consider the problem of computing the prime numbers less than a given value n. The basic idea of the algorithm can be described as follows: "start with the set of values from 2 to n and remove from this set any element which is the multiple of another element". So the Gamma program is built as the sequential composition of $iota$ which computes the set of values from 2 to n and rem which removes multiples. The program $iota$ itself is made of two reactions: the first one splits an interval (x, y) with $x \neq y$ in two parts and the second one replaces any interval (x, x) by the value x.

$$
\begin{aligned}
primes(n) &= rem(iota(\{(2, n)\})) \\
iota \quad &= (x, y) \rightarrow (x, \lfloor (x+y)/2 \rfloor), \ (\lfloor (x+y)/2 \rfloor + 1, y) \Leftarrow x \neq y \\
&\quad (x, y) \rightarrow x \Leftarrow x = y \\
rem \quad &= x, \ y \rightarrow y \Leftarrow multiple(x, y)
\end{aligned}
$$

The first reaction increases the size of the multiset, the second one keeps it constant and the third one makes the multiset shrink. In contrast with the usual sequential or parallel solutions to this problem (usually based on the successive application of sieves [9]), the Gamma program proceeds through a collection of atomic actions applying on individual and independent pieces of data.

Another program exhibiting these expansion and shrinking phases is the Gamma version of the Fibonacci function:

$$
\begin{aligned}
fib(n) &= add(dec_1(n)) \\
dec_1 \quad &= x \rightarrow x - 1, \ x - 2 \Leftarrow x > 1 \\
&\quad x \rightarrow 1 \Leftarrow x = 0 \\
add \quad &= x, \ y \rightarrow x + y \Leftarrow True
\end{aligned}
$$

The initial value is decomposed by dec_1 into a number of ones which are then summed up by add to produce the result. The first phase corresponds to the recursive descent in the usual functional definition

$$
fib(x) = if \ x \leq 1 \ then \ 1 \ else \ fib(x - 1) + fib(x - 2)
$$

while the reduction phase is the counterpart of the recursive ascent. However the Gamma program does not introduce any constraint on the way the additions are carried out, which contrasts with the functional version in which additions must be performed following the order imposed by the recursion tree of the execution.

As a last example of this introduction, let us consider the "maximum segment sum" problem. The input parameter is a sequence of integers. A segment is a subsequence of consecutive elements and the sum of a segment is the sum of its values. The program returns the maximum segment sum of the initial sequence. The elements of the multiset are triples (i, x, s) where i is the position of value x in the sequence and s is the maximum sum (computed so far) of segments ending at position i. The s field of each triple is originally set to the x field. The program max_l computes local maxima and max_g returns the global maximum.

$$
\begin{aligned}
maxss(M) &= max_g(max_l(M)) \\
max_l \quad &= (i, x, s), \ (i', x', s') \rightarrow (i, x, s), \ (i', x', s + x') \\
&\quad \Leftarrow (i' = i + 1) \ and \ (s + x' > s') \\
max_g \quad &= (i, x, s), \ (i', x', s') \rightarrow (i', x', s') \Leftarrow s' > s
\end{aligned}
$$

2.2   The Tropes: Five Basic Programming Schemes

The reader may have noticed a number a recurrent programming patterns in the small examples presented in the previous section. After some experience in writing Gamma programs, we came to the conclusion that a very small number of program schemes were indeed necessary to write most applications. Five schemes (basic reactions), called *tropes* (for *t*ransmuter, *r*educer, *o*ptimiser, *e*xpander, *s*elector) are particularly useful. We present only three of them here:

- Transmuter.
$$\mathcal{T}(C, f) = x \to f(x) \Leftarrow C(x)$$

  The transmuter applies the same operation to all the elements of the multiset until no element satisfies the condition.
- Reducer.
$$\mathcal{R}(C, f) = x, y \to f(x, y) \Leftarrow C(x, y)$$

  This trope reduces the size of the multiset by applying a function to pairs of elements satisfying a given condition. The counterpart of the traditional functional *reduce* operator can be obtained with an always true reaction condition.
- Expander.
$$\mathcal{E}(C, f_1, f_2) = x \to f_1(x), f_2(x) \Leftarrow C(x)$$

  The expander is used to decompose the elements of a multiset into a collection of basic values.

The Fibonacci function can be expressed as the following combination of tropes:

$$
\begin{aligned}
fib(n) &= add\ (\ zero\ (\ dec\ (\{n\}))) \\
dec\ \ &= \mathcal{E}(C, f_1, f_2)\ \ \textbf{where} \\
&\quad C(x)\ =\ x > 1,\ \ f_1(x)\ =\ x - 1,\ \ f_2(x)\ =\ x - 2 \\
zero\ &= \mathcal{T}(C, f)\ \ \textbf{where} \\
&\quad C(x)\ =\ (x\ =\ 0),\ f(x)\ =\ 1 \\
add\ \ &= \mathcal{R}(C, f)\ \ \textbf{where} \\
&\quad C(x, y)\ =\ True,\ f(x, y)\ =\ x + y
\end{aligned}
$$

The maximum segment sum program presented in the previous section can be defined in terms of tropes in a very similar way. Further details about tropes may be found in [35].

2.3   Larger Applications

The interested reader can find in [9] a longer series of examples chosen from a wider range of domains (string processing problems, graph problems, geometric problems). We just sketch in this section a small selection of applications that we consider more significant, either because of their size or because of their target domain.

Image processing application. Gamma has been used in a project aiming at experimenting high-level programming languages for prototyping image processing applications [24]. The application was the recognition of the tridimensional topography of the vascular cerebral network from two radiographies. A version of this application written in PL/1 was in use before the start of the experiment but it was getting huge and quite difficult to master. One of the benefits of rewriting the application in Gamma has been a better understanding of the key steps of the application and the discovery of a number of bugs in the original software. So Gamma has been used in this context as an executable specification language and it turned out to be very well suited to the description of this class of algorithms. The basic reason is probably that many treatments in image processing are naturally expressed as collections of local applications of specific rules.

Another example of application of Gamma to image processing is reported in [47]. The aim of this application is to generate fractals to model the growth of biological objects. Again, the terseness and the facility of the expression of the problem is Gamma was seen as a great advantage. In both experiences however, the lack of efficient general purpose implementation of Gamma was mentioned as a serious drawback because it prevented any test on large examples.

Reactive programming. In [58], an operating system kernel is defined in Gamma and proven correct in a framework inspired by the Unity logic [14]. An important result of this work is the definition of a temporal logic for Gamma (extended with a fairness assumption) and the derivation of the kernel of a file management system by successive refinements from a temporal logic specification. Each refinement step results in a greater level of detail in the definition of the network of processes. We are not aware of comparable attempts in the area of operating systems.

2.4   Implementations

A property of Gamma which is often presented as an advantage is its potential for concurrent interpretation. In principle, due to the locality property, each tuple of elements fulfilling the reaction condition can be handled simultaneously. It should be clear however that managing all this parallelism efficiently can be a difficult task and complex choices have to be made in order to map the chemical model on parallel architectures. The major problems to be solved are:

1. The detection of the tuples which may react.
2. The transformation of the multiset by application of reactions.
3. The detection of the termination.

This section sketches several attempts to provide parallel implementations for Gamma programs.

Distributed memory implementations. Two protocols have been proposed [6,7] for the implementation of Gamma on network of communicating machines. They differ in the way rewritings are controlled:

- *Centralized control.* The elements of the multiset are distributed over the local memories of the processors. A central controller is connected to all the processors and monitors information exchanges. This protocol has been implemented on a Connexion Machine [23]. Other experiments have been conducted on the Maspar 1 SIMD machine: [46] describes an implementation of Gamma which results in a very good speed-up and a good exploitation of parallel resources. [45] shows how Higher-Order Gamma programs can be refined for an efficient execution on a parallel machine.
- *Distributed control.* Information transfers are managed in a fully asynchronous way. The values of the multiset are spread over a chain of m processors. There is no central controller in the system and each processor knows only its two neighbors. The termination detection algorithm is fully distributed over the chain of processors; however, the cost of this detection can be high compared with the cost of the computation itself [6,7]. This solution has been implemented on an Intel iPSC2 machine [6,7] and on a Connexion Machine [23]. The results show a good exploitation of the processing power and speedup.

Shared memory implementations. The Gamma model can also be seen as a shared memory model: the multiset is the unique data structure from which elements are extracted and where elements resulting from the reaction are stored. Shared memory multiprocessors are good candidates for parallel implementations of Gamma. A specific software architecture has been developed in [33] in order to provide an efficient Gamma implementation on a Sequent multiprocessor machine. Several techniques have been experimented in order to improve significantly the overall performances. A kernel operating system has also been developed in order to cope with various traditional problems and in particular with the synchronization required by Gamma (a multiset element cannot participate in more than one reaction at a time).

Hardware implementation. The tropes defined in section 2.2 have been used as a basis for the design of a specialized architecture [62]. A hardware skeleton is associated with each trope and these skeletons are parameterized and combined according to the program to be implemented. A circuit can then be produced from a program description. The hardware platform was the PRL-DEC Perle 1 board which is built around a large array of bit-level configurable logic cells [11].

2.5   Linguistic Extensions

The Gamma programs that we have presented so far are made from a single block of reaction rules. In this section, we review several linguistic extensions for structuring programs or multisets.

Composition operators for Gamma. For the sake of modularity, it is desirable that a language offers a rich set of operators for combining programs. It is also fundamental that these operators enjoy a useful collection of algebraic laws in order to make it possible to reason about programs. Several proposals which have been made to extend Gamma with facilities for building complex programs from simple ones.

[34] presents of a set of operators for Gamma and studies their semantics and the corresponding calculus of programs. The two basic operators considered in this paper are the sequential composition $P_1 \circ P_2$ and the parallel composition $P_1 + P_2$. The intuition behind $P_1 \circ P_2$ is that the stable multiset reached after the execution of P_2 is given as argument to P_1. On the other hand, the result of $P_1 + P_2$ is obtained (roughly speaking) by executing the reactions of P_1 and P_2 (in any order, possibly in parallel), terminating only when neither can proceed further. The termination condition is particularly significant and heavily influences the choice of semantics for parallel composition. As an example of sequential composition of Gamma programs, let us consider another version of sort.

$$sort' : match \circ init$$
$$\textsf{where} \quad init : (x \to (1, x)) \Leftarrow integer(x))$$
$$match : ((i, x), (j, y) \to (i, x), (i + 1, y) \Leftarrow (x \leq y \ and \ i = j))$$

The program $sort'$ takes a multiset of integers and returns an increasing list encoded as a multiset of pairs $(index, value)$. The reaction $init$ gives each integer an initial rank of one. When this has been completed, $match$ takes any two elements of the same rank and increases the rank of the larger.

The case for parallel composition is slightly more involved. In fact $sort'$ could have been defined as well as:

$$sort' : match + init$$

because the reactions of $match$ can be executed in parallel with the reactions of $init$ (provided they apply on disjoint subsets, but this is implied by the fact that their respective reaction conditions are exclusive). As far as the semantics of parallel composition is concerned, the key point is that a *synchronized termination* of P_1 and P_2 is required for $P_1 + P_2$ to terminate. It may be the case that, at some stage of the computation, none of the reaction conditions of, P_1 (resp. P_2) holds; but some reactions by P_2 (resp. P_1) may create new values which will then be able to take part in reactions by P_1 (resp. P_2). This situation precisely occurs in the above example where no reaction of $match$ can take place in the initial multiset; but $init$ transforms the multiset and triggers subsequent reactions by $match$. Thus the termination condition of $P_1 + P_2$ indicates that neither P_1 nor P_2 can terminate unless both terminate and the composition as well.

This new vision of parallel composition and its combination with the sequential composition creates interesting semantical problems. [34] defines a set of program refinement and equivalence laws for parallel and sequential composition, by considering the input-output behavior induced by an operational semantics. Particular attention is paid on conditions under which $P_1 \circ P_2$ can be transformed into $P_1 + P_2$ and vice-versa. These transformations are useful to improve the efficiency of a program with respect to some particular machine and implementation strategy.

Several compositional semantics of the language have also been proposed [59,20]. Not all the laws established in the operational semantics remain valid in these semantics; this is because they distinguish programs with identical input/output behavior but which behave differently in different contexts. It is shown however that most interesting properties still hold, the great advantage of a compositional semantics being that laws can be used in a modular way to prove properties of large programs. Other semantics of Gamma have been proposed, including [28] which defines a congruence based on transition assertions and [55] which describes the behavior of Gamma programs using Lamport's Temporal Logic of Actions.

Composition operators have been studied in a more general framework called *reduction systems* [60] which are sets equipped with some collection of binary rewrite relations. This work has led to a new graph representation of Gamma programs which forms a better basis for the study of compositional semantics and refinement laws.

Higher-order Gamma. Another approach for introduction of composition operators in a language consists in providing a way for the programmer to define them as higher-order programs. This is the traditional view in the functional programming area and it requires to be able to manipulate programs as ordinary data. This is the approach followed in [41] which proposes a higher-order version of Gamma. The definition of Gamma used so far involves two different kinds of terms: the programs and the multisets. The multiset is the only data structure and programs are described as collections of pairs *(Reaction Condition, Action)*. The main extension of higher-order Gamma consists in unifying these two categories of expressions into a single notion of configuration. One important consequence of this approach is that active configurations may now occur inside multisets and reactions can take place (simultaneously) at different levels. Thus two conditions must be satisfied for a simple program to terminate: no tuple of elements satisfies the reaction condition and the multiset does not contain active elements.

A configuration is denoted:

$$[Prog, Var_1 = Multexp_1, \ldots, Var_n = Multexp_n].$$

It consists of a (possibly empty) program *Prog* and a record of named multisets Var_i. A configuration with an empty program component is called passive, otherwise it is active. The record component of the configuration can be seen as the environment of the program. Each component of the environment is a typed multiset. Simple programs extract elements from these multisets and produce new elements. A stable component $Multexp_i$ of a configuration C can be obtained as the result of $C.Var_i$.

The operational semantics is essentially extended with the following rules to capture the higher-order features:

$$\frac{X \to X'}{\{X\} \oplus M \to \{X'\} \oplus M}$$

$$\frac{M_k \to M'_k}{[P, \ldots Var_k = M_k, \ldots] \to [P, \ldots Var_k = M'_k, \ldots]}$$

The first and the second rule respectively account for the computation of active configurations inside multisets and for the transformation of multisets containing active configurations inside a configuration. Note that these rules are very similar to the chemical law and the membrane law of the Cham (section 4.1).

Let us take one example to illustrate the expressive power provided by this extension. The application of the sequential composition operator to simple programs can be defined in higher-order Gamma, and thus does not need to be included as a primitive. $(P_2 \circ P_1)(M_0)$ is defined by the following configuration:

$$[Q, E_1 = \{[P_1, M = M_0]\}, E_2 = \emptyset].E_2$$
$$\textbf{where} \quad Q = [\emptyset, M = M_1] : E_1 \to [P_2, M = M_1] : E_2$$

E_1 is a multiset containing the active configuration $[P_1, M = M_0]$ initialy. Note that Q reactions only apply to passive values of E_1 which means that M_1 must be a stable state for P_1. Then the new active configuration $[P_2, M = M_1]$ is inserted into E_2 and the computation of P_2 can start. When a stable state is obtained, it is extracted from the top-level configuration through the access operation denoted by $.E_2$.

[41] shows how other useful combining forms can be defined in higher-order Gamma (including the chemical abstract machine). It is also possible to express more sophisticated control strategies such as the *scan vector model* suitable for execution on fine-grained parallel machines. Another generalization of the chemical model to higher-order is presented in [21].

Structured Gamma. The choice of the multiset as the unique data constructor is central in the design of Gamma. However, this may lead to programs which are unnecessary complex when the programmer needs to encode specific data structures. For example, it was necessary to resort to pairs *(index,value)* to represent sequences in the *sort* program. Trees or graphs have to be encoded in a similar way. This lack of structuring is detrimental both for reasoning about programs and for implementing them. The proposal made in [31] is an attempt to solve this problem without jeopardizing the basic qualities of the language. It would not be acceptable to take the usual view of recursive type definitions because this would lead to a recursive style of programming and ruin the fundamental locality principle (the data structure would then be manipulated as a whole).

The solution proposed in [31] is based on a notion of *structured multiset* which can be seen as a set of addresses satisfying specific relations and associated with a value. As an example, the list [5; 2; 7] can be represented by a structured multiset whose set of addresses is $\{a_1, a_2, a_3\}$ and associated values (written $\overline{a_i}$) are $\overline{a_1} = 5$, $\overline{a_2} = 2$, $\overline{a_3} = 7$. Let **next** be a binary relation and **end** a unary relation; the addresses satisfy

$$\textbf{next } a_1 \, a_2, \ \textbf{next } a_2 \, a_3, \ \textbf{end } a_3$$

A new notion of type is introduced in order to characterize precisely the structure of the multiset. A type is defined in terms of a graph grammar (or rewrite rules). A structured multiset belongs to a type T if its underlying set of addresses satisfies the invariant expressed by the grammar defining T. As an example, the list type can be defined by the following context-free graph grammar:

$$List = L \ x$$
$$L \ x = \textbf{next} \ x \ y, \ L \ y$$
$$L \ x = \textbf{end} \ x$$

Any multiset which can be produced by this grammar belongs the the $List$ type. Reading the grammar rules from right to left gives the underlying rewrite system. So alternatively, any multiset which can be reduced using this rewrite rules to the singleton $List$ belongs to the $List$ type. The variables in the rules are instantiated with addresses in the multiset. $L \ x$ can be seen as a non-terminal standing for a list starting at address x and $\textbf{end} \ x$ is a one element list. A circular list can be defined as follows:

$$Circular = L \ x \ x$$
$$L \ x \ y \quad = L \ x \ z, \ L \ z \ y$$
$$L \ x \ y \quad = \textbf{next} \ x \ y$$

Note that the use of different variable names in a rule is significant: two variables are instantiated with the same address if and only if the variables have the same name. In this definition, $L \ x \ y$ is the non terminal for a list starting at position x and ending at position y.

A reaction in Structured Gamma can:

- test and modify the relations on addresses,
- test and modify the values associated with addresses.

Here are some examples of programs operating on lists:

$$Sort \ : List = \textbf{next} \ a \ b \qquad\qquad \rightarrow \textbf{next} \ a \ b, \ a \ := \ \overline{b}, b \ := \ \overline{a} \quad \Leftarrow \overline{a} < \overline{b}$$
$$Mult : List = \textbf{next} \ a \ b, \ \textbf{next} \ b \ c \rightarrow \textbf{next} \ a \ c, a \ := \ \overline{a} * \overline{b}$$
$$Iota \ : List = \textbf{end} \ a \qquad\qquad\quad \rightarrow \textbf{next} \ a \ b, \ \textbf{end} \ b, b \ := \ \overline{a} - 1 \Leftarrow \overline{a} > 1$$

Actions are now described as assignments to given addresses. A consumed address which does not occur in the result of the action disappears from the multiset: this is the case for b in the $Mult$ program. On the other hand, new addresses can be added to the multiset with their value, like b in the $Iota$ program. Actions must also state explicitly how the relations are modified. For instance, the $Sort$ does not modify the $\textbf{next}$ relation, but $Mult$ shrinks the list by removing the intermediate element b.

The natural question following the introduction of a new type system concerns the design of an associated type checking algorithm. In the context of Structured Gamma, type checking must ensure that a program maintains the underlying structure defined by a type. It amounts to the proof of an invariant property. There exists a sound checking algorithm based on the construction of an abstract reduction graph. For a reaction $C \ \rightarrow \ A$ and type T, the algorithm computes the possible contexts X such that $X + C$ reduces to $\{T\}$ (i.e. belongs to $\{T\}$). It is then sufficient to check that $X + A$ (i.e. the multiset after the reaction) reduces to $\{T\}$.

Structured Gamma allows the programmer to define his own types and have his programs checked according to the type definitions. For example, it is possible to check that the three programs above manipulate multisets of type *List*: in other words, the list property is an invariant of the programs. Applications of this approach to imperative programming and the analysis of software architectures are described in sections 4.2 and 4.3. It is important to notice that this new structuring possibility is obtained without sacrificing the fundamental qualities of the language. Gamma programs are just particular cases of Structured Gamma programs and Structured Gamma programs can be translated in a straightforward way into Gamma.

3   Gamma as a Bridge between Specifications and Implementations

In the previous section, we presented Gamma as a programming language and tried to convey the programming style entailed by the chemical reaction model through a series of examples. Gamma can be also seen as a very high-level language bridging the gap between specification languages and low-level (implementation oriented) languages.

3.1   From Specifications to Gamma Programs

We first present the techniques that can be used to prove properties of Gamma programs. Then, we suggest how they can be used to derive programs from specifications in a systematic way.

In order to prove the correctness of a program in an imperative language, a common practice consists in splitting the property into two parts: the *invariant* which holds during the whole computation, and the *variant* which is required to hold only at the end of the computation. In the case of total correctness, it is also necessary to prove that the program must terminate. The important observation concerning the variant property is that a Gamma program terminates when no more reaction can take place, which means that no tuples of elements satisfy the reaction condition. So the obtain the variant of the program by taking the negation of the reaction condition. Let us consider as an example the *sort* program introduced in the introduction. The reaction condition corresponds to the property:

$$\exists (i, x) \in M. \; \exists (j, y) \in M. \; (i > j) \; and \; (x < y)$$

and its negation

$$\forall (i, x) \in M. \; \forall (j, y) \in M. \; (i > j) \Rightarrow x \geq y$$

This variant is very informative indeed since it is the well-ordering property. The invariant of the program must ensure that the set of indexes and the multiset of values are constant. This can be checked by a simple inspection of the action

$$A((i,x),(j,y)) \;=\; \{(i,y),\ (j,x)\}$$

It is easy to see that the global invariance follows from the local invariance. In order to prove the termination of the program, we have to provide a well-founded ordering (an ordering such that there is no infinite descending sequences of elements) and to show that the application of an action decreases the multiset according to this ordering. To this aim, we can resort to a result from [26] allowing the derivation of a well-founded ordering on multisets from a well-founded ordering on elements of the multiset. Let $\succ$ be an ordering on V and $\gg$ be the ordering on $Multisets(V)$ defined in the following way:

$$M \gg M' \Leftrightarrow$$

$$\exists X, Y \in Multisets(V). \; X \neq \emptyset \;\; and$$

$$X \subseteq M \;\; and \;\; M' = (M - X) + Y \;\; and \;\; (\forall y \in Y. \; \exists x \in X. \; x \succ y)$$

The ordering $\gg$ on $Multisets(V)$ is well-founded if and only if the ordering $\succ$ on V is well-founded. This result is fortunate because the definition of $\gg$ precisely mimics the behavior of Gamma (removing elements from the multiset and inserting new elements). The significance of this result is that it allows us to reduce the proof of termination, which is essentially a global property, to a local condition. In order to prove the termination of the *sort* program, we can use the following ordering on the elements of the multiset:

$$(i,x) \sqsubseteq (i',x') \Leftrightarrow (i \geq i' \;\; and \;\; x' \geq x)$$

It is easy to see that this ordering is well-founded (the set of indexes and the multiset of values are finite), so the corresponding multiset ordering is also well-founded. We are left with the proof that for each value produced by the action, we can find a consumed value which is strictly greater. To prove this we observe that:

$$(i,y) \sqsubset (j,y) \;\; and \;\; (j,x) \sqsubset (j,y)$$

This concludes the correctness proof of the sort program.

Rather than proving a program *a posteriori*, it may be more appropriate to start from a specification and try to construct the program systematically. The derived program is then correct by construction. A method for the derivation of Gamma programs from specifications in first order logic is proposed in [8]. The basic strategy consists in splitting the specification into a conjunction of two properties which will play the rôles of the invariant and the variant of the program to be derived. The invariant is chosen as the part of the specification which is satisfied by the input multiset (or that can be established by an initialization program). If the variant involves only $\forall$ quantifiers than its negation yields the reaction condition of the program directly. The technique for deriving the action consists in validating the variant locally while maintaining the invariant. These two constraints are very often strong enough to guide the construction of the action. Let us consider as an example the *rem* program in the definition of *primes* in section 2.1. The input multiset is $\{2, \ldots, n\}$ and a possible specification of the result M is the following:

$$M \subseteq \{2, \ldots, n\} \tag{1}$$
$$\forall x \in \{2, \ldots, n\}. \ (\forall y \in \{2, \ldots, n\}. \ \neg multiple(x, y)) \Rightarrow x \in M \tag{2}$$
$$\forall x, y \in M. \ \neg multiple(x, y) \tag{3}$$

Both properties (1) and (2) are satisfied by the input multiset $\{2, \ldots, n\}$, so the invariant is defined as $I = (1) \wedge (2)$ and the variant is $V = (3)$. The negation of the variant is

$$\exists x, y \in M. \ multiple(x, y)$$

which yields to the reaction condition $multiple(x, y)$. The action must satisfy the invariant which means that no value outside $\{2, \ldots, n\}$ can be added to the multiset and no value should be removed from the multiset unless it is the multiple of another value. On the other hand, the action should establish the variant locally which means that the returned values should not contain any multiples. So the action cannot return both x and y and it cannot remove y: the only possibility is to return y; this action satisfies all the conditions and the derived program is:

$$rem = x, \ y \rightarrow y \Leftarrow multiple(x, y)$$

The interested reader can find a more complete treatment of several examples in [8]. A slightly different approach is taken in [52] which introduces a very general form of specification and the derivation of the corresponding Gamma program. It is then shown that a number of classical and apparently unrelated problems (the knapsack, the shortest paths, the maximum segment sum and the longest up-sequences problems) turn out to be instances of the generic specification. The generic derivation can then be instantiated to these applications, yielding the corresponding Gamma programs.

Let us stress the pervasive influence of the locality principle (stated in the introduction) in the correctness proofs and the derivations. Each part of the correctness proof of the *sort* program sketched above exploits this feature by reducing the global reasoning (manipulation of properties of the whole multiset) to a local reasoning (on the elements involved in a single reaction).

3.2   From Gamma Programs to Efficient Implementations

As mentioned earlier, the philosophy of Gamma is to introduce a clear separation between correctness issues and efficiency issues in program design. In particular, Gamma can be seen as a specification language which does not introduce unnecessary sequentiality. As a consequence, designing a reasonably efficient implementation of the language is not straightforward. This section outlines several optimizations allowing Gamma programs to be refined into efficient, sequential programs.

Consider a very simple form of Gamma program:

$$x_1, \ldots, x_n \to f(x_1, \ldots, x_n) \Leftarrow R(x_1, \ldots, x_n)$$

A straightforward implementation of this program can be described by the following imperative program:

While tuples remain to be processed
    do
        choose a tuple $(x_1, \ldots, x_n)$ not yet processed;
        if $R(x_1, \ldots, x_n)$ **then**
            (1) **remove** $x_1, \ldots, x_n$ **from** M
            (2) **replace** them by $f(x_1, \ldots, x_n)$
    end

This very naïve implementation puts forward most of the problems which have to be tackled in order to produce a Gamma implementation with a realistic complexity. The hardest problem concerns the construction of all tuples to be checked for reaction. A blind approach to this problem leads to an untractable complexity but a thorough analysis of the possible relationships between the elements of the multiset and the shape of the reaction condition may lead to improvements which highly optimize the execution and produce acceptable performances. In his thesis, C. Creveuil [22] studied several optimizations which are summarized here.

One important source of inefficiency comes from useless (redundant or deemed to fail) checks of the reaction condition. Three optimizations can dramatically reduce the overhead resulting from this redundancy:

1. **Decomposition of the reaction condition**: instead of considering R as a whole, one may decompose it as a conjunction of simpler conditions like:

$$R(x_1, \ldots, x_n) = R_1(x_1) \wedge R_2(x_1, x_2) \wedge \ldots \wedge R_n(x_1, x_2, \ldots, x_n)$$

   The test for condition R is done incrementally, avoiding the construction of tuples whose prefix does not satisfy one R_i.
2. **Detection of neighborhood relationships**: the analysis of the reaction condition may provide information which can be used to limit the search space. For example, it may happen from the reaction condition that only adjacent values can react, or that only values possessing a common "flag" can be confronted. These properties can be detected at compile time and, in some situations (some sorting examples, pattern detection in image processing applications), the run-time improvement is considerable.
3. **Control of the non-determinism:** The Gamma paradigm imposes no constraint on the way tuples are formed. [22] shows that limiting non-determinism by imposing an ordering in the choice of values to be checked against the reaction condition can be very fruitful.

An interesting conclusion of the work described in [22] is that well-known efficient versions of sequential algorithms (shortest path for instance) can be "rediscovered" and justified as the result of several optimizations of a naïve implementation of Gamma. It is very often the case that the most drastic optimizations rely on structural properties of the values belonging to the multiset (neighborhood relationship, ordering in the choice of values ...).

Such properties are difficult to find automatically and the optimizations described above can be seen as further refinement steps rather than compilation techniques. Several proposals have been made to enrich Gamma with features which could be exploited by a compiler to reduce the overhead associated with the "magic stirring" process. For example, the language of *schedules* [17,18] provides extra information about control in Gamma programs, and *local linear logic* [48] as well as Structured Gamma [31] structure the multiset.

4   Gamma as of Source of Inspiration

The chemical reaction model has served as the basis of a number of works in various, often unexpected, research directions. In particular, ideas borrowed from Gamma have been applied to process calculi, imperative programming and software architectures. We describe these three developments in turn and conclude with a sketch of a few other proposals. Most of these works are quite significant and open new research directions but it is important to note that none of them jeopardizes the fundamental characteristics of the model which is the expression of computation as "the global result of the successive applications of local, independent, atomic reactions".

4.1   The Chemical Abstract Machine

The chemical abstract machine (or *Cham*) was proposed by Berry and Boudol [12] to describe the operational semantics of process calculi. The most important additions to Gamma are the notions of *membrane* and *airlock mechanism*. Membranes are used to encapsulate solutions and to force reactions to occur locally. In terms of multisets, a membrane can be used to introduce multiset of molecules inside a multiset that is to say *"to transform a solution into a single molecule"* [13]. The airlock mechanism is used to describe communications between an encapsulated solution and its environment. The reversible airlock operator $\triangleleft$ extracts a element m of a solution $\{m, m_1, \ldots, m_n\}$:

$$\{m, m_1, \ldots, m_n\} \rightleftharpoons \{m \triangleleft \{m_1, \ldots, m_n\}\}$$

The new molecule can react as a whole while the sub-solution $\{m_1, \ldots, m_n\}$ is allowed to continue its internal reactions. So the main rôle of the airlock is to allow one molecule to be visible from outside the membrane and thus to take part in a reaction in the embedding solution. The need for membranes and airlocks emerged from the description of CCS [50] in Cham and especially the treatment of the restriction operation (which restricts the communication capabilities of a process to labels different from a particular value a). The computation rules of the Cham are classified into general laws and two classes of rules:

- The general laws include the chemical law and the membrane law:

$$\frac{S \to S'}{S + S" \to S' + S"}$$

$$\frac{S \to S'}{\{C[S]\} \to \{C[S']\}}$$

The former shows that reactions can be performed freely within any solution, which captures the locality principle. The latter allows reactions to take place within a membrane ($C[S]$ denotes any context of a solution S).
- The first class of rules corresponds to the proper reaction rules similar to the rules presented so far in the paper. The definition of a specific Cham requires the specification of a syntax for molecules and the associated reaction rules. As an example, molecules can be CCS processes and the rule corresponding to communication in CCS would be:

$$\alpha.P \; , \; \overline{a}.Q \mapsto P \; , \; Q$$

- The second kind of rules are called structural and they are reversible. They can be decomposed into two inverse relations $\rightharpoonup$ and $\rightharpoondown$ called respectively heating and cooling rules. The first ones break complex molecules into smaller ones, preparing them for future reactions, and the second ones rebuild heavy molecules from light ones. Continuing the CCS example, we have the structural rule:

$$(P \mid Q) \rightleftharpoons P \; , \; Q$$

where $\mid$ is the CCS parallel composition operator.

The Cham was used in [12] to define the semantics of various process calculi (TCCS, Milner's π-calculus of mobile processes) and a concurrent lambda calculus. A Cham for the call-by-need reduction strategy of λ-calculus is defined in [13]. The Cham has inspired a number of other contributions. Let us mention some of them:

- [1] uses a *linear Cham* to describe the operational semantics of proof expressions for the classical linear logic.
- [51] defines an operational semantics of the π-calculus in a Cham style.
- [39] describes a graph reduction in terms of a Cham.
- [43] applies the Cham in the context of the Facile implementation.

The Cham approach illustrates the significance of multisets and their connection with concurrency. The fact that multisets are inherently unordered makes them suitable as a basis for modeling concurrency which is an essentially associative and commutative notion. As stated in [12]: *"In the SOS style of semantics, labeled transitions are necessary to overcome the rigidity of syntax when performing communications between two syntactically distant agents. ... On the contrary, in the Cham, we just make the syntactic distance vanish by putting molecules into contact when they want to communicate, and their communication is direct."* As a consequence, this makes it possible to bring the semantics of concurrent systems closer to the execution process of sequential languages, or the evaluation mechanism of functional languages [13].

4.2   Shape Types

Type systems currently available for imperative languages are too weak to detect a significant class of programming errors. The main reason is that they fail to capture properties about the sharing which is inherent in many data structures used in efficient imperative programs. As an illustration, it is impossible to express the property that a list is doubly-linked or circular in existing type systems. The work around Structured Gamma (section 2.5) showed that many data structures could be described as graph grammars and manipulated by reactions. Furthermore, a static algorithm can be used to check that a reaction preserve the structure specified by the grammar. These ideas and techniques have been adapted in order to extend the type system of C and make pointer manipulation safer [30].

Shape-C is an extension of C which integrates the notion of types as graph grammars (called here *shapes*) and reactions. The notion of graph grammars is powerful enough to describe most complex data structures (see [30] for a description of skip lists, red-black trees, left-child-right-sibling trees in terms of graph grammars).

The design of Shape-C was guided by the following criteria:

– the extensions should be blended with other C features and be natural enough for C programmers,
– the result of the translation of Shape-C into pure C should be efficient,
– the checking algorithm of section 2.5 should be applicable to ensure shape invariance.

We present Shape-C through an example: the Josephus program. This program, borrowed from [56], first builds a circular list of n integers; then it proceeds through the list, counting through $m - 1$ items and deleting the next one, until only one is left (which points to itself). Figure 1 displays the program in Shape-C. The Josephus program first declares a shape `cir` denoting a circular list of integers with a pointer `pt`. Besides cosmetic differences, the definition of shapes is similar to the context free grammars presented in Section 2.5. The variables are interpreted as addresses. They possess a value whose type must be declared (here `int`). Values can be tested or updated but cannot refer to addresses. They do not have any impact on shape types.

Intuitively, unary relations (here `pt`) correspond to roots whereas binary relations (here `next`) represent pointer fields. Shapes can be translated into C structures with a value field and as many fields (of pointer type) as the shape has binary relations.

```
/* Integer circular list                                              */
shape int cir { pt x, L x x;
                L x y = L x z, L z y;
                L x y = next x y;      };
main()
{int i, n, m;
/*initialization to a one element circular list                       */
  cir s = [| => pt x; next x x; $x=1; |];
  scanf("%d%d", &n, &m);
/* Building the circular list 1->2->...->n->1                         */
  for (i = n; i > 1; i--)
      s:[| pt x; next x y; => pt x; next x z; next z y; $z=i; |];
/* Printing and deleting the m th element until only one is left      */
  while (s:[| pt x; next x y; x != y; => |])
  {
    for (i = 1; i < m-1; ++i)
        s:[| pt x; next x y; => pt y; next x y; |];
      s:[| pt x; next x y; next y z; => pt z; next x z; printf("%d ‘‘,$y); |];
  }
/* Printing the last element                                          */
  s:[| pt x => pt x;   printf("%d\n",$x); |];
}
```

Fig. 1. Josephus Program

Shape-C uses only a subset of graph grammars that corresponds to the rooted pointer structures manipulated in imperative languages. This subset is defined by the following properties:

(S1) *Relations are either unary or binary.*
(S2) *Each unary relation is satisfied by exactly one address in the shape.*
(S3) *Binary relations are functions.*
(S4) *The whole shape can be traversed starting from its roots.*

These conditions allow shapes to be implemented by simple C structures (with a value and pointer fields). They can be enforced by analyzing the definition of grammars.

The reaction, written [| C => A |], is the main operation on shapes. Two specialized versions of reactions are also provided: initializers, with only an action, written [| => A |] and tests, with only a condition, written [| C => |].

The Josephus program declares a local variable s of shape cir and initializes it to a one element circular list.

```
    cir s =  [| => pt x; next x x; $x = 1; |];
```

The value of address x is written $x and is initialized to 1. In general, actions may include arbitrary C-expressions involving values. The for-loop builds a n element circular list using the reaction

```
    s:[| pt x; next x y; => pt x; next x z; next z y; $z=i; |];
```

The condition selects the address x pointed to by pt and its successor. The action inserts a new address z and initializes it to i. The translation in pure C is local and applied to each shape operation of the program. Shape-C enforces a few simple restrictions on reactions so that the translation is both direct and efficient.

Shape checking amounts to verify that initializations and reactions preserve the shape of objects. The checking algorithm is directly based on the algorithm outlined in section 2.5. Note that values and expressions on values are not relevant for shape checking purposes. Using this algorithm, it is easy to ensure that the list s is cyclic throughout the Josephus program.

Due to their precise characterization of data structures, shape types are a very useful facility for the construction of safe programs. Most efficient versions of algorithms are based on complex data structures which must be maintained throughout the execution of the program [16,56]. The manipulation of these structures is an error-prone activity. Shape types permits to describe invariants of their representation in a natural way and have them automatically verified.

4.3 Software Architectures

Another related area of application which has attracted a great amount of interest during the last decade is the formal definition of software architectures. As stated in [2], *"Software systems become more complex and the overall system structure - or software architecture - becomes a central design problem. An important step towards an engineering discipline of software is a formal basis for describing and analyzing these designs"*. Typical examples of software architectures are the "client-server organization", "layered systems", "blackboard architecture". Despite the popularity of this topic, little attention has focused on methods for comparing software architectures or proving that they satisfy certain properties. One major reason which makes these tasks difficult is the lack of common and formally based language for describing software architectures. These descriptions are typically expressed informally with box and lines drawings indicating the global organization of computational entities and the interactions between them [2]. The chemical reaction model has been used for specifying software architectures [38] and architecture styles [42].

Software architecture specification. The application considered in [38] is a multi-phase compiler and two architectures are defined using the "chemical abstract machine" [12,13]. The different phases of the compiler are called *lexer, parser, semantor, optimiser and generator*. An initial phase called *text* generates the source text. The types of the data elements circulating in the architecture are *char, tok, phr, cophr, obj*. The elements of the multiset have one of the following forms:

$$i(t_1) \; \diamond \; o(t_2) \; \diamond \; phase$$

$$o(t_1) \; \diamond \; phase \; \diamond \; i(t_2)$$

$$phase \; \diamond \; i(t_1) \; \diamond \; o(t_2)$$

where $\diamond$ is a free constructor, t_1 and t_2 represent data types and *phase* is one of the phases mentioned above. An element starting with $i(t_1)$ (resp. $o(t_1)$) corresponds to a phase which is consuming inputs (resp. producing outputs). An element starting with *phase* is not ready to interact. So $i(t_1)$ and $o(t_1)$ can be seen as ports defining the communications which can take place in a given state.

The following is a typical reaction in the definition of an architecture:

$$i(d_1) \diamond o(d_2) \diamond m_1, o(d_1) \diamond m_2 \diamond i(d_3)$$
$$\rightarrow$$
$$o(d_2) \diamond m_1 \diamond i(d_1), m_2 \diamond i(d_3) \diamond o(d_1)$$

This rule describes pairwise communication between processing elements: m_1 consumes input d_1 produced as output by another processing element m_2. For example, the reaction:

$$i(tok) \diamond o(phr) \diamond parser, o(tok) \diamond lexer \diamond i(char)$$
$$\rightarrow$$
$$o(phr) \diamond parser \diamond i(tok), lexer \diamond i(char) \diamond o(tok)$$

represents the consumption by the parser of tokens produced by the lexer. At the end of the reaction, the parser is ready to produce its output and the lexer is inert because it has completed its job. In fact another reaction may be applied later to make it active again to process another piece of text.

One major benefit of the approach is that it makes it possible to define several architectures for a given application and compare them in a formal way. As an example, [38] defines a correspondence between multisets generated by two versions of the multi-phase compiler and establishes a form of bisimulation between the two architectures. They also prove normalization properties of the programs.

Software architecture styles. The approach described in [42] focuses on the interconnection between individual components of the software architecture. The main goal is to describe architecture styles (i.e. classes of architectures) and to check that the dynamic evolution of an architecture preserves the constraints imposed by the style. Techniques developed for Structured Gamma (graph grammars and the associated checking algorithm) can be applied to this problem.

Structured Gamma allows connections to become "first class" objects and to prove invariance properties on the structure of the network. For example, a client-server architecture style can be defined as the graph grammar

$$
\begin{aligned}
ClientServer &= CS\ m \\
CS\ m\ \ \ \ &= \mathbf{cr}\ c\ m, \mathbf{ca}\ m\ c,\ \mathbf{C}\ c,\ CS\ m \\
CS\ m\ \ \ \ &= \mathbf{sr}\ m\ s, \mathbf{sa}\ s\ m,\ \mathbf{S}\ s,\ CS\ m \\
CS\ m\ \ \ \ &= \mathbf{M}\ m,\ \mathbf{X}\ x
\end{aligned}
$$

The unary relations $\mathbf{C}$, $\mathbf{S}$, $\mathbf{M}$ and $\mathbf{X}$ correspond respectively to client, server, manager and external entities. The external entity stands for the external world; it records requests for new clients wanting to be registered in the system. The binary relations **cr** and **ca** correspond to client request links and client answer links respectively (**sr** and **sa** are the dual links for servers). For example, the architecture

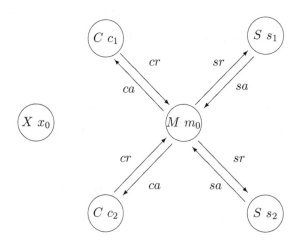

involves two clients c_1 and c_2, two servers s_1 and s_2, a manager m_0 and the external entity x_0. It belongs to the client-server class (grammar) *ClientServer*.

It is often the case that the architecture of an application should be able to evolve dynamically. For instance, a client-server organization must allow for the introduction of new clients or their departure, a pipeline may grow or shrink, facilities for dealing with mobile computing may be required. In this framework, the evolution of the architecture is defined by a *coordinator*. The task of the coordinator is expressed by reaction rules. As an illustration, the following co-ordinator applies to a client-server architecture:

$$\mathbf{X}\ x,\ \mathbf{M}\ m \to \mathbf{X}\ x',\ \mathbf{M}\ m,\ \mathbf{cr}\ c\ m, \mathbf{ca}\ m\ c,\ \mathbf{C}\ c$$

$$\mathbf{cr}\ c\ m, \mathbf{ca}\ m\ c,\ \mathbf{C}\ c\ \to \varnothing$$

The two rules[1] describe respectively the introduction of a new client in the architecture and its departure. The main benefit of this approach is that the algorithm of section 2.5 can be used to ensure that a coordinator does not break the constraints of the architecture style. This makes it possible to reconcile a dynamic view of the architecture with the possibility of static checking. For example, had we forgotten, say **cr** $c\ m$ in the right-hand side of the first rule, then the coordinator would have been able to transform a client-server architecture into an architecture which would not belong any longer to the class defined by *ClientServer*.

The interested reader can find in [37] the treatment of an industrial case study proposed by the Signaal company [25] using a multiple views extension

---

[1] In fact, these rules are completed with side conditions on the states of the entities otherwise, the coordinator could add or remove entities without any consideration of the current state of the system.

of the formalism presented above. The goal of the work was the specification of a railway network system. The static verification of coordination rules with respect to grammars was deemed the most attractive feature of the formalism [37].

These applications provide evidence that (Structured) Gamma is an intuitive and formally based formalism to describe and analyze software architectures.

4.4   Other Works

Influences of the chemical reaction model can be found in other domains such as visual languages [36], protocols for shared virtual memories [49] or logic programming [19,61]. We just sketch here works around coherence protocols and logic programming where Gamma played a key rôle.

Formalization of coherence protocols. Coherence protocols for shared virtual memories have been formalized as Gamma programs in [49]. The multiset, whose elements are the protocol entities or events, represents a global view of the system. The protocol itself is described as a Gamma program (i.e. a collection of reaction rules). Besides, a fragment of first-order logic is used to specify properties that the protocol is expected to satisfy. This formalization made it possible to design an algorithm checking that properties are indeed invariants of the protocol. [49] presents a Gamma formalization of the Li and Hudak protocol [44] as well as the automatic verification of a collection of invariants.

This approach has been applied to software architectures in [53]. The dynamic evolution of the architecture is described as a Gamma program (as in section 4.3). Instead of checking membership to a given style (as in section 4.3), the algorithm of [49] is reused to check that the evolutions of the architecture respect some logical properties specified in a separate language.

Gamma and logic programming. Several proposals have been made for integrating Gamma and logic programming languages. A first approach, followed in [19], uses multisets of terms and describes conditions and actions as predicates. This model is implemented as an extension of Gödel, a strongly typed logic programming language with a rich module system. It involves a definition of multiset unification and a careful integration with the operational semantics of Gamma, in which the "choices" made by reaction conditions are not backtrackable. This extension, called Gammalög, includes the sequential and parallel composition operators introduced in [34]. Another approach is followed in [61] where the objects in the multisets are goal formulas and the *(condition,action)* pairs are goal-directed deduction rules. This results in λLO, an extension of LO [3] which can itself be seen as an elaboration on the basic chemical reaction model. λLO can be seen as a higher-order extension of LO in the same way as λProlog is a higher-order extension of Prolog. Implication in goals provides the ability to construct (or augment) the program at run-time and the use of multisets leads to a uniform treatment of programs and data.

5   Conclusion

A number of languages and formalisms bearing similarities with the chemical reaction paradigm have been proposed in the literature. Let us briefly review the most significant ones:

- A programming notation called *associons* is introduced in [54]. Essentially an associon is a tuple of names defining a relation between entities. The state can be changed by the creation of new associons representing new relations derived from the existing ones. In contrast with Gamma, the model is deterministic and does not satisfy the locality properties (dues to the presence of ∀ properties).

- A Unity program [14] is basically a set of multiple-assignment statements. Program execution consists in selecting non deterministically (but following a fairness condition) some assignment statement, executing it and repeating forever. [14] defines a temporal logic for the language and the associated proof system is used for the systematic development of parallel programs. Some Unity programs look very much like Gamma programs (an example is the exchange sort program presented in the introduction). The main departures from Gamma is the use of the array as the basic data structure and the absence of locality property. On the other hand, Unity allows the programmer to distinguish between synchronous and asynchronous computations which makes it more suitable as an effective programming languages for parallel machines. In the same vein as Unity, the *action systems* presented in [5] are *do-od* programs consisting of a collection of guarded atomic actions, which are executed nondeterministically so long as some guard remains true.

- Linda [32,15] contains a few simple commands operating on a tuple space. A producer can add a value to the tuple space; a consumer can read (destructively or not) a value from the tuple space. Linda is a very elegant communication model which can easily be incorporated into existing programming languages.

- LO (for Linear Objects) was originally proposed as an integration of logic programming and object-oriented programming [3]. It can be seen as an extension of Prolog with formulae having multiple heads. From an object-oriented point of view, such formulae are used to implement methods. A method can be selected if its head matches the goal corresponding to the object in its current state. The head of a formula can also be seen as the set of resources consumed by the application of the method (and the tail is the set of resources produced by the method). In [4], LO is used as a foundation for *interaction abstract machines*, extending the chemical reaction metaphor with a notion of broadcast communication: sub-solutions (or "agents") can be created dynamically and reactions can have the extra effect of broadcasting a value to all the agents.

Taking a dual perspective, it is interesting to note that the physical modeling community has borrowed concepts from computer science leading to formalisms which bear similarities with higher-order Gamma. An example of this trend of activity is the "Turing gas" [29] where molecules float at random in a solution, reacting if they come into contact with each other. A language akin to lambda-calculus is used to express the symbolic computation involved in the reactions.

As a conclusion, we hope that this paper has shown that the chemical reaction model is a particularly simple and fruitful paradigm. No doubt that new surprising developments are yet to come.

Acknowledgments

We would like to express our thanks to all the people who have contributed to the development of Gamma during these years.

References

1. S. Abramsky, *Computational interpretations of linear logic*, Theoretical Computer Science, Vol. 111, pp. 3-57, 1993.
2. R. Allen and D. Garlan, *Formalising architectural connection*, Proceedings of the IEEE 16th International Conference on Software Engineering, pp. 71-80, 1994.
3. J.-M. Andreoli and R. Pareschi, *Linear Objects: logical processes with built-in inheritence*, New Generation Computing, Vol. 9, pp. 445-473, 1991.
4. J.-M. Andreoli, P. Ciancarini and R. Pareschi, *Interaction abstract machines*, in *Proc. of the workshop Research Directions in Concurrent Object Oriented Programming*, 1992.
5. R. Back, *Refinement calculus, part II: parallel and reactive programs*, in *Proc. of the workshop on Stepwise Refinement of Distributed Systems: Models, Formalisms, Correctness*, 1989, Springer Verlag, LNCS 430.
6. J.-P. Banâtre, A. Coutant and D. Le Métayer, *A parallel machine for multiset transformation and its programming style*, Future Generation Computer Systems, pp. 133-144, 1988.
7. J.-P. Banâtre, A. Coutant and D. Le Métayer, *Parallel machines for multiset transformation and their programming style*, Informationstechnik, Oldenburg Verlag, Vol. 2/88, pp. 99-109, 1988.
8. J.-P. Banâtre and D. Le Métayer, *The Gamma model and its discipline of programming*, Science of Computer Programming, Vol. 15, pp. 55-77, 1990.
9. J.-P. Banâtre and D. Le Métayer, *Programming by multiset transformation*, Communications of the ACM, Vol. 36-1, pp. 98-111, January 1993.
10. J.-P. Banâtre and D. Le Métayer, *Gamma and the chemical reaction model: ten years after*, in Coordination Programming: Mechanisms, Models and Semantics, Imperial College Press, 1996.
11. P. Bertin, D. Roncin and J. Vuillemin, *Programmable active memories: a performance assessment*, in *Proc. of the workshop on Parallel architectures and their efficient use*, 1992, Springer Verlag, LNCS , pp. 119-130.
12. G. Berry and G. Boudol, *The chemical abstract machine*, Theoretical Computer Science, Vol. 96, pp. 217-248, 1992.
13. G. Boudol, *Some chemical abstract machines*, in *Proc. of the workshop on A decade of concurrency*, 1994, Springer Verlag, LNCS 803, pp. 92-123.
14. Chandy M. and Misra J., *Parallel program design: a foundation*, Addison-Wesley, 1988.

15. N. Carriero and D. Gelernter, *Linda in context*, Communications of the ACM, Vol. 32-4, pp. 444-458, April 1989.

16. T. H. Cormen, C. E. Leiserson and R. L. Rivest, *Introduction to algorithms*, MIT Press, 1990.

17. M. Chaudron and E. de Jong, *Schedules for multiset transformer programs*, in Coordination Programming: Mechanisms, Models and Semantics, Imperial College Press, 1996.

18. M. Chaudron and E. de Jong, *Towards a compositional method for coordinating Gamma programs*, in *Proc. Coordination'96 Conference*, Lecture Notes in Computer Science, Vol. 1061, pp. 107-123, 1996.

19. P. Ciancarini, D. Fogli and M. Gaspari, *A logic language based on multiset rewriting*, in Coordination Programming: Mechanisms, Models and Semantics, Imperial College Press, 1996.

20. P. Ciancarini, R. Gorrieri and G. Zavattaro, *An alternate semantics for the calculus of Gamma programs*, in Coordination Programming: Mechanisms, Models and Semantics, Imperial College Press, 1996.

21. D. Cohen and J. Muylaert-Filho, *Introducing a calculus for higher-order multiset programming*, in *Proc. Coordination'96 Conference*, Lecture Notes in Computer Science, Vol. 1061, pp. 124-141, 1996.

22. C. Creveuil, *Techniques d'analyse et de mise en œuvre des programmes Gamma*, Thesis, University of Rennes, 1991.

23. C. Creveuil, *Implementation of Gamma on the Connection Machine*, in *Proc. of the workshop on Research Directions in High-Level Parallel Programming Languages*, Mont-Saint Michel, 1991, Springer Verlag, LNCS 574, pp. 219-230.

24. C. Creveuil and G. Moguérou, *Développement systématique d'un algorithme de segmentation d'images à l'aide de Gamma*, Techniques et Sciences Informatiques, Vol. 10, No 2, pp. 125-137, 1991.

25. E. de Jong, *An industrial case study: a railway control system*, Proc. Second int. Conf. on Coordination Models, Languages and Applications, Springer Verlag, LNCS 1282, 1997.

26. Dershowitz N. and Manna Z., *Proving termination with multiset ordering*, Communications of the ACM, Vol. 22-8, pp. 465-476, August 1979.

27. Dijkstra E. W., *The humble programmer*, Communications of the ACM, Vol. 15-10, pp. 859-866, October 1972.

28. L. Errington, C. Hankin and T. Jensen, *A congruence for Gamma programs*, in *Proc. of WSA conference*, 1993.

29. W. Fontana, *Algorithmic chemistry*, *Proc. of the workshop on Artificial Life*, Santa Fe (New Mexico), Addison-Wesley, 1991, pp. 159-209.

30. P. Fradet and D. Le Métayer, *Shape types*, in Proc. of Principles of Programming Languages, POPL'97, ACM Press, pp. 27-39, 1997.

31. P. Fradet and D. Le Métayer, *Structured Gamma*, Science of Computer Programming, 31, pp. 263-289, 1998.

32. Gelernter D., *Generative communication in Linda*, ACM Transactions on Programming Languages and Systems, Vol. 7,1, pp. 80-112, January 1985.

33. K. Gladitz and H. Kuchen, *Parallel implementation of the Gamma-operation on bags*, *Proc. of the PASCO conference*, Linz, Austria, 1994.

34. C. Hankin, D. Le Métayer and D. Sands, *A calculus of Gamma programs*, in *Proc. of the 5th workshop on Languages and Compilers for Parallel Computing*, Yale, 1992, Springer Verlag, LNCS 757.

35. C. Hankin, D. Le Métayer and D. Sands, *A parallel programming style and its algebra of programs*, in *Proc. of the PARLE conference*, LNCS 694, pp. 367-378, 1993.
36. B. Hoffmann. *Shapely Hierarchical Graph Transformation*, Symposium on Visual Languages and Formal Methods (VL FM'01) in the IEEE Symposia on Human-Centric Computing Languages and Environments (HCC'01), IEEE Press, 2001.
37. A. A. Holzbacher, M. Périn and M. Südholt, *Modeling railway control systems using graph grammars: a case study*, Proc. Second int. Conf. on Coordination Models, Languages and Applications, Springer Verlag, LNCS 1282, 1997.
38. P. Inverardi and A. Wolf, *Formal specification and analysis of software architectures using the chemical abstract machine model*, IEEE Transactions on Software Engineering, Vol. 21, No. 4, pp. 373-386, April 1995.
39. A. Jeffrey, *A chemical abstract machine for graph reduction*, TR 3/92, University of Sussex, 1992.
40. H. Kuchen and K. Gladitz, *Parallel implementation of bags*, in *Proc. ACM Conf. on Functional Programming and Computer Architecture*, ACM, pp. 299-307, 1993.
41. D. Le Métayer, *Higher-order multiset programming*, in *Proc. of the DIMACS workshop on specifications of parallel algorithms*, American Mathematical Society, Dimacs series in Discrete Mathematics, Vol. 18, 1994.
42. D. Le Métayer, *Describing software architecture styles using graph grammars*, IEEE Transactions on Software Engineering (TSE), Vol. 24(7), pp. 521-533, 1998.
43. L. Leth and B. Thomsen, *Some Facile chemistry*, TR 92/14, ECRC, 1992.
44. K. Li, P. Hudak, *Memory Coherence in Shared Virtual Memory Systems*, in Proc. of ACM Symposium on Principles of Distributed Computing, pp. 229-239, 1986.
45. Lin Peng Huan, Kam Wing Ng and Yong Qiang Sun, *Implementing higher-order Gamma on MasPar: a case study*, Journal of Systems Engineering and Electronics, Vol. 16(4), 1995.
46. Lin Peng Huan, Kam Wing Ng and Yong Qiang Sun, *Implementing Gamma on MasPar MP-1*, Journal of Computer Science and Technology.
47. H. McEvoy, *Gamma, chromatic typing and vegetation*, in Coordination Programming: Mechanisms, Models and Semantics, Imperial College Press, 1996.
48. H. McEvoy and P.H. Hartel, *Local linear logic for locality consciousness in multiset transformation*, Proc. Programming Languages: Implementations, Logics and Programs, PLILP'95, LNCS 982, pp. 357-379, 1995.
49. D. Mentré, D. Le Métayer, T. Priol, *Formalization and Verification of Coherence Protocols with the Gamma Framework*, in Proc. of the 5th Int. Symp. on Software Engineering for Parallel and Distributed Systems (PDSE-2000), ACM, 2000.
50. R. Milner, *Communication and concurrency*, International Series in Computer Science, Prentice Hall, Englewood Cliffs, NJ, 1989.
51. R. Milner, *Functions as processes*, Mathematical Structures in Computer Science, Vol. 2, pp. 119-141, 1992.
52. L. Mussat, *Parallel programming with bags*, in *Proc. of the workshop on Research Directions in High-Level Parallel Programming Languages*, Mont-Saint Michel, 1991, Springer Verlag, LNCS 574, pp. 203-218.
53. M. Périn, *Spécifications graphiques multi-vues : formalisation et vérification de cohérence*, PhD thesis, Université de Rennes 1, 2000.
54. M. Rem, *Associons: a program notation with tuples instead of variables*, ACM Trans. on Programming Languages and Systems, Vol. 3,3, pp. 251-261, 1981.
55. M. Reynolds, *Temporal semantics for Gamma*, in Coordination Programming: Mechanisms, Models and Semantics, Imperial College Press, 1996.

56. R. Sedgewick, *Algorithms in C*, Addison-Wesley publishing company, 1990.
57. W.-P. de Roever, *Why formal methods are a must for real-time system specification*, in *Proc. Euromicro'92*, Panel discussion, June 1992, Athens.
58. H. Ruiz Barradas, *Une approche à la dérivation formelle de systèmes en Gamma*, Thesis, University of Rennes 1, July 1993.
59. D. Sands, *A compositional semantics of combining forms for Gamma programs*, in *Proc. of the Formal Methods in Programming and their Applications conference*, Novosibirsk, 1993, Springer Verlag, LNCS 735, pp. 43-56.
60. D. Sands, *Composed reduction systems*, in Coordination Programming: Mechanisms, Models and Semantics, Imperial College Press, 1996.
61. L. Van Aertryck and O. Ridoux, *Gammalog as goal-directed proofs*, internal report.
62. M. Vieillot, *Synthèse de programmes Gamma en logique reconfigurable*, Technique et Science Informatiques, Vol. 14, pp. 567-584, 1995.

Visual Multiset Rewriting: Applications to Diagram Parsing and Reasoning

Paolo Bottoni[1], Bernd Meyer[2], and Francesco Parisi Presicce[1]

[1] Dipartimento di Scienze dell' Informazione
Università La Sapienza di Roma
[2] School of Computer Science & Software Engineering
Monash University, Australia
[bottoni | parisi]@dsi.uniroma1.it
bernd.meyer@acm.org

Abstract. Diagrammatic notations, such as Venn diagrams, Petri-Nets and finite state automata, are in common use in mathematics and computer science. While the semantic domain of such systems is usually well formalized, the visual notation itself seldom is, so that they cannot be used as valid devices of formal reasoning. A complete formalization of such notations requires the construction of diagram systems with rigorously defined syntax and semantics. We discuss how diagram specification can be interpreted as multiset rewriting and, based on this, how it can be formalized in linear logic. We discuss the power of our approach through an illustration of its possible extension with reflective capabilities to manage negative conditions, and through the identification of a class of diagrammatic transformations which can be directly expressed in our framework.

1 Introduction

A common way of modelling the world is as a collection of entities and relations. Such a paradigm, though adequate to build static definitions of a domain, presents difficulties in modelling dynamics. In particular, the *frame problem* arises when modelling syntactic transformation of diagrams.

Syntactic diagram transformations constitute the fundamental tool for performing resoning with diagrammatic representations and visual simulation of processes or systems. The main problem is that, graphical sentences being two-dimensional, they cannot be constrained to support only a privileged relation, such as the prefix relation of one dimensional representations, e.g. lists, strings, etc. Nor can they be restricted to support only partial orders, such as the ancestor-descendent relation in a tree. Therefore, the direct extension to the two dimensional world of images of techniques developed for one dimensional algebraic structures falls short in describing dynamic aspects of visual sentences.

Graphs have been proposed as a data type adequately supporting the representation of entities (nodes) and relations (edges), and able to express the abstract syntax and semantics of diagrams [BMST99]. Several approaches to

C.S. Calude et al. (Eds.): Multiset Processing, LNCS 2235, pp. 45–67, 2001.

graph transformation have been developed. However, these approaches are not free from problems, as will be discussed later, and are not adequate to model the insurgence of new relations as a consequence of the modification of a part of a diagram.

What is needed is a formalism which does not commit to any specific interpretation or reading convention of images, i.e. that does not fix the set of spatial relations that are considered once and for all, and that supports a uniform way for the management of different spatial relations between significant structures in an image. We argue that multisets are an adequate formalism in which to set the management of visual entities and that linear logic provides an adequate calculus for diagram transformation.

In particular, we demonstrate that a small fragment of linear logic is adequate for expressing the parsing of diagrammatic languages, through which we can build the meaning of a visual sentence. We argue that this same fragment characterises a class of diagram transformations encompassing several types of executable diagrammatic specifications employed in Software Engineering, such as Petri Nets or Transition Systems, as well as typical transformations employed in diagrammatic reasoning.

The identification of such a fragment provides an advantage in that it allows the expression in a uniform way of several calculi developed ad hoc for the different types of visual language.

2 Representing Diagrams as Multisets

The first step to formalizing diagram transformation systems is to find a proper representation for static diagrams. Diagrams can be considered as collections of graphical tokens such as *lines*, *circles*, etc. A fundamental distinction between diagrammatic languages and string languages is that no single linear order is inherently defined on diagram elements. Hence, diagrammatic expressions must be considered as non-linear structures. Typical algebraic non-linear structures used to support diagram management are graphs and multisets, rather than sequences of tokens. Diagram transformation can then be interpreted as rewriting of such structures. In this paper we describe a multiset-based approach, i.e. we view the tokens as resources that can be produced, consumed, queried and changed. Linear logic presents itself naturally as a framework for studying diagram transformations based upon such a model.

The objects in a multiset that represents a diagram must contain a representation of the diagram geometry. Moreover, a proper typing mechanism is also desirable. For example, on the most elementary level, the square of Figure 1 could be represented as the multiset $\{line(140, 5), line(140, 85), line(220, 85), line(220, 5)\}$. However, we might want to model objects at a higher level of abstraction, so that the entire square in the figure is represented as a single object $\{square(140, 5, 80)\}$. The construction of this new type from the primitives could be defined by a simple rule and parsing could be used to construct the higher-level presentation.

For the purpose of this article, we adopt a view in which meaningful types of graphical entities are given by specific graphical data types. We regard their construction as encapsulated in abstract constructors so that the process by which they can be recognized is hidden. From a technical perspective, the recognition of these data types from a diagram can be done in various forms, for example by transformation of prototypes [BCM99], by parsing [MMW98], with algebraic specifications [BMST99] or using description logic [Haa98].

A strong requirement to be imposed on the graphical data type system is that type assignment must be unambiguous. For example, in a system where both a *rectangle* and a *square* data type exist, rectangles with equal sides must consistently be considered as squares. Inheritance can be used to model the relation between these types [Haa98].

An important requirement for the transformation mechanism is implied: A transformation which modifies the geometry of an element so that it is incompatible with its original data type must also change the type of this element accordingly. In the rest of this paper we assume that these conditions are met by the type systems and the transformation rules adopted in the examples. For discussions of the limits of these assumptions see [BCM99,BMST99,MMW98].

Fig. 1. A simple diagram.

The second question to be addressed is how to represent the spatial relations between the diagrammatic objects. Several alternatives for this have been considered in the literature:

Explicit representation of relations: The relations existing between tokens are reified and inserted in the multiset as tokens in themselves. In this case, the diagram of Figure 1 could be represented as $\{letter_1(\text{``}a\text{''}, 35, 45, 10, 10), square_1(0, 0, 80), circle_1(40, 40, 20), inside(letter_1, circle_1), inside(circle_1, square_1)\}$.

Use of constraints: The multiset M is combined with a set of constraints C that describes the relevant spatial relations expressed as (usually arithmetic) constraints on attributes of the objects. The pair (M, C) representing the diagram of Figure 1 could be described by $\{letter(\text{``}a\text{''}, \boldsymbol{x}_1), circle(\boldsymbol{x}_2), square(\boldsymbol{x}_3)\}$ together with the constraints $\{inside(\boldsymbol{x}_1, \boldsymbol{x}_2) \wedge inside(\boldsymbol{x}_2, \boldsymbol{x}_3)\}$. Since attributes are used to represent the geometries, spatial relations can be queried by inspection of this geometry, e.g. by testing constraint entailment or consistency. These functions are realized by a constraint solver for the underlying domain.

Use of concrete geometries: For this representation, tokens with geometric attributes are used as in the case of the constraint representation. However, attributes are always given concrete values. In this case, the representation of Figure 1 would be $\{letter(\text{``}a\text{''}, 35, 45, 10, 10), square(0, 0, 80), circle(40, 40, 20)\}$.

The choice of representation has some important implications related to the frame problem. Consider the transformation rule depicted in Figure 2 which expresses the removal of a *circle* object. It corresponds to a multiset transformation rule which could informally be written as:

$$circle_1(\_,\_,\_), square_1(X,Y,Z), inside(circle_1, square_1) \rightarrow square_1(X,Y,Z)$$

Applying this rule to Figure 1 we are left with the problem of what to do with the *inside* relation between the letter and the circle. Since the circle no longer exists, it should be removed, but now an *inside* relation should hold between the letter and the square. This causes problems with the first two representations.

The fundamental difference in the way these representations handle the problem is this: In the first representation, the *inside* relation is represented explicitly by an uninterpreted, symbolic relation. Any transformation must therefore explicitly handle such relations and relevant change in the spatial relations must be propagated explicitly. Essentially, one needs to re-implement the relevant fragment of geometry, which makes the specification cumbersome and error-prone.

In the constraint-based representation, the *inside* relation is managed by mapping the geometry to the underlying arithmetic domain and by having a constraint solver handle the arithmetic constraint theory. The propagation of spatial constraints happens for

Fig. 2. Simple Transformation.

free, because the corresponding arithmetic constraints are propagated automatically by this solver. However, in the context of *transformation* we are now facing a new problem, which stems from the fact that an appropriate constraint solver must work incrementally. The full set of constraints is only evaluated once for the initial diagram. After this, the constraint store should be adapted *incrementally* during the transformation, i.e. only those spatial constraints that change should be updated, added or deleted. The problem arises, because constraints in the store have to be kept in a solved form, so that it is difficult to perform a meaningful constraint deletion. Even if a single constraint can explicitly be removed, it is difficult to remove all implied constraints. As a simple example, consider the existence of three attributes a, b, c. When the constraints $a = b \wedge b = c$ are asserted, $a = c$ is automatically derived. Even if we remove $b = c$ from the constraint store, $a = c$ would still be in the store, but it would no longer be justified. Transformation of constraint diagram representations therefore requires a re-evaluation of the existing spatial relations, which is inefficient. These problems have been explored in [Mey97].

These kinds of problem make it difficult to provide a linear logic characterization of diagram transformation based on either of these representations. For this reason, we adopt the representation based on concrete geometry here, and we assume the existence of a specialized *geometry agent* which is able to answer queries regarding spatial relations in a given diagram by performing arithmetic calculations based on its concrete geometry.

3   Diagram Transformation as Multiset Rewriting

We now move from discussing the representation of diagrams as multisets to investigating diagrammatic transformations through multiset transformations. In general, such diagram transformations detail how one diagram can *syntactically* be transformed into another diagram that directly and explicitly exhibits additional information that is either not present or only implicit and hidden in the prior diagram. This transformation can be understood as multiset transformation or rewriting. This is not a completely new idea, since whereas the specification of diagram transformation has often been based on graph grammars [BMST99], other approaches have also used *type-0* variants of attributed multiset grammars [MMW98,MM00].

Even though different formal theories for multi-dimensional grammars have been developed, there is no grammar calculus that would allow us to deduce soundness or completeness results for diagrammatic calculi on the basis of their grammar specifications. It is for these reasons that logic has been explored as an alternative tool for diagram rewriting. The first obvious choice to explore is classical first-order logic. Roughly speaking, two different embeddings are possible: Either the objects of the diagram and their relations are represented as predicates or, alternatively, they can be modelled as term structures. For diagram rewriting, the first type of embedding has been demonstrated in [HM91,Mar94,Mey00]. The second approach is closely related to modelling string language grammars in logic programming by Definite Clause Grammars and to their extension in the form of Definite Clause Set Grammars [Tan91]. Variations for the use of diagram parsing and/or rewriting have been demonstrated and discussed in [HMO91,Mey97,MMW98].

Both embeddings of diagram rewriting into first-order logic have drawbacks that make their universal utility questionable. In the first embedding (graphical entities as predicates) a typical rewrite step needs to add as well as to remove objects, which amounts to deriving new predicates (conclusions) and deleting old conclusions. This is not possible in classical first order logic due to its monotonicity. In the second embedding there are no restrictions on how the diagram could be rewritten, but the modelling does not leverage from the structure of the underlying logic anymore, since, essentially, this is "abused" as a rule-based rewrite mechanism. In contrast, what is really desirable is that the logical derivation relation can directly deal with terms representing the graphical elements, so that a direct correspondence between logical derivations and those in the diagram system exists.

A detailed analysis of these problems has recently been given for the first time in [MM00], where the use of linear logic [Gir87,Gir91] is advocated instead. The major advantage of linear logic over classical logic in the context of diagram rewriting is that it is a logic of resources and actions, and therefore adequately models a non-monotonous rewrite process. Linear implication models the fact that the left-hand side resources (the antecedent) are used and consumed (*sic*) in the process of producing the right-hand side resources (the consequent). This is exactly the process of multiset rewriting that we have to model.

The question arises which fragment of linear logic we should use. One needs to use multiplicative connectives to express the simultaneous existence of elements in the multiset of diagram elements, while linear implication is adequate to express the rewriting as such. Since *all* the elements in a multiset of objects representing a diagram have to exist simultaneously, a natural choice is to use multiplicative conjunction ($\otimes$) to model their union in a multiset. Therefore, we might decide to express, for instance, the transformation rule in Figure 2 as

$$circle(\boldsymbol{x_1}) \otimes square(\boldsymbol{x_2}) \otimes inside(\boldsymbol{x_1}, \boldsymbol{x_2}) \multimap square(\boldsymbol{x_2})$$

This choice, advocated in [MM00], seems natural and conceptually correct. However, we are not interested in modelling diagram rewriting in linear logic for its theoretical properties alone, but we are also interested in linear logic as a well-founded theoretical framework for declarative computational implementations of diagram transformation. Therefore, we have to pay attention to whether the chosen fragment is adequate as the basis of implementations.

3.1   LO and Interaction Abstract Machines

The basic idea for an implementation of our framework is that it should be directly transformable into a linear logic program. Unfortunately, current linear logic programming languages [Mil95] do not offer multiplicative conjunction in the rule head. It would therefore be advantageous to find a different fragment of linear logic that directly corresponds to a linear logic programming language. The fragment we introduce in this paper is a slight extension of the one used in the linear logic programming language LO.

A benefit of LO is that it has an interpretation as a system of interacting collaborating systems of agents, called Interaction Abstract Machine (IAM). This can later be used as the basis of integrating interaction specifications with our approach. We will now give a very brief introduction to the IAM and its interpretation in LO. For a full introduction the interested reader is referred to [ACP93,AP91].

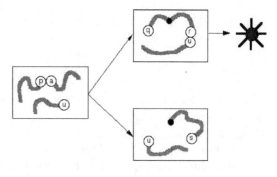

Fig. 3. IAM Evolution

The IAM is a model of interacting agents in which an agent's state is fully described as a multiset of resources. The agents' behavior is described by a set of rules called "methods". A method takes the form $A_1 \,\bindnasrepma\, \cdots \,\bindnasrepma\, A_n \multimap B$, meaning that a multiset containing the elements $A_1, \cdots, A_n$ (the *head*) can be transformed by eliminating these elements and producing effects as specified by the form of B (the *body*). Each resource A_i and each resource in B is syntactically given as a first-order term $f(X_1, \ldots, X_n)$ like in standard logic programming. A method can fire if the appropriate head elements are found among the agent's resources.

If more than one rule can fire at a given time, e.g. if the agent's resource set is $\{a, p, q\}$ and its rules are $p \,\bindnasrepma\, a \circ\!\!-\, r$ and $q \,\bindnasrepma\, a \circ\!\!-\, s$, these rules compete, i.e. one of them will be chosen, committed and applied. If the body could only contain the connective $\bindnasrepma$, the rules would only allow us to describe the behavior of a single agent, but IAMs also use a second kind of connective (in the body of a rule) which effectively spawns another agent. All the resources of the spawning agent not consumed by the rule application are copied into each spawned agent. This connective is the *additive conjunction* &, read "with". & takes priority on $\bindnasrepma$. A final IAM operation is to terminate an agent. This is denoted by a $\top$ in the body of a method. This behavior is illustrated in Figure 3 (redrawn from [ACP93]), depicting a possible development of an agent with state $\{p, a, u\}$ whose behavior is specified by the rules $p \,\bindnasrepma\, a \circ\!\!-\, (q \,\bindnasrepma\, r) \;\&\; s$ and $r \,\bindnasrepma\, u \circ\!\!-\, \top$.

The complete behavior of a method in a multiagent IAM is summarized by the following rules: (1) A rule is applicable to an agent only if all the rule head atoms occur in the agent's state. (2) If a rule is applicable and is selected for application to an agent, the rule's head atoms are first removed from the agent's state; (3) The new configuration of the IAM is defined according to the form of the body: (a) If the body is the symbol $\top$, the agent is terminated; (b) If the body is the symbol $\bot$, no atom is added to the state of the agent; (c) If the body does not contain any &, then the body elements are added to the agent state; (d) If the body consists of several conjuncts connected by &, then for each occurrence of & a new agent containing a copy of the original agent's resources is spawned. For all the resulting agents (including the original one) the atoms in the corresponding conjunct are added to the agent's state.

The IAM, even when restricted to the single agent case, defines a form of multiset rewriting that is applicable to the rewriting of diagrams as outlined above.

The advantage gained from using the IAM as our basic model is that its interpretation can be given in a small fragment of linear logic which only consists of the connectives *par* ($\bindnasrepma$), *with* (&) and linear implication ($\circ\!\!-$), as implemented by the linear logic programming language LO [AP91]. The most important idea of this interpretation is that a terminating evolution of an IAM directly corresponds to a successful proof according in linear logic.[1]

3.2   Diagram Parsing as Linear Logic Programming

Parsing can be considered as the most basic task in diagram transformation. First, it seems of fundamental importance to be able to analyze the correctness of a diagram and to interpret its structure. Secondly, context-free parsing according to a multiset grammar corresponds to a particularly elementary form

---

[1] IAMs and LO also use another kind of connective that emulates broadcasting among different agents. We do not need this connective for our tasks and will therefore exclude it.

of diagram transformation in which each transformation step replaces a multi-set of diagram objects with some non-terminal object. We will therefore first look at diagram parsing in linear logic, before proceeding to arbitrary diagram transformation. Diagram parsing has been studied by a number of different researchers before. The interested reader is referred to [MMW98,BMST99] for comprehensive surveys and to [MM00] for diagram parsing in linear logic. Here we review a particular type of attributed multiset grammars, termed constraint multiset grammars (CMGs), which have been used by a number of researchers for reasonably complex tasks such as the interpretation of state diagrams and mathematical equations. In [Mar94] a precise formal treatment is given, and we review only the basic notions here. CMG productions rewrite multisets of attributed tokens and have the form

$$U ::= U_1, \ldots, U_n \ where \ (C) \ \{E\} \tag{1}$$

indicating that the non-terminal symbol U can be recognized from the symbols $U_1, \ldots, U_n$ whenever the attributes of $U_1, \ldots, U_n$ satisfy the constraints C. The attributes of U are computed using the assignment expression E. The constraints enable information about spatial layout and relationships to be naturally encoded in the grammar. The terms *terminal* and *non-terminal* are used analogously to the case in string languages. The only difference lies in the fact that terminal types in CMGs refer to graphic primitives, such as *line* and *circle*, instead of textual tokens and each of these *symbol types* has a set of one or more attributes, typically used to describe its geometric properties. A *symbol* is an instance of a symbol type. In each grammar, there is a distinguished non-terminal symbol type called the *start* type.

CMGs also include context-sensitive productions. Context symbols, i.e. symbols that are not consumed when a production is applied, are existentially quantified in a production. As an example, the following context-sensitive production from a grammar for state transition diagrams recognizes transitions:

> *T:transition ::= A:arc exists S1:state,S2:state where (*
> *OnCircle(A.start,S1.mid,S1.radius) and OnCircle(A.end,S2.mid,S2.radius))*
> *{T.start = A.start and T.tran = A.label and T.end = A.end }*

A diagrammatic sentence to be parsed by a CMG is just an attributed multiset of graphical tokens. Therefore we can view a sentential form as the resources of an IAM agent. Intuitively, it is clear that the application of a CMG production corresponds closely to the firing of IAM methods and that a successful parse consists of rewriting the original set of terminal symbols into a multiset that only contains the single non-terminal symbol which is the start symbol.

We can map a CMG production to an LO rule (IAM method, respectively) and hence to a linear logic implication in the following way: For a CMG production

$$U ::= U_1, \ldots, U_n \ exists \ U_{n+1}, \ldots, U_m \ where \ (C) \ \{E\} \tag{2}$$

we use the equivalent LO method:

$$\tau(u_1) \ \mathbf{\otimes} \ldots \mathbf{\otimes} \ \tau(u_m) \circ\!\!-\{C\} \ \& \ \{E\} \ \& \ \tau(u) \ \mathbf{\otimes} \ \tau(u_{n+1}) \ \mathbf{\otimes} \ldots \mathbf{\otimes} \ \tau(u_m) \tag{3}$$

In LO, each CMG terminal and non-terminal object u_i will be represented by a first order term $\tau(u_i)$ which has the token type of u_i as the main functor and contains the attributes in some fixed order. We extend this mapping function canonically so that we use $\tau(u_1, \ldots, u_n)$ to denote $\tau(u_1) \,\vartheta \ldots \vartheta\, \tau(u_n)$. In the same way, $\tau(p)$, for a CMG production p, will denote its mapping to an LO rule and $\tau(G) = \{\tau(p) \mid p \in P\}$ denotes the complete mapping of a CMG G to an LO program.

The linear logic reading of such a method $\tau(p)$ is its exponential universal closure:

$$! \widetilde{\forall}\, \tau(u_1) \,\vartheta \ldots \vartheta\, \tau(u_m) \circ\!\!-\{C\} \,\&\, \{E\} \,\&\, \tau(u) \,\vartheta\, \tau(u_{n+1}) \,\vartheta \ldots \vartheta\, \tau(u_m) \quad (4)$$

To evaluate the constraints in the grammar and to compute the attribute assignments we assume that the geometric (arithmetic) theory is available as a first-order theory Γ_g in linear logic. Obviously, geometry cannot be completely axiomatized in LO, because its fragment of linear logic is too small. However, we can encapsulate a more powerful "geometry machine" (and arithmetic evaluator) in a separate agent and give evaluation requests to this agent. This is what we do by using "with" ($\&$) to spawn agents for these computations in the above LO translation.

From an operational point of view, this requires us to adopt a proactive interpretation of LO in which we can spawn an agent and wait for it to return a result before proceeding with the rest of the computation. A different implementation of a proactive LO interpretation, by sending requests from a coordinator to registered participants, is provided by the Coordination Language Facility [AFP96].

Each rule $\tau(p)$ emulates exactly one production p. To emulate parsing fully, we also need a rule which declares that a parse is successful if and only if the initial diagram is reduced to the start symbol and no other symbols are left. For a CMG G with start symbol s, we could do this in linear logic by adding $\tau(s)$ as an axiom to $\tau(G)$. Unfortunately, from an implementation point of view, we cannot formulate true linear axioms in LO. It is more consistent with the LO model to extend the language with the linear goal $\mathbf{1}$, which terminates an agent if and only if this agent does not have any resources left (i.e. $\mathbf{1}$ succeeds if and only if the linear proof context is empty). We will call this extension of the *LO* language LO_1. Instead of the axiom $\tau(s)$ we can then add the method $\tau(s) \circ\!\!-\mathbf{1}$ to the LO_1 program. The complete set of LO_1 rules that implement a grammar G is: $\Pi = \tau(G) \cup \{(\tau(s) \circ\!\!-\mathbf{1})\}$

Operationally, a successful parse of a diagram D now corresponds to an IAM evolution with method set Π starting from a single agent with the resource set $\tau(D)$ in which all agents eventually terminate. Logically, it corresponds to a proof of $\Gamma_g, \Pi \vdash \tau(D)$. This linear logic embedding of CMGs is sound and complete.

Theorem 1. $D \in \mathcal{L}(G) \Leftrightarrow \Gamma_g, \Pi \vdash \tau(D)$

The proof is given in [BMMP01].

LO_1 is only a minor extension of LO and a proper subset of the linear logic programming language Lygon [HPW96]. Thus we still have an executable logic as the basis of our model. In fact, it is Lygon's sequent calculus [HP94] that we will use in the remainder of this article. In contrast to LO which applies rules by committed choice, Lygon actually performs a search for a proper proof tree. Therefore, if there is a proof for $\Gamma_g, \Pi \vdash \tau(D)$, i.e. if $D \in \mathcal{L}(G)$, Lygon will find this proof. This is in contrast to LO, which, even disregarding the extension with 1, can only be guaranteed to find the proof if G is confluent.

3.3 Negative Application Conditions

We have shown that context-sensitive rules are expressed in the adopted fragment of linear logic by simply using the contextual resources in the head and in the body of a rule. It is interesting to investigate how negative context conditions, stating that something must not exist in the diagram, can be expressed. Though negative application contexts are not adopted in all grammar-based approaches to diagram parsing, there are many visual languages that can only be defined with negative conditions. Therefore, some grammar frameworks, such as Constraint Multiset Grammars [MMW98] and several types of graph grammars, have adopted negative contexts. In particular, in the Single Push-Out Approach (SPO) [EHK+97], negative application conditions (NAC) are expressed as an addition to the left-hand side (LHS) of a rule, so that for a rule to be applied, the mapping m from the LHS to the host graph G must not be extendable with a mapping n from the NAC to G.

Since negation in linear logic is constructive, it is not possible to directly state such a negative condition in a form such as $a^\perp ℘ b\circ{-}c$ to express that b is transformed into c if a is not present.

Given the importance of negative application conditions for diagram specification and transformation we allow ourselves some speculation how an extension of the framework presented above could allow us to handle negative contexts.

In linear logic the notation $a^\perp$, in fact, does not represent the absence of a in the multiset, which is what we really want to express. On the contrary, $a^\perp$ is used to represent a *dual* resource with respect to a. Intuitively, we can say that $a^\perp$ can be seen as the answer to a request of type a, so that they can disappear together in a context where a has been requested and $a^\perp$ has been produced. In the linear logic programming language Lygon [HP94] indeed, negated resources are used to represent facts, while positive resources in the right-hand side of a rule are goals to be satisfied.

Let us formulate negative application conditions in the form $b\circ{-}c\,\&\,neg(a)$, to indicate that we do not want a to be present or, in other words, that we want the proof not to terminate successfully in the branch created by the & if a is present. We might want to achieve this by using a rule of the form $X ℘ neg(X)\circ{-}\#$, where $\#$ is a resource which is not used anywhere else and therefore cannot be removed. However, this is not valid, since we can not guarantee that this rule is really applied. In fact, there is no guarantee that X will not be consumed by some other rule or that no rule of type $c ℘ neg(a)\circ{-}d$ could undo the effect of producing the negated resource. Moreover, this approach does not allow us to obtain a guard on the application of the rule.

We have to look for an approach that allows us to model the committed choice application of a rule based on negated guards, i.e. which ensures that the element was not present at the time of its usage. The problem is that negative application conditions essentially express conditions on the global configuration of the multiset. Hence, to be sure that a rule is applicable we need to maintain a global representation of the state and check it in the rule itself. This can be achieved if we keep book of all the resources we produce and consume so that the check for non-existence is reduced to asessing membership in a given set in the resource records. Since the condition may involve checking values of the attributes of a resource, we use resources which include a representation of the actual resource. This approach obviously amounts to introducing a limited form of metalevel resources in the multiset and its augmentation with reflective capabilities. Such an extension has been discussed in [BBMP97].

It would also be possible to use a representation purely based on the counting of resources to achieve the same effect, but this would lead to a much more complex rule set.

To achieve this, we could instantiate in the initial multiset a collection of resources of type $recX(SetOfX)$ for each type X of atoms in the alphabet of the grammar. The idea is that $SetOfX$ is simply a (multi)set of all resources of the type X that are currently present. It is easy (though complex) to re-write a rule set to use these reflective resources. In the above approach a rule of the form $a(X) \circ\!\!-\!\! f(X,Y)$ & $b(Y)$ where $f(X,Y)$ is some function on the attributes of X and Y would then be transformed into the form:

$recA(SetOfA)$ $\wp$ $recB(SetOfB)$ $\wp$ $a(X)$
   $\circ\!\!-$ $f(X,Y)$ & $diff(SetOfA, a(X), NSA)$ &
      $diff(NSB, b(Y), SetOfB)$ & $b(Y)$ $\wp$ $recA(NSA)$ $\wp$ $recB(NSB)$

A proof for the operator *diff* succeeds if and only if the third argument is the multiset obtained by removing the resource denoted by the second argument from the multiset in the first argument. In our treatment, we assume that this is available through a theory Γ_g which contains geometric, arithmetic and elementary set theory facts as axioms. Tests on the attributes of C are admissible as long as their negation can be formalized in Γ_g. For all practical implementation purposes *diff* can also be implemented via the host language.

Now negative conditions can be checked as non-membership in the suitable record of a resource with given properties. A rule of the form $a(X)$ $\wp$ $neg(c(Z))$ $\circ\!\!-$ $f_1(X,Y)$ & $f_2(X,Z)$ & $b(Y)$ involving a negative condition on the presence of $c(Z)$ will then be written as:

$recA(SetOfA)$ $\wp$ $recB(SetOfB)$ $\wp$ $recC(SetOfC)$ $\wp$ $a(X)$
   $\circ\!\!-$ $f_1(X,Y)$ & $f_2(X,Z)$ & $diff(SetOfC,c(Z),SetOfC)$ &
      $diff(SetOfA, a(X), NSA)$ & $diff(NSB, b(Y), SetOfB)$ &
      $b(Y)$ $\wp$ $recA(NSA)$ $\wp$ $recB(NSB)$ $\wp$ $recC(SetOfC)$

The price to pay in this approach is obviously the management of reflective resources and a much more complex rule-set, which, however, could be generated automatically. It is also obvious that this approach can only handle a limited type of negation as failure.

4   Applications

In the previous sections we have presented a general approach to diagram transformation based on linear logic. In this section we present two simple applications of this approach: One in which the diagram transformation corresponds to the execution of a computation and one in which diagram transformations are used to reason in some underlying domain represented by the diagram.

4.1   Executable Diagrams

By executable diagrams we refer to diagram notations that are used to specify the configurations of some system. Transformation of such diagrams can be used to simulate and animate the transformation of these configurations. Typical examples of such systems are Petri nets, finite state machines and a number of domain specific visual languages.

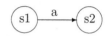

Fig. 4. FSA Transition

For these systems, a multiset representation is intuitively apt and the transformation rules for the multiset closely correspond to the diagram transformation rules. As an example, consider a transition in a finite state diagram, such as the one in Figure 4.

Let us adopt a set of data types corresponding to the definition of *states*, with a geometry, a name and a couple of attributes denoting whether the state is initial and/or final; *transitions*, defined by their geometry and an input symbol; and *input* labels, which are positioned under the current state, to be read one symbol at a time.

A straightforward translation of the partial diagram of Figure 4 results in a rule which corresponds exactly to the semantics of the depicted transition, assuming that in a diagrammatic animation of the transformation the input string is always placed under the current state:

$state((109,24),s1,nonfinal,noninitial)$ $⅋$ $transition((129,24),(189,24),"a")$ $⅋$
  $state((209,24),s2,nonfinal,noninitial)$ $⅋$ $input((109,40),["a"|Rest])$
  $\circ\!-$ $state((109,24),s1,nonfinal,noninitial)$ $⅋$ $transition((129,24),(189,24),"a")$ $⅋$
  $state((209,24),s2,nonfinal,noninitial)$ $⅋$ $input((209,40),Rest)$

The whole diagram is translated into such a set of rules, one for each transition, and its execution can be started by placing the input string under the initial state.

We can, however, have a more general view of this process and define the behavior of such animations independently of any concrete diagram:

$state(Geom1, Name1, F1, I1)$ $⅋$ $state(Geom2, Name2, F2, I2)$ $⅋$
  $transition(Geom3, Lab)$ $⅋$ $input(Geom4, [Lab|Rest])$
  $\circ\!-$ $startsAt(Geom3, Geom1)$ $\&$ $endsAt(Geom3, Geom2)$ $\&$
  $below(Geom4, Geom1)$ $\&$ $below(Geom5, Geom2)$ $\&$
  $state(Geom1, Name1, F1, I1)$ $⅋$ $state(Geom2, Name2, F2, I2)$ $⅋$
  $transition(Geom3, Lab)$ $⅋$ $input(Geom5, Rest)$

where *startsAt, endsAt* and *below* are suitable predicates that check the corresponding spatial relations, possibly instantiating the *geom* attribute appropriately.

In the LO_1 setting, a rule for expressing acceptance of the input would be:

$$state(Geom1, \_, final, \_) \ \text{⅋} \ input(Geom2, []) \circ\!\!-\ \top \ \& \ below(Geom2, Geom1).$$

Note that the termination of an agent indicates the success of a branch in the corresponding proof.

It is easy to see how the two alternative approaches can both provide an operational semantics for executable diagrams. In both cases, the actual execution of the transformations occurs uniformly according to the LO_1 proof system.

4.2   Diagrammatic Reasoning

Often we are using diagram transformations not so much to define the configuration of a computational system, but to reason about some abstract domain. A typical case of the use of diagrams to perform such reasoning are Venn Diagrams. A variant of these, developed by Shin [Shi95], provides a formal syntax, semantics and a sound and complete system of visual inference rules.

In these diagrams, sets are represented by regions, shaded regions imply that the corresponding set is empty, and a chain of X implies that at least one of the regions marked by an X in the chain must be non-empty.

As an example, Figure 5 says that A and B are not both empty (expressed by the chain of X) and nothing is both in A and in B (expressed by shading). By inference we can obtain that the elements in B and in A must belong to the symmetric difference $A \bigtriangleup B = A - B \cup B - A$. This diagram is equivalent to one in which the X in the shaded region is removed. Such an equivalence is expressed

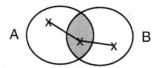

Fig. 5. A Venn diagram

by the "Erasure of Links" inference rule. This can be stated as "an X in a shaded region may be removed from an X-chain provided the chain is kept connected." We reformulate this textual rule as a set of linear logic rules, defined on the following graphical data types: (1) *chain*, associated with an attribute *setOfX* which stores the locations of the X elements in the chain; (2) *x*, with an attribute *pt* giving its position and an attribute *num*, giving the number of lines attached to it; (3) *line*, with an attribute *ends* giving the positions of its two ends; (4) *region*, with an attribute *geom*, allowing the reconstruction of the geometry of the region, and an attribute *shading*, indicating whether the region is shaded or not. We assume that these elements are correctly and consistently identified in the diagram. Additionally, some synchronization resources are used to indicate that the transformation is performing some not yet completed process. Link erasure is defined by the following actions:

(1) A point inside a shaded region is eliminated and the set in the chain is updated accordingly. A synchronization resource is used to ensure that all the elements previously connected to it will be considered:

$chain(SofX) \,⅋\, x(Pt, Num) \,⅋\, region(Geom, shaded)$
  $∘{-}\; chain(G, Pt, Num) \,⅋\, region(Geom, shaded)$ &
      $inside(Pt, Geom)$ & $diff(SofX, Pt, G)$

(2) Points previously connected to the removed element are marked and the connecting lines are removed:

$chain(G, Pt, Num) \,⅋\, line(Ends) \,⅋\, x(Pt1, Num1)$
  $∘{-}\; equal(Ends, \{Pt, Pt1\})$ & $minus(Num1, 1, Num11)$ &
      $chain(G, Pt, Num) \,⅋\, x(Pt1, cand, Num11)$

(3) If the removed point was inside the chain, its two neighbors are connected by a new line. Synchronization resources are removed and a consistent state is restored:

$chain(G, Pt, 2) \,⅋\, x(Pt1, cand, Num1) \,⅋\, x(Pt2, cand, Num2)$
  $∘{-}\; equal(Ends, \{Pt1, Pt2\})$ & $plus(Num1, 1, Num11)$ &
      $plus(Num2, 1, Num21)$ & $chain(G) \,⅋\, x(Pt1, Num11) \,⅋\,$
      $x(Pt2, Num21) \,⅋\, line(Ends)$

(4) If the removed point was at an end of the chain, its neighbor is now at an end or it is isolated.

$chain(G, Pt, 1) \,⅋\, x(Pt1, cand, 1) \; ∘{-} \; chain(G) \,⅋\, x(Pt1, 1)$
$chain(G, Pt, 1) \,⅋\, x(Pt1, cand, 0) \; ∘{-} \; x(Pt1, 0)$

(5) If the removed point was an isolated element, the diagram was inconsistent, and the chain is removed altogether:

$chain(G, Pt, 0) \; ∘{-} \; \perp$

The erasure process goes through intermediate steps in which the diagram is not a Venn diagram (for instance, dangling edges appear in the chain). Such inconsistent diagrams correspond to states in which synchronization resources appear in the multiset. The process is, however, guaranteed to terminate with a consistent diagram. Such situations often occur in diagrammatic transformations, where a complex step is broken up to produce several intermediate diagrams, of which only the start and final diagram belong to the language of interest. The problem of deciding whether a diagram produced during the transformation process belongs to the language or is just an intermediate diagram can in many cases be solved without resorting to a new full analysis of the diagram. In fact, knowing that the starting diagram was in the language and knowing the possible transformations, we can usually define some simple test for the validity of a transformed diagram. For example, among all the multisets produced during the link erasure process, all and only those which do not contain any synchronization resource represent a Venn diagram.

In general, a language L can be specified by a triple $(L_0, \Rightarrow^*, L_f)$, where L_0 is an initial language, $\Rightarrow^*$ is the reflexive and transitive closure of the *yield* relation, and L_f is a final language acting as a filter for sentences produced starting from L_0 according to $\Rightarrow^*$, i.e. $L = \{s \mid \exists s' \in L_0 : s' \Rightarrow^* s\} \cap L_f$.

This view was proposed for string languages in [Man98] and independently adopted for the diagrammatic case in [BCM99,BPPS00]. This suggests a line of attack for typical problems in diagram transformations related to the possible production of inconsistent diagrams. In our approach, the filter language can be characterized by a set of LO_1 rules. A valid state in a diagram transformation process is one for which there exists an LO_1 proof of the filter property.

As an example, consider the *dangling edge* problem which is typical of graph transformation systems. The double-pushout approach to algebraic graph transformation [CMR$^+$97] faces this problem by not allowing deletion of a node if its elimination would leave dangling edges after rule application. From our perspective, this could be modelled by giving a simple set of LO_1 filter rules. Indeed, a proof using the following program terminates successfully (i.e. with **1**) only if all edges are eliminated.

$edge(G1) \ \otimes \ node(G2) \ \otimes \ node(G3)$
    $\circ\!\!-\ touches(G1, G2) \ \& \ touches(G1, G3) \ \& \ node(G2) \ \otimes node(G3)$
$node(\_) \circ\!\!-\ \bot$
$node(\_) \circ\!\!-\ \mathbf{1}$

Again, this "filtering" set of rules is seen as an ex-post check that the application of rules was correct. The implementation of checks to avoid dangling edges would require the encoding and maintaining in the nodes of a record of the touching edges and a check that all touching edges have been removed before removing the node. As seen in Section 3.3, guarded rule application would require the use of metalevel representations.

5   On a Class of Local Transformations

We have seen that the semantics of graphs defining finite state machines can be given in two alternative ways: the first consists of generating a set of rules from the analysis of the graph and providing the initial configuration of the process (with the string to be analysed in a position corresponding to the initial state); the second consists of giving the general law for transitions in the form of a set of rules and providing the whole description of the diagram, in the form of a multiset of resources together with the initial position of the string. In both cases the inference calculus realises the computation. In a sense we could say that in the first case we directly compile a concrete diagram into the rule set whereas in the second case the rule set represents a general interpreter for such diagrams.

This is possible because a notion of *transition* results explicitly defined in the visual representation, (depicted as arrows in the finite state machines), such that the pre-and post-conditions of the transition are directly inferrable from the diagram. A process corresponds to a sequence of markings of the diagram, according to the specifications of the pre- and post-conditions of the fired transitions.

In general a visual transformation s can be defined by specifying a law for transforming a diagram d into a diagram d' in the same visual language. We indicate with $pre(d, s)$ the set of elements in d involved in the precondition for s and with $post(s, d')$, the analogous set of elements involved in the postcondition. A couple of partial functions $\mu : pre(d, s) \mapsto post(s, d')$ and $\mu' : post(s, d') \mapsto pre(d, s)$ exist, relating elements in the two diagrams.

Visual languages providing explicit representation of the transition enjoy some properties, that we now consider independently to investigate which is the class of languages for which such a dual approach can be taken. To start this research we first introduce the property of *locality*, based on a notion of *reachability* between elements. For the purpose of this discussion we consider only *connectedness* and *containment* as significant relations between visual elements. We say that an element A is *reachable* from another element B if there is a navigation path composed of only relations of connectedness and containment leading from B to A. Reachability is here taken as a symmetric relation, in the sense that the navigation does not have to take into account the possible existence of directions in the connections. This is to say that if A is contained into B, we consider that A and B are reachable from one another, and similarly if they are connected by a directed edge. If we extend the relation to admit that any element is reachable from itself, reachability becomes an equivalence relation, as it is obviously transitive. We can now establish a notion of *distance* between elements in the following way. Let $P(A, B)$ be the set of paths between A and B.[2]

$$d(A, B) = \begin{cases} min - length(P(A, B)) & \text{if } reachable(A, B) \\ \infty & \text{otherwise} \end{cases}$$

We can now formalise the notion of explicit representation of a transition as follows: given a transition step s, there exist in d exactly one graphical element t, a collection of graphical elements $P = \{p_1, \ldots, p_n\}$ representing the elements affected by the transformation and such that $reachable(t, p_i)$ holds for $i = 1, \ldots, n$ and there exist two collections of graphical elements $M_1 = \{m_{1,1}, \ldots, m_{1,n_1}\}$ and $M_2 = \{m_{2,1}, \ldots, m_{2,n_2}\}$. Then there are two mappings $pre : P \mapsto M_1 \cup \{blank\}$ and $post : P \mapsto M_2 \cup \{blank\}$, such that M_1 is a marking for P iff the system is able to perform s, and M_2 is a marking for P iff the system can have just performed s.

Such a definition appears as a rephrasing of the firing condition for Petri nets, but one can see that with the suitable restrictions on P, it occurs also for finite state machines. For example, a transition from a state s_1 to a state s_2, consuming a symbol X from an input $X \circ R$ is visualised by considering $P = \{s_1, s_2\}$, $M1 = string(X \circ R)$ $M2 = string(R)$, $pre(s_1) = string(X \circ R)$, $pre(s_2) = blank$, $post(s_1) = blank$, $post(s_2) = string(R)$.

In languages supporting an explicit representation of transformations, μ and μ' constitute a bijection and $pre(d, s) = \mu'(post(s, d')) = P$. Hence a diagram

---

2 Essentially $d(A, B)$ is the path length between A and B in the underlying relationship graph.

depicting an initial state may vary only for its markings. This observation may be generalised to define a class of local transformations.

We say that a transformation is *local* if and only if there exist an integer K_1 such that for any pair of diagrams d, d' and transition step s transforming d into d', the following hold:

1. $\forall x \in post(s, d') \ \exists y \in pre(d, s)$ s.t. $d(x, \mu(y)) \leq K_1$
2. there is an integer K_d such that no sequence of transformation steps may affect more than K_d elements.

In practice, this notion of locality rules out transformations involving sets of elements of arbitrary cardinality or at an arbitrary distance from elements mentioned in the precondition of a transition.

Observation 1 *If a diagrammatic system supports an explicit representation of transitions, then all the transformations realisable on the system are local.*

Reasoning. If transitions are represented explicitly, then for any transition there is a set of elements in the pre- or post-conditions that are reachable from the element representing the transition. Moreover, since the diagram is modified only for its marking, an inspection of the initial diagram provides the constants K_1 and K_d.

One can observe, however, that the notion of locality goes beyond that of explicit representation of transitions. In particular, it enlists all transformation laws that can be expressed in the form of before-after rules, where the left-hand side enumerates all the elements in the preconditions and the right-hand side all the elements in the postconditions, without any form of side effects. Typical examples are Agentsheets [RS95] or Kidsim [SCS94] rules. Such a notion can be formalised by constraining μ and μ' to be injections such that if $\mu'(\mu(e))$ is defined for some $e \in pre(d, s)$, then $\mu'(\mu(e)) = e$ and if $\mu(\mu'(e'))$ is defined for some $e' \in post(s, d')$, then $\mu(\mu'(e')) = e'$.

We can now relate the general notion of locality to our linear logic framework.

Observation 2 *If a transformation law is local, then it is possible to translate a visual sentence defining a dynamic system into a collection of rules in the considered fragment of linear logic, such that any sequence of applications of these rules corresponds to a possible dynamics of the system.*

Reasoning. In a local transformation law, the effect of a transition step can be described through a rule which presents all elements in the pre-condition in the left-hand side, and all elements in the post-condition in the right-hand side. Since all elements in the pre-condition must be present for the rule to succeed, we can use the connective $\mathrm{⅋}$ to state this fact. The same reasoning justifies the use of $\mathrm{⅋}$ in the right-hand side. Conditions on attributes can be mapped to constraints involving elements in the right-hand side, and can be used as conjuncts for the connective &. Since elements in local transformations are finite, a finite set of rules is derived.

The above suggests the general form of the rules directly derivable from a diagram:

$$originators(X) \circ\!\!-\, transf(X, Y) \,\&\, affected(Y)$$

where we have slightly abused the notation, by collecting into the terms *originators* and *destinations* the (fixed) set of terms respectively in $pre(d, s)$ and $post(s, d')$. The resource *transf* summarises the operators defining the modification of the marking and the check that additional preconditions for the transition are met.

As noted before, the types of transformations considered above have also another characteristic: they do not modify the topology of the diagram along transformations.

We say that a visual transformation is *topology preserving* if for any diagram d and transition step s, for any two elements x, y appearing either in $D = pre(d, s) \cup \mu'(post(s, d'))$ or in $D' = post(s, d') \cup \mu(pre(d, s))$, the relations between x and y are the same in D and D'.

Observation 3 *If a diagrammatic system supports an explicit representation of transitions, then it describes a topology preserving transformation.*

Reasoning. Indeed, since transitions are defined once and for all, their application can only modify the marking of the diagram, and not the topology.

A weaker condition is for a transition to be *reachability preserving*, holding if for any diagram d and transition step s, for any two elements $x, y \in pre(d, s))$ such that $reachable(x, y)$ holds in d, if $x, y \in post(s, d')$, then $reachable(x, y)$ holds in $post(s, d')$. In words, if two elements are present in a diagram both before and after a transition step, then they have to remain reachable.

Observation 4 *If a transformation law is topology preserving, then it can be expressed by a collection of rules in the considered fragment of linear logic.*

Reasoning. In a topology preserving transformation law, the set of elements to be checked for any transition can be obtained from navigation along connections. The application of the rule requires the identification of the elements to be checked in the precondition for any transition, the checking of the precondition, the identification from the postcondition of the elements to be transformed, and their actual transformation. All these steps can be performed through the adopted connectives, namely $\otimes$ for identifying elements and $\&$ for checking conditions.

Also in this case, the argument suggests the form of the set of rules representing the general transformation law. In this case, the set of elements involved in the transformation has to be constructed.

The first rule starts the process of collecting the originators for a given transition Dep, assuming that some trigger occurs from outside and is inserted in the multiset and that a resource $transf(Dep)$ is used to represent a transition.

$$trigger(Dep) \,\otimes\, transf(Dep) \circ\!\!-\, transf(Dep, collect, nil)$$

The second rule is applied as long as there are originators to collect.

transf(*Dep, collect, Store*) $\otimes$ *originator*(*X*) $\circ\!\!-$
  include(*Dep, X, Store, NewStore*) & *nonComplete*(*Dep, Store*) &
  transf(*Dep, collect, NewStore*)

The third rule evaluates the actual transformation and starts the process of distributing the new configuration elements.

transf(*Dep, collect, Store*) $\circ\!\!-$
  complete(*Dep, Store*) & *eval*(*Store, Result*) & *transf*(*Dep, distrib, Result*)

The fourth rule is applied as long as there are results to distribute.

transf(*Dep, distrib, Store*) $\circ\!\!-$
  take(*Dep, Store, Y, NewStore*) & *nonempty*(*Store*) &
  destination(*Y*) $\otimes$ *transf*(*Dep, distrib, NewStore*)

The final rule allows starting the transition again if a *trigger* is generated.

transf(*Dep, distrib, nil*) $\circ\!\!-$ *transf*(*Dep*)

Again, we have used generic resources *originator* and *destination* to indicate elements which are affected, either in pre- or in post-conditions by the transition.

The use of *transf*(*Dep*) prevents a transition to be started again before the completion of the procedure, while allowing concurrent execution of several transitions. It might be the case that conflicting transitions are triggered together, so that some form of mutual exclusion must be realised. This can be achieved by having a global *available* resource, of which at most one copy may exist in the multiset, removed any time a procedure starts and placed back each time it finishes. In this way consistent states of the diagram correspond to possible states of the system according to an interleaving semantics.

The procedure here described relies on some suitable reflective encoding of the connectedness between elements, i.e. an element must have access to a description of its neighbour set. Such an encoding, namely a counting of neighbours, has been used in the procedure for link erasure in the previous section. Such a counting is in general sufficient and can be consistently updated if we can overview side effects. For example, if the containment relation is important, then rules must be defined which allow removal of containers only if they are empty. In general the encoding must be restricted to immediately reachable elements and not to transitive closures, so that updating of topological information can be performed on a local basis. In this way, we avoid problems connected with the representation of spatial relations described in Section 2.

More general metalevel capabilities are required if rule application depends on global properties of the multiset, such as in transformations which can occur only if the collection of all instances of a given type enjoys some properties.

In this case, a record of the object level resources of type X must be kept in metalevel resources of type $recX$.

Combining the two conditions above, we obtain the following result:

Observation 5 *If a transformation law is both local and topology preserving, then the system dynamics can be equivalently expressed as a set of rules containing at least an instantiated attribute for each term and an initial multiset, or as a set of rules containing only free variables and an initial multiset.*

For systems such as FSA, Petri Nets or dataflow diagrams, the set of originators and destination elements coincide. For the case of graph-like diagrams, the locality requirement, and dually the topology preserving one, rules out transformations which involve the updating of arbitrary embeddings. Basically, the requirements establish a sort of context-freeness property. This is not meant to say that our embedding only accounts for such context-free transformations, but that the duality between the two representations is only possible in this case.

These patterns of rule sets are definable not only for graph-like diagrams, but also for other types of diagrams. For example, consider a diagrammatic language in which atomic elements are located into container elements and transformations may affect only elements within a same container (an example of such a system was described in [BBMP96]). In this case the requirements for context-freeness are satisfied only if the allowed transformations are restricted to the following types:

1. Elements can be individually created within or removed from a container
2. Transformations involve a fixed set of elements
3. Containers are created empty
4. Containers can be removed only if they are empty.

6   Conclusions

We have shown how diagram transformation can be formalized in linear logic and we have discussed interpretations in multiset rewriting. Many important kinds of diagrammatic reasoning, which can be understood as syntactic diagram transformation, can be formalized in this way.

We have discussed a small fragment of linear logic that is expressive enough to model diagrammatic transformations, yet small enough to directly correspond to a calculus of linear logic programming. Our formalism therefore is a directly executable specification language. We have also proven equivalence of our model with attributed multiset grammar approaches and correctness of the corresponding mapping.

The paper also described how to manage a restricted form of negative context conditions checking the non-existence of certain resources or ensuring exhaustive rule application. Such conditions cannot be directly stated on the resources present in the multiset, but require the use of metalevel resources maintaining a record of the object-level resources.

Ultimately, we are interested in specification languages for diagram notations in which the rules themselves are visual. The idea is that this can be formalized by an additional mapping between linear logic rules and visual rules. Such an approach necessarily raises the question if and when visual rules are adequate to describe a transformation system. We hope that the ability to formalize the transformation as well as the embedding conditions and the underlying geometric theory within the unifying framework of linear logic will allow us to develop formal criteria that help to answer this important question.

Acknowledgments

Work partially supported by the European Community TMR network GET-GRATS and by the Italian Ministry of Research under the PRIN project on Specification, Design and Development of Visual Interactive Systems

Appendix A: Linear Sequent Calculus

This appendix shows the relevant rules of the sequent calculus presented in [HP94].

$$\frac{}{\phi \vdash \phi} \ (ax) \qquad\qquad \frac{\Gamma \vdash \phi, \Delta \quad \Gamma', \phi \vdash \Delta'}{\Gamma, \Gamma' \vdash \Delta, \Delta'} \ (cut)$$

$$\frac{\Gamma, \phi, \psi, \Gamma' \vdash \Delta}{\Gamma, \psi, \phi, \Gamma' \vdash \Delta} \ (X-L) \qquad \frac{\Gamma \vdash \Delta, \phi, \psi, \Delta'}{\Gamma \vdash \Delta, \psi, \phi, \Delta'} \ (X-R)$$

$$\frac{\Gamma, \phi \vdash \Delta}{\Gamma, \psi \,\&\, \phi \vdash \Delta} \qquad \frac{\Gamma, \psi \vdash \Delta}{\Gamma, \psi \,\&\, \phi \vdash \Delta} \ (\&-L) \quad \frac{\Gamma \vdash \phi, \Delta \quad \Gamma \vdash \psi, \Delta}{\Gamma \vdash \phi \,\&\, \psi, \Delta} \ (\&-R)$$

$$\frac{\Gamma, \phi \vdash \Delta \quad \Gamma', \psi \vdash \Delta'}{\Gamma, \Gamma', \psi \,\mathbin{\rotatebox[origin=c]{180}{\&}}\, \phi \vdash \Delta, \Delta'} \ (\mathbin{\rotatebox[origin=c]{180}{\&}}-L) \qquad \frac{\Gamma \vdash \phi, \psi, \Delta}{\Gamma \vdash \phi \,\mathbin{\rotatebox[origin=c]{180}{\&}}\, \psi, \Delta} \ (\mathbin{\rotatebox[origin=c]{180}{\&}}-R)$$

$$\frac{\Gamma \vdash \phi, \Delta \quad \Gamma', \psi \vdash \Delta'}{\Gamma, \Gamma', \phi \multimap \psi \vdash \Delta, \Delta'} \ (\multimap-L) \qquad \frac{\Gamma, \phi[t/x] \vdash \Delta}{\Gamma, \forall x.\phi \vdash \Delta} \ (\forall-L)$$

$$\frac{\Gamma \vdash \Delta}{\Gamma, !\phi \vdash \Delta} \ (W!-L) \qquad \frac{\Gamma, !\phi, !\phi \vdash \Delta}{\Gamma, !\phi \vdash \Delta} \ (C!-L)$$

$$\frac{\Gamma, \phi \vdash \Delta}{\Gamma, !\phi \vdash \Delta} \ (!-L) \qquad \frac{}{\vdash 1} \ (1-R)$$

References

[ACP93] J.-M. Andreoli, P. Ciancarini, and R. Pareschi. Interaction abstract machines. In G. Agha, P. Wegner, and A. Yonezawa, editors, *Research Directions in Concurrent Object-Oriented Programming*, pages 257–280. MIT Press, Cambridge, MA, 1993.

[AFP96]   J.-M. Andreoli, S. Freeman, and R. Pareschi. The coordination language
           facility: Coordination of distributed objects. *Theory and Practice of Object
           Systems*, 2:77–94, 1996.
[AP91]   J.-M. Andreoli and R. Pareschi. Linear objects: Logical processes with
           built-in inheritance. *New Generation Computing*, 9:445–473, 1991.
[BBMP96]   U.M. Borghoff, P. Bottoni, P. Mussio, and R. Pareschi. A systemic
           metaphor of multi-agent coordination in living systems. In *Proc. Euro-
           pean Simulation Multiconference (ESM'96)*, pages 245–253, 1996.
[BBMP97]   U.M. Borghoff, P. Bottoni, P. Mussio, and R. Pareschi. Reflective agents
           for adaptive workflows. In *Proc. 2nd Int. Conf. on the Practical Application
           of Intelligent Agents and Multi-Agent Technology (PAAM '97)*, pages 405
           – 420, 1997.
[BCM99]   P. Bottoni, M.F. Costabile, and P. Mussio. Specification and dialogue con-
           trol of visual interaction through visual rewriting systems. *ACM Transac-
           tions on Programming Languages and Systems*, 21:1077–1136, 1999.
[BMMP01]   P. Bottoni, B. Meyer, K. Marriott, and F.Parisi Presicce. Deductive parsing
           of visual languages. In *Int. Conf. on Logical Aspects of Computational
           Linguistics*, Le Croisic, France, June 2001.
[BMST99]   R. Bardohl, M. Minas, A. Schürr, and G. Taentzer. Application of graph
           transformation to visual languages. In H. Ehrig, G. Engels, H.-J. Kreowski,
           and G. Rozenberg, editors, *Handbook of Graph Grammars and Computing
           by Graph Transformation*, volume 2, pages 105–180. World Scientific, 1999.
[BPPS00]   P. Bottoni, F. Parisi Presicce, and M. Simeoni. From formulae to rewrit-
           ing systems. In H. Ehrig, G. Engels, H.-J. Kreowsky, and G. Rozenberg,
           editors, *Theory and Application of Graph Transformations*, pages 267–280.
           Springer, Berlin, 2000.
[CMR+97]   A. Corradini, U. Montanari, F. Rossi, H. Ehrig, R. Heckel, and M. Lowe.
           Algebraic approaches to graph transformation - Part I: basic concepts and
           double pushout approach. In G. Rozenberg, editor, *Handbook of Graph
           Grammars and Computing by Graph Transformation*, volume 1, pages 163–
           245. World Scientific, 1997.
[EHK+97]   H. Ehrig, R. Heckel, M. Korff, M. Löwe, L. Ribeiro, A. Wagner, and Cor-
           radini. Algebraic approaches to graph transformation II: Single pushout
           approach and comparison with double pushout approach. In G. Rozenberg,
           editor, *Handbook of Graph Grammars and Computing by Graph Transfor-
           mation, Volume 1: Foundations*, chapter 4. World Scientific, 1997.
[Gir87]   J.-Y. Girard. Linear logic. *Theoretical Computer Science*, 50:1–102, 1987.
[Gir91]   J.-Y. Girard. Linear logic: A survey. Technical report, Int. Summer School
           on Logic and Algebra of Specification, 1991.
[Haa98]   V. Haarslev. A fully formalized theory for describing visual notations. In K.
           Marriott and B. Meyer, editors, *Visual Language Theory*, pages 261–292.
           Springer, New York, 1998.
[HM91]   R. Helm and K. Marriott. A declarative specification and semantics for
           visual languages. *Journal of Visual Languages and Computing*, 2:311–331,
           1991.
[HMO91]   R. Helm, K. Marriott, and M. Odersky. Building visual language parsers.
           In *ACM Conf. Human Factors in Computing*, pages 118–125, 1991.
[HP94]   J. Harland and D. Pym. A uniform proof-theoretic investigation of linear
           logic programming. *Journal of Logic and Computation*, 4(2):175–207, April
           1994.

[HPW96]   J. Harland, D. Pym, and M. Winikoff. Programming in Lygon: An overview. In *Algebraic Methodology and Software Technology*, LNCS 1101, pages 391–405. Springer, July 1996.

[Man98]   V. Manca. String rewriting and metabolism: A logical perspective. In G. Paun, editor, *Computing with Bio-Molecules*, pages 36–60. Springer-Verlag, Singapore, 1998.

[Mar94]   K. Marriott. Constraint multiset grammars. In *IEEE Symposium on Visual Languages*, pages 118–125. IEEE Computer Society Press, 1994.

[Mey97]   B. Meyer. Formalization of visual mathematical notations. In M. Anderson, editor, *AAAI Symposium on Diagrammatic Reasoning (DR-II)*, pages 58–68, Boston/MA, November 1997. AAAI Press, AAAI Technical Report FS-97-02.

[Mey00]   B. Meyer. A constraint-based framework for diagrammatic reasoning. *Applied Artificial Intelligence*, 14(4):327–344, 2000.

[Mil95]   D. Miller. A survey of linear logic programming. *Computational Logic*, 2(2):63–67, December 1995.

[MM00]   K. Marriott and B. Meyer. Non-standard logics for diagram interpretation. In *Diagrams 2000: International Conference on Theory and Application of Diagrams*, Edinburgh, Scotland, September 2000. Springer. To appear.

[MMW98]   K. Marriott, B. Meyer, and K:B. Wittenburg. A survey of visual language specification and recognition. In K. Marriott and B. Meyer, editors, *Visual Language Theory*, pages 5–85. Springer, 1998.

[RS95]   A. Repenning and T. Sumner. Agentsheets: A medium for creating domain-oriented visual languages. *IEEE Computer*, 28(3):17–26, March 1995.

[SCS94]   D.C. Smith, A. Cypher, and J. Sporer. Kidsim: Programming agents without a programming language. *Comm. ACM*, 37(7):55–67, July 1994.

[Shi95]   S.-J. Shin. *The Logical Status of Diagrams*. Cambridge University Press, Cambridge, 1995.

[Tan91]   T. Tanaka. Definite clause set grammars: A formalism for problem solving. *Journal of Logic Programming*, 10:1–17, 1991.

Multiset Automata*

Erzsébet Csuhaj-Varjú[1], Carlos Martín-Vide[2], and Victor Mitrana[3]

[1] Computer and Automation Research Institute
Hungarian Academy of Sciences
Kende u. 13-17, H-1111 Budapest, Hungary
csuhaj@sztaki.hu

[2] Research Group on Mathematical Linguistics
Rovira i Virgili University
Pça. Imperial Tàrraco 1, 43005 Tarragona, Spain
cmv@astor.urv.es

[3] University of Bucharest, Faculty of Mathematics
Str. Academiei 14, 70109 Bucharest, Romania
mitrana@funinf.math.unibuc.ro

Abstract. We propose a characterization of a Chomsky-like hierarchy of multiset grammars in terms of multiset automata. We also present the deterministic variants of all the automata investigated and prove that, unlike the string case, most of them are strictly less powerful than the nondeterministic variants. Some open problems and further directions of research are briefly discussed.

1 Introduction

In [8], a formalism based on multiset rewriting, called *constraint multiset grammar*, is introduced with the aim of providing a high-level framework for the definition of visual languages. They have been investigated in [11] from the parsing complexity of visual languages point of view. These grammars may be viewed as a bridge between the usual string grammars, dealing with string rewriting, and constraint logic programs, dealing with set rewriting. Other devices based on multiset rewriting have been reported in [16,17] where *abstract rewriting multiset systems* were used for modelling some features of the population development in artificial cell systems. Furthermore, a Chomsky-like hierarchy of grammars rewriting multisets was proposed in [9]. In all these approaces, the multisets are processed in a sequential, non-distributed way.

On the other hand, other authors have considered parallel or distributed ways of processing multisets for defining computational models inspired from biochemistry [1,2,3] or microbiology [12,13,14]. Different properties of commutative languages (macrosets in our terminology) have been considered in a series of papers, see, e.g., [5,10] and their references.

* Research supported by the Hungarian Scientific Research Fund "OTKA" Grant No. T 029615 and by the Dirección General de Enseñanza Superior e Investigación Cientifica, SB 97–00110508

C.S. Calude et al. (Eds.): Multiset Processing, LNCS 2235, pp. 69–83, 2001.

We address in this paper the accepting problem of multisets by means of some devices similar to automata from the formal language theory.

We shall assume the reader familiar with the fundamental concepts of formal language theory and automata theory; all notions we use can be found in [15]. We denote by $[1..n]$ the set of natural numbers $1, 2, \ldots, n$. For a finite set A we denote by $card(A)$ the cardinality of A. An *alphabet* is always a finite set whose elements are called *symbols*. A *string* over an alphabet V is a mapping $s : [1..n] \longrightarrow V$, for some n which is called the *length* of the string s, denoted by $|s|$. By convention, we shall consider also the *empty string* of length 0 which is written e. Moreover, for each $a \in V$ we set $|s|_a = card(\{i \mid s(i) = a\})$. The set of all strings over an alphabet V is denoted by V^* (the free monoid generated by V with the concatenation operation); moreover, $V^+ = V^* - \{e\}$. A subset of V^* is called *language*. For an alphabet $V = \{a_1, a_2, \ldots, a_k\}$, the Parikh mapping associated with V is a homomorphism ψ_V from V^* into the monoid of vector addition on $\mathbf{N}^k$, defined by $\Psi_V(s) = (|s|_{a_1}, |s|_{a_2}, \ldots, |s|_{a_k})$; moreover, given a language L over V, we define $\psi_V(L) = \{\Psi_V(x) \mid x \in L\}$.

A finite *multiset* over an alphabet V is a mapping $\sigma : V \longrightarrow \mathbf{N}$; $\sigma(a)$ expresses the number of copies of the symbol a in the multiset σ. The *empty multiset* is denoted by ε, that is $\varepsilon(a) = 0$ for all $a \in V$. The set of all multisets over V is denoted by V° while the set of all non-empty multisets is denoted by $V^\#$. A subset of V° is called *macroset*. The *weight* of a multiset σ is $\| \sigma \| = \sum_{a \in V} \sigma(a)$, and $\| \sigma \|_U = \sum_{a \in U} \sigma(a)$ for any subset U of V.

One can naturally associate with each multiset σ the finite language $L_\sigma = \{w \mid |w|_a = \sigma(a) \text{ for all } a \in V\}$, and conversely, with each string $w \in V^*$ one can associate the multiset $\mu_w : V \longrightarrow \mathbf{N}$ defined by $\mu_w(a) = |w|_a$. Clearly, the language formed by the empty string only is associated with the empty multiset and the empty multiset is associated with the empty string. Therefore, each multiset σ over V can be represented, in a one-to-one manner, by the vector $\Psi_V(L_\sigma)$, therefore, we do not distinguish between macrosets over an alphabet of cardinality k and sets of vectors from $\mathbf{N}^k$.

Normally, we use lower case letters from the end of the Latin alphabet for strings and small Greek letters for multisets. Furthermore, the Greek letter μ is used with an index only, for the multiset associated with the string denoted by the index.

For two multisets σ, τ over the same alphabet V we define:

- the inclusion relation $\sigma \subseteq \tau$ iff $\sigma(a) \leq \tau(a)$ for all $a \in V$;
- the addition multiset $\sigma + \tau$ defined by $\sigma + \tau(a) = \sigma(a) + \tau(a)$ for all $a \in V$;
- the difference multiset $\sigma - \tau$ defined by $(\sigma - \tau)(a) = \sigma(a) - \tau(a)$ for each $a \in V$, provided that $\tau \subseteq \sigma$.

2   Multiset Grammars

We recall from [9] the definition of a multiset grammar. A *multiset grammar* is a construct $G = (N, T, S, P)$, where N, T are disjoint alphabets, referred as the *nonterminal* and the *terminal* one, respectively, S is a finite macroset over

$N \cup T$, called the *starting* macroset, P is a finite set of *multiset rewriting rules* of the form $\sigma \to \tau$, where σ, τ are multisets over $N \cup T$ and $\sum_{A \in N} \sigma(A) \geq 1$.

For two multisets α_1, α_2 over $N \cup T$, we write $\alpha_1 \Longrightarrow \alpha_2$ if there exists a multiset rule $r : \sigma \to \tau \in P$ such that $\sigma \subseteq \alpha_1$ and $\alpha_2 = \alpha_1 - \sigma + \tau$. We denote by $\Longrightarrow^*$ the reflexive and transitive closure of the relation $\Longrightarrow$. The macroset *generated* by G is defined by

$$Gen(G) = \{\alpha \in T^\# \mid S \Longrightarrow^* \alpha\}.$$

Following [9], a natural Chomsky-like classification of such grammars is given below:

1. Grammars G as above (without any restriction on the multiset rules) are said to be *arbitrary*.
2. If $\| \sigma \| \leq \| \tau \|$ for all rules $\sigma \to \tau$ in P, then G is said to be *monotone*.
3. If $\| \sigma \| = 1$ for all rules $\sigma \to \tau$ in P, then G is said to be *context-free*.
4. If $\| \sigma \| = 1, \| \tau \| \leq 2$, and $\| \tau \|_N \leq 1$ for all rules $\sigma \to \tau$ in P, then G is said to be *regular*.

We denote by *mARB, mMON, mCF, mREG* the families of macrosets generated by arbitrary, monotone, context-free, and regular multiset grammars, respectively. By *REG, CF, MON, ARB* we denote the families of regular, context-free, context-sensitive, and recursively enumerable languages, respectively. For a family of languages F, we denote by PsF the family of Parikh sets of vectors associated with the languages in F. The family of all semilinear sets is denoted by *SLin* while the family of all linear sets is denoted by *Lin*. Furthermore, *mFIN* stands for the class of all finite macrosets while *PsFIN* stands for the class of all finite Parikh sets of vectors.

Following [9], the aforementioned classes of macrosets are related as in the diagram from Figure 1, where a pair of arrows indicates a strict inclusion of the lower class into the upper class, and a single arrow indicates an inclusion for which the properness is open.

The proper inclusions $mMON \subset PsMON$ and $mARB \subset PsARB$ appear to be of interest. The first relation is no longer unexpected if we remember that all variants of scattered context grammars generate only context-sensitive languages [7,4,6].

3   Multiset Finite Automata

Roughly speaking, a multiset finite automaton consists of an input bag in which a multiset is placed, and a detecting head which can detect whether or not a given symbol appears in the bag (there is at least one copy of that symbol in the multiset). The automaton works as follows: it starts in the initial state having a multiset in its bag and changes its current state depending on the former state and the detection of a symbol in the bag. If a symbol has been detected, it is automatically removed from the bag. The automaton stops when no further

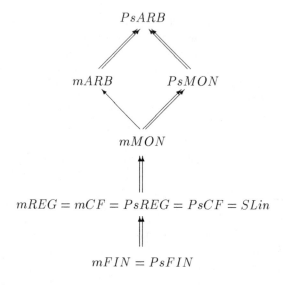

$PsARB$

$mARB$ $PsMON$

$mMON$

$mREG = mCF = PsREG = PsCF = SLin$

$mFIN = PsFIN$

Fig. 1.

move is possible or the bag is empty. If the bag is empty and the current state is a final state, then the automaton accepts the input multiset. In all the other situations, the initial multiset is rejected.

Formally, a *multiset finite automaton* (MFA, for short) is a structure

$$\mathcal{A} = (Q, V, f, q_0, F),$$

where Q is a finite non-empty set of states, V is the input alphabet, q_0 is the initial state, F is the set of final states, and f is the transition mapping

$$f : Q \times V \longrightarrow 2^Q.$$

A configuration of $\mathcal{A}$ is a pair (q, τ), where q is a state (the current state) and τ is a multiset (the current content of the bag). We define the following relation on the set of all configurations:

$$(q, \tau) \vdash (s, \rho) \quad \text{iff there exists } a \in V \text{ such that } s \in f(q, a) \text{ and } \tau - \rho = \mu_a.$$

The reflexive and transitive closure of this operation is denoted by $\vdash^*$. The macroset accepted by $\mathcal{A}$ is defined by

$$Rec(\mathcal{A}) = \{\tau \mid (q_0, \tau) \vdash^* (q, \varepsilon) \text{ for some } q \in F\}.$$

Otherwise stated, the macroset accepted by the finite automaton $\mathcal{A}$ is the set of all multisets τ for which the automaton $\mathcal{A}$ starting in q_0 and having τ in its bag reaches a final state with an empty bag.

Example 1 *We consider the multiset finite automaton* $\mathcal{A} = (\{q_0, q_1, q_2\}, \{a, b, c\},$ $f, q_0, \{q_0\})$, *where the transition mapping f is defined as in the next diagram:*

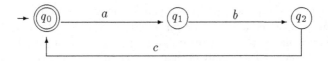

Fig. 2.

It is easy to notice that $Rec(\mathcal{A}) = \{\tau \mid \tau(a) = \tau(b) = \tau(c)\}$.

The well-known proof for showing the equivalence of finite automata and regular string grammars, with respect to their computational power, can be easily handled in order to prove the equivalence of multiset finite automata and multiset regular grammars with respect to the same criterion. We denote by $\mathcal{L}(MFA)$ the family of all macrosets accepted by multiset finite automata.

Proposition 1 $\mathcal{L}(MFA) = mREG = mCF = PsREG = PsCF = SLin.$

A MFA $\mathcal{A} = (Q, V, f, q_0, F)$ is said to be *deterministic* (DMFA for short) if the following two conditions are satisfied:

(i) $card(f(q, a)) \leq 1$ for all $q \in Q$ and $a \in V$.
(ii) For each state q, if $card(f(q, a)) \neq 0$, then $card(f(q, b)) = 0$ for all $b \neq a$.

The equivalence relation between nondeterministic and deterministic finite automata on strings naturally suggests the same question regarding the DMFAs and MFAs. For it is easy to notice that there exist finite macrosets which cannot be accepted by any DMFA, the answer is immediate. However, the next result states more, namely

Proposition 2 *1. $\mathcal{L}(DMFA)$ is incomparable with Lin and the family of finite macrosets, with respect to the set inclusion. 2. $\mathcal{L}(DMFA) \subset \mathcal{L}(MFA)$.*

Proof. Clearly, the second assertion follows immediately from the first one. We shall prove only the relation between $\mathcal{L}(DMFA)$ and Lin. To this aim, we consider the linear set $L = \{x(1, 2) + y(2, 1) \mid x, y \in \mathbf{N}\}$. Assume that L is accepted by a DMFA $\mathcal{A} = (Q, \{a, b\}, f, q_0, F)$. Since both multisets $(2n+4, n+5)$ and $(2n + 4, 4n + 2)$, for a given arbitrary large n, belong to L, it follows that there exist two final states $q_1, q_2 \in Q$ such that

$$(q_0, (2n + 4, 4n + 2)) \vdash^* (q_1, (0, 3n - 3)) \vdash^* (q_2, (0, 0)).$$

Because there are repeated states in the computation $(q_1, (0, 3n-3)) \vdash^* (q_2, (0,0))$, one infers that vectors of the form $(2n + 4, m)$, with arbitrary large m, have to be in L, a contradiction.

Conversely, the DMFA defined by the transition mapping

$$g(s_0, a) = s_1, \quad g(s_1, a) = s_2, \quad g(s_2, b) = s_3, \quad g(s_3, c) = s_1,$$

having the initial state s_0 and the final states s_1 and s_3, accepts the macroset

$$\{(n + 1, n, n) \mid n \geq 0\} \cup \{(n + 2, n + 1, n) \mid n \geq 0\}$$

which is not linear, and the proof is complete.

We now consider a slight extension of multiset finite automata in the following way. This automaton, called *multiset finite automata with detection* (MFAD, for short) is able to change its state not only when some symbols are detected in its bag but also when some symbols do not appear in the bag. More precisely, an MFAD is a structure $\mathcal{A} = (Q, V, f, q_0, F)$, where all parameters are defined as for a MFA, excepting the transition mapping which is defined on $Q \times (V \cup \bar{V})\}$ into Q. By $\bar{V}$ we mean an alphabet containing all barred copies of the symbols in V. A move defined by the function f applied to a pair (q, a), $a \in V$, is the same move defined by the transition function of a MFA applied to the same pair, while a move defined by the function f applied to a pair $(q, \bar{a})$, $a \in V$, may determine the automaton to enter a state from $f(q, \bar{a})$ providing that no copy of a is in the bag. More precisely,

$$(q, \tau) \vdash (s, \rho)$$

if and only if

(i) there exists $a \in V$ such that $s \in f(q, a)$ and $\tau - \rho = \mu_a$, or
(ii) there exists $a \in V$ such that $s \in f(q, \bar{a}), \tau(a) = 0$ and $\rho = \tau$.

A MFAD $\mathcal{A} = (Q, V, f, q_0, F)$ is said to be *deterministic* (DMFAD for short) if the following two conditions are satisfied:

(i) $card(f(q, x)) \le 1$ for all $q \in Q$ and $x \in V \cup \bar{V}$.
(ii) For each state q and $a \in V$, if $card(f(q, a)) \ne 0$ or $card(f(q, \bar{a})) \ne 0$, then $card(f(q, y)) = 0$ for all $y \in (V \setminus \{a\}) \cup (\bar{V} \setminus \{\bar{a}\})$.

This extension does not induce any increase of the computational power of nondeterministic multiset finite automata but it does it for the deterministic variants. Denoting by $\mathcal{L}(MFAD)$ and $\mathcal{L}(DMFAD)$ the families of macrosets accepted by MFADs and DMFADs, respectively, we state:

Proposition 3 *1. $\mathcal{L}(MFA) = \mathcal{L}(MFAD)$. 2. $\mathcal{L}(DMFAD)$ contains all finite macrosets. 3. Lin and $\mathcal{L}(DMFAD)$ are incomparable with respect to the set inclusion.*

Proof. 1. Let $\mathcal{A} = (Q, V, f, q_0, F)$ be a MFAD; we assume that $V = \{a_1, a_2, \ldots, a_n\}$ for some positive integer n. Construct the regular multiset grammar $G = (N, V, S, P)$, where

$$N = \{\langle q, x \rangle \mid q \in Q, x \in V^*, |x| < n, |x|_{a_i} \le 1, i \in [1..n]\},$$
$$S = \{\mu_{\langle q_0, x \rangle} \mid q \in Q, x \in V^*, |x| < n, |x|_{a_i} \le 1, i \in [1..n]\},$$

and P contains all rules defined as follows:

- For each transition $q \in f(s, a)$, add the rules $\mu_{\langle s, x \rangle} \to \mu_{a\langle q, x \rangle}$ for all nonterminals $\langle s, x \rangle, \langle q, x \rangle \in N$, providing $|x|_a = 0$.
- For each transition $q \in f(s, \bar{a})$, add the rules $\mu_{\langle s, x \rangle} \to \mu_{\langle q, y \rangle}$, for all nonterminals $\langle s, x \rangle, \langle q, x \rangle \in N$, where $y = xa$, if $|x|_a = 0$, or $y = x$, otherwise.
- For each final state $q \in F$, add the rules $\mu_{\langle q, x \rangle} \to \varepsilon$, for all nonterminals $\langle q, x \rangle \in N$.

Let

$$(q_0, (k_1^{(0)}, k_2^{(0)}, \ldots, k_n^{(0)})) \vdash^* (s_1, (k_1^{(1)}, k_2^{(1)}, \ldots, k_n^{(1)})) \vdash$$
$$(q_1, (k_1^{(1)}, k_2^{(1)}, \ldots, k_n^{(1)})) \vdash^* (s_2, (k_1^{(2)}, k_2^{(2)}, \ldots, k_n^{(2)})) \vdash$$
$$(q_2, (k_1^{(2)}, k_2^{(2)}, \ldots, k_n^{(2)})) \vdash^* \ldots \vdash^* (q_{m-1}, (k_1^{(m-1)}, k_2^{(m-1)}, \ldots, k_n^{(m-1)})) \vdash^*$$
$$(s_m, (k_1^{(m)}, k_2^{(m)}, \ldots, k_n^{(m)})) = (s_m, (0, 0, \ldots, 0))$$

be a computation in $\mathcal{A}$, which accepts the multiset $(k_1^{(0)}, k_2^{(0)}, \ldots, k_n^{(0)})$, such that the following conditions are satisfied:

(i) $q_i \in f(s_i, \bar{a}_{j_i}), j_i \in [1..n], i \in [1..m-1]$.

(ii) Each computation $(q_i, (k_1^{(i)}, k_2^{(i)}, \ldots, k_n^{(i)})) \vdash^* (s_{i+1}, (k_1^{(i+1)}, k_2^{(i+1)}, \ldots, k_n^{(i+1)})), i \in [0..m-1]$, contains no transition based on checking the presence of a symbol in the current bag.

One can construct the following derivation in G:

$$\rho_0 \Longrightarrow^* \eta_1 \Longrightarrow \rho_1 \Longrightarrow^* \eta_2 \ldots \rho_{m-1} \Longrightarrow^* \eta_m \Longrightarrow \eta_{m+1},$$

where

$\rho_0(\langle q_0, e \rangle) = 1$,

$\eta_i(\langle s_i, y_i \rangle) = 1$, where y_i is the shortest string containing
all symbols $a_{j_1}, a_{j_2}, \ldots, a_{j_{i-1}}, \eta_i(a_p) = k_p^{(0)} - k_p^{(i)}, p \in [1..n], i \in [1..m]$,

$\rho_i(\langle q_i, y_i \rangle) = 1$, where y_i is the shortest string containing
all symbols $a_{j_1}, a_{j_2}, \ldots, a_{j_i}, \rho_i(a_p) = k_p^{(0)} - k_p^{(i)}, p \in [1..n], i \in [1..m-1]$,

$\eta_{m+1}(a_p) = k_p^{(0)}, p \in [1..n]$.

Note that each symbol in $N \cup V$ which does not appear in the definition of some multiset from above has no occurrence in the respective multiset.

Therefore, G can generate all the multisets accepted by $\mathcal{A}$. The converse inclusion follows in the same way, and this concludes the first assertion.

2. Let $L = \{(n_1^{(i)}, n_2^{(i)}, \ldots, n_k^{(i)}) \mid i \in [1..p]\}$ be an arbitrary finite macroset over the alphabet $V = \{a_1, a_2, \ldots, a_k\}$, having its vectors listed in the lexicographical order. We prove by induction on the cardinality of V that L belongs to $\mathcal{L}(DMFAD)$. If $k = 1$, then a DMFAD which accepts L is presented in Figure 3.

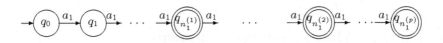

Fig. 3.

We assume that

$$n_1^{(1)} = n_1^{(2)} = \ldots = n_1^{(t_1)} < n_1^{(t_1+1)} = \ldots = n_1^{(t_2)} < \ldots < n_1^{(t_{s-1}+1)} = \ldots$$
$$= n_1^{(t_s)} = n_1^{(p)}.$$

By the induction hypothesis, for each $i \in [1..s]$, let $\mathcal{A}_i = (Q_i, V \setminus \{a_1\}, q_0^{(i)}, f_i, F_i)$ be the DMFAD which accepts the language

$$L_{t_i} = \{(n_2^{(j)}, n_3^{(j)}, \ldots, n_k^{(j)}) \mid j \in [t_{i-1} + 1..t_i]\},$$

where $t_0 = 0$. Moreover, the sets Q_i are pairwise disjoint.

We construct the DMFAD

$$\mathcal{A} = (\cup_{i=1}^{s} Q_i \cup \{q_0, q_1, \ldots, q_{n_1^{(p)}}\}, V, q_0, f, \cup_{i=1}^{s} F_i)$$

as shown in the next figure.

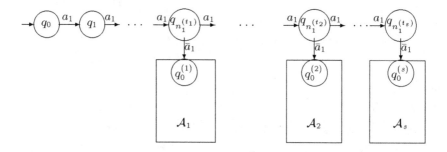

Fig. 4.

More formally, the transition mapping is defined by

$$f(q_j, a_1) = q_{j+1}, j \in [0..n_1^{(p)} - 1],$$
$$f(q_{n_1^{(t_j)}}, \bar{a}_1) = q_0^{(j)}, j \in [1..s],$$
$$f(q, a_l) = f_j(q, a_l), l \in [2..k], \text{ provided that } q \in Q_j.$$

Cleary, the automaton is deterministic and it accepts the given macroset L.

3. The linear macroset considered in the proof of Proposition 2 cannot be recognized by any DMFAD – a similar reasoning to that used in the aforementioned proof is valid – which completes the proof.

4   Multiset Linear Bounded Automata

A multiset linear bounded automaton has a bag able to store a finite multiset and a read-write head which can pick up (read) a symbol from the bag and add at most one symbol to the contents of the bag. Adding a symbol is meant as increasing the number of copies of that symbol by 1. Being in a state, the multiset linear bounded automaton reads a symbol which occurs in the bag, changes its state and adds or not a symbol to the bag contents. When the machine stops (no move is possible anymore) in a final state with an empty bag, we say that the input multiset is accepted, otherwise it is rejected.

Formally, a *multiset linear bounded automaton* (shortly, MLBA) is a construct

$$\mathcal{M} = (Q, V, U, f, q_0, F),$$

where Q and F are the finite sets of states and final states, respectively, V and U are the input and the bag alphabets, respectively, $V \subseteq U$, q_0 is the initial state, and f is the transition mapping from $Q \times U$ into the set of all subsets of $Q \times (U \cup \{e\})$. As in the case of MFAs, a configuration is a pair (q, τ), where q is the current state and τ is the content of the bag. We write

$$(q, \tau) \models (s, \rho)$$

if and only if there exists $a \in V$ such that

(i) $(s, b) \in f(q, a), b \neq e, \tau(a) \geq 1, \rho(a) = \tau(a) - 1, \rho(b) = \tau(b) + 1,$
    and $\rho(c) = \tau(c)$ for all $c \in V, c \notin \{a, b\}$, or

(ii) $(s, e) \in f(q, a), \tau(a) \geq 1, \rho(a) = \tau(a) - 1,$
    and $\rho(c) = \tau(c)$ for all $c \in V, c \neq a.$

The reflexive and transitive closure of this operation is denoted by $\models^*$. The macroset accepted by $\mathcal{A}$ is defined by

$$Rec(\mathcal{M}) = \{\tau \mid (q_0, \tau) \models^* (q, \varepsilon) \text{ for some } q \in F\}.$$

The deterministic MLBAs are defined as usual. In [9], a result corresponding to the workspace theorem in formal language theory was proved for multisets.

For an arbitrary multiset grammar $G = (N, T, A, P)$, let

$$D : \alpha_1 \Longrightarrow \alpha_2 \Longrightarrow \ldots \Longrightarrow \alpha_k$$

be a derivation with $\alpha_1 \in A$. The *workspace* of the derivation D, denoted by $WS(D)$, is the maximal weight of the multisets appearing in D, namely $\max\{\| \alpha_i \| \mid i \in [1..k]\}$. Then, for $\beta \in Gen(G)$ we put

$$WS_G(\beta) = \inf\{WS(D) \mid D : \alpha \Longrightarrow^* \beta, \alpha \in A\}.$$

We say that G has a linearly bounded workspace if there is a constant k such that for each non-empty $\beta \in Gen(G)$ we have $WS_G(\beta) \leq k|\beta|$.

Theorem 1 (Workspace Theorem) *The class of macrosets generated by grammars with linearly bounded workspace equals the class of macrosets generated by monotone grammars.*

Based on this result, it is easy to prove the next result. Note that the second assertion, which can be easily proved, corresponds to one of the most important open problems in the formal language theory.

Proposition 4 *1. mMON equals the class of macrosets accepted by MLBAs.*
2. Deterministic MLBAs are strictly weaker than MLBAs.

Proof. Let $G = (N, T, A, P)$ be a monotone grammar; without loss of generality
(see [9]), we may assume that A contains only one multiset μ_S over $N \cup T$
(remember that S has to be a nonterminal and μ_S is the multiset associated
with the string S) and μ_S is not a submultiset of any righthand side of the rules
in P. We construct the MLBA $\mathcal{A} = (Q, T, N \cup T, f, q_0, F)$, where

$$Q = \{q_0, q_f\} \cup \{\langle \beta, \alpha \rangle \mid \text{ there exists a rule } \eta \to \tau \in P \text{ such that}$$
$$\beta \text{ and } \alpha \text{ are submultisets of } \tau \text{ and } \eta, \text{ respectively}\}$$

and the transition mapping f is defined as follows:

$$f(q_0, a) = f(\langle \varepsilon, \varepsilon \rangle, a) = \{(\langle \beta, \alpha \rangle, b) \mid \text{ there exists a rule } \eta \to \tau \in P$$
$$\text{such that } \tau(a) - \beta(a) = \eta(b) - \alpha(b) = 1 \text{ and}$$
$$\tau(c) - \beta(c) = \eta(c) - \alpha(c) = 0 \text{ for all } c \in (N \cup T) \setminus \{a, b, S\}\},$$
$$\text{for all } a \in (N \cup T) \setminus \{S\}$$
$$f(\langle \varepsilon, \varepsilon \rangle, S) = \{(q_f, e)\}$$
$$f(\langle \beta, \alpha \rangle, a) = \{(\langle \beta', \alpha' \rangle, b) \mid \beta(a) - \beta'(a) = \alpha(b) - \alpha'(b) = 1 \text{ and}$$
$$\beta(c) - \beta'(c) = \alpha(c) - \alpha'(c) = 0 \text{ for all } c \in (N \cup T) \setminus \{a, b, S\}\},$$
$$\text{for all } a \in (N \cup T) \setminus \{S\}$$
$$f(\langle \beta, \varepsilon \rangle, a) = \{(\langle \beta', \varepsilon \rangle, e) \mid \beta(a) - \beta'(a) = 1 \text{ and } \beta(c) - \beta'(c) = 0$$
$$\text{for all } c \in (N \cup T) \setminus \{a, b, S\}\},$$
$$\text{for all } a \in (N \cup T) \setminus \{S\}.$$

The set of final states contains the state q_f; if the empty multiset is generated
by G, then we add q_0 to the set of final states.

It is easy to infer that, on each input multiset, the automaton simulates the
application of the rules of G in the reverse order – the righthand side of a rule,
which is identified in the bag, is replaced by the lefthand side of that rule - and
accepts (enters q_f) if and only if the contents of the bag is exactly μ_S. Since G
is monotone, the weight of the bag at any step is bounded by the weight of the
input multiset, hence the automaton is linearly bounded.

Conversely, let $\mathcal{A} = (Q, V, U, f, q_0, F)$ be a MLBA; we construct the arbitrary
multiset grammar $G = (N, V, \mu_S, P)$, where $N = \{S\} \cup V \cup V' \cup Q$. We denote
by V' the set $\{a' \mid a \in V\}$. The set of multiset productions P consists of the
following rules:

1. $\mu_S \to \mu_{aa'}S$ for all $a \in V$,
2. $\mu_S \to \mu_{aa'}q_0$ for all $a \in V$,
3. $\mu_{qa'} \to \mu_{sb'}$ if $(s, b) \in f(q, a)$,
4. $\mu_{qa'} \to \mu_s$ if $(s, e) \in f(q, a)$,
5. $\mu_q \to \varepsilon$ if $q \in F$.

The above grammar works as follows. In the first phase, by using the rules from the first two groups, itgenerates a multiset having just one copy of q_0 and the same number of copies of a and a' for all $a \in V$, respectively. Roughly speaking, we may say that the sentential form comprises two multisets, of equal weight, one of them being the primed copy of the other. The grammar now simulates the automaton $\mathcal{A}$ on the primed multiset. If the automaton accepts, the final state is erased from the sentential form and the result of this derivation is the non-primed multiset. Thus, a multiset over V is generated by G if and only if $\mathcal{A}$ accepts that multiset. We conclude the proof by mentioning that the workspace used by G for any derivation generating a multiset ρ is bounded by $2 \parallel \rho \parallel +1$, hence G has a linearly bounded workspace.

2. The finite macroset τ over the binary alphabet $\{a, b\}$, containing the two multisets μ_a and μ_b, cannot be accepted by any $DMLBA$.

Remember that the extension considered in the previous section had no effect on the computational power of nondeterministic MFAs. The situation changes when considering the same extension for MLBAs. By [9], all the macrosets over one letter alphabet in $mMON$ are semilinear. The next example provides a deterministic MLBAD (multiset linear bounded automaton with detection) which is able to recognize the non-semilinear macroset $\{2^n \mid n \geq 1\}$. We list the transition mapping only and indicate the set of final states, the other parameters can be deduced easily.

Example 2

$$f(q_0, a) = \{(q_1, e)\} \quad f(q_1, a) = \{(q_0, b)\}$$
$$f(q_0, \bar{a}) = \{(q_2, e)\} \quad f(q_1, \bar{a}) = \{(q_r, e)\}$$
$$f(q_r, b) = \{(q_e, e)\} \quad f(q_r, \bar{b}) = \{(q_f, e)\}$$
$$f(q_2, b) = \{(q_3, e)\} \quad f(q_3, b) = \{(q_2, a)\}$$
$$f(q_2, \bar{b}) = \{(q_0, e)\} \quad f(q_3, \bar{b}) = \{(q'_r, e)\}$$
$$f(q'_r, a) = \{(q_e, e)\} \quad f(q'_r, \bar{a}) = \{(q_f, e)\}$$

The set of final states consists of the state q_f only. One can easily check that the automaton iteratively halves the number of copies of a or b as long as this is possible. Then, it checks whether or not the previous odd number of copies was one, in which case it accepts, or reject, otherwise.

Clearly, from the previous example it immediately follows that

Proposition 5 *1. MLBAs are strictly weaker than MLBADs. 2. Deterministic MLBAs are strictly weaker than deterministic MLBADs.*

In order to get more information concerning the accepting power of ML-BADs, we consider here multiset random context grammars. The definition of a multiset random context grammar is a direct extension of the definition of a string random context grammar, see [15]. We mention that the multiset random context grammars contains only non-erasing context-free rules. The class of macrosets accepted by multiset random context grammars is denoted by mRC.

Proposition 6 *Every macroset in mRC is accepted by an MLBAD.*

Proof. Let $G = (N, T, \mu_S, P)$ be a multiset random context grammar. We construct the MLBAD $\mathcal{A} = (Q, T, T \cup N \cup N', f, q_0, \{q_f\})$, where N' is the alphabet containing the primed copies of all the letters in N,

$$Q = \{q_0, q_f\} \cup \{[X, Y, Z, \rho] \mid X, Y, Z \subset T \cup N\} \cup$$
$$\{[X, Y, Z, \rho] \mid X, Y, Z \subset T \cup N, \rho \subseteq \tau \text{ for some } \sigma \to \tau \in P\}.$$

and the transition mapping f is defined as follows:

- For each rule $(P, F; \sigma \to \tau) \in P$ we have

  - $([P, P, F, \tau - \mu_a], e) \in f(q_0, a)$, $a \in T \cup N$, if $\| \tau \| \geq 2$,
  - $([P, P, F], A) \in f(q_0, a)$, if $\tau = \mu_a$ and $\sigma = \mu_A$,
  - $([P, P, F, \rho - \mu_a], e) \in f([P, P, F, \rho], a)$, $a \in T \cup N$, if $\| \rho \| \geq 2$,
  - $([P, P, F], A) \in f([P, P, F, \rho], a)$, if $\rho = \mu_a$ and $\sigma = \mu_A.$.
  - For each $a \in X$ we have

  * $f([X, Y, Z], a) = \{([X \setminus \{a\}, Y, Z], a')\}$,
  * $f([\emptyset, X, Y], a') = \{([\emptyset, X \setminus \{a\}, Y], a)\}$,
  * $f([\emptyset, \emptyset, X], \bar{a}) = \{([\emptyset, \emptyset, X \setminus \{a\}], e)\}$,

  for all $X, Y, Z \subseteq T \cup N$.

- $f([\emptyset, \emptyset, \emptyset], a) = \{(q_0, a)\}$
- $f(q_0, S) = \{(q_f, e)\}$.

It is rather easy to observe that

$$\mu_s \Longrightarrow^* \eta \text{ if and only if } (q_0, \eta) \models^* (q_f, \varepsilon)$$

for any multiset η over $N \cup T$. It follows immediately that $Gen(G) = Rec(\mathcal{A})$ holds which concludes the proof.

We point out here some open problems:

1. What is the relation between $mARB$ and the family of macrosets accepted by MLBADs?

2. The same question for $PsMON$ and the family of macrosets accepted by MLBADs.

3. Are deterministic MLBADs strictly weaker than MLBADs?

4. What is the relation between the macroset classes recognized by deterministic MLBADs and MLBAs, respectively?

5   Multiset Turing Machines

Multiset Turing machines, unlike MLBAs, have a bag able to store an infinite multiset and a read-write head which can pick up a symbol from the bag and add at most one symbol to the content of the bag. The working mode is defined exactly as for MLBAs. Such a machine accepts a multiset when it enters a final state having in its bag an infinite number of B's and nothing else. Let us denote by β the multiset $\beta(B) = \infty$ and $\beta(a) = 0$ for any other symbol a.

Formally, a *multiset Turing machine* (shortly, MTM) is a construct

$$\mathcal{M} = (Q, V, U, f, q_0, B, F),$$

where Q and F are the finite sets of states and final states, respectively, V and U are the input and the bag alphabets, respectively, $V \subset U$, q_0 is the initial state, B is a special symbol which occurs infinitely many times in the bag, and f is the transition mapping from $Q \times U$ into the set of all subsets of $Q \times ((U \cup \{e\}) \setminus \{B\})$. A configuration is a pair (q, τ), where q is the current state and τ is the content of the bag. We write

$$(q, \tau) \models (s, \rho)$$

if and only if there exists $a \in V$ such that

$$(i) \quad (s, b) \in f(q, a), b \neq e, \ \tau(a) \geq 1, \rho(a) = \tau(a) - 1, \rho(b) = \tau(b) + 1,$$
$$\text{and } \rho(c) = \tau(c) \text{ for all } c \in V, c \notin \{a, b\}, \text{ or}$$
$$(ii) \quad (s, e) \in f(q, a), \tau(a) \geq 1, \rho(a) = \tau(a) - 1, \rho(b) = \tau(b) + 1,$$
$$\text{and } \rho(c) = \tau(c) \text{ for all } c \in V, c \neq a.$$

The reflexive and transitive closure of this operation is denoted by $\models^*$. The macroset accepted by $\mathcal{A}$ is defined by

$$Rec(\mathcal{M}) = \{\tau \mid (q_0, \tau) \models^* (q, \beta) \text{ for some } q \in F\}.$$

The proof of the next result is left to the reader who can be successful by using the proof of Proposition 4 with some minor changes.

Proposition 7 *1. mARB equal the accepting power of MTMs. 2. Deterministic MTMs are strictly weaker than MTMs.*

The same extension defined in the previous two sections can be considered also for MTMs. Thus, one gets a modified version of a register machine. One can prove that:

Proposition 8 *1. MTMDs are stronger than MTMs. 2. PsARB equals the class of macrosets accepted by MTMDs.*

Proof. The first statement is a consequence of the second one which follows from the fact that MTMDs are equivalent to register machines.

It is not known whether or not deterministic MTMs are strictly weaker than deterministic MTMDs, neither whether deterministic MTMDs are strictly weaker than MTMDs.

6   Final Remarks and Further Directions of Research

We have considered some variants of multiset automata whose working mode is based on the multiset addition and subtraction. These automata turned out to be suitable for an automata characterization of the Chomsky-like hierarchy of multiset grammars introduced in [9].

Several interesting problems listed in the previous sections remain to be solved. Besides them, there are lots of problems concerning multiset automata, coming from different possible directions of further research. We briefly discuss below some of them.

We start with a remark which could make an essential distinction among MLBADs as well as MTMDs: In the definition of an MLBAD, for instance, there is no restriction when checking the existence of a given symbol in its bag; it may add or not a symbol to the content of the bag. It appears natural to consider that whenever it checks the existence of a symbol in its bag it cannot add anything to the bag. Does this restriction diminish the computational power of these variants?

It seems to be of interest to apply the program we used in this paper to other types of automata: multi-head automata, multi-tape automata, contraction automata, etc.

Another attractive direction of investigation is to consider the complexity of the automata introduced in this paper. This study may develop in two directions: descriptional complexity of these automata with respect to different measures (states, transitions, etc.) and computational complexity of them with respect to time/space resources they need.

References

1. J.P. Banâtre, A. Coutant, D. Le Metayer, A parallel machine for multiset transformation and its programming style, *Future Generation Computer Systems*, 4 (1988), 133–144.
2. J. P. Banâtre, D. Le Metayer, Gamma and chemical reaction model: ten years after, *Coordination Programming: Mechanisms, Models and Semantics* (C. Hankin, ed.), Imperial College Press, 1996.
3. G. Berry, G. Boudol, The chemical abstract machine, *Theoretical Computer Sci.*, 96 (1992), 217–248.
4. A.B. Cremers, Normal forms for context-sensitive grammars, *Acta Informatica*, 3 (1973), 59–73.
5. S. Crespi-Reghizzi, D. Mandrioli, Commutative grammars, *Calcolo*, 13, 2, (1976), 173–189.
6. J. Gonczarowski, M.K. Warmuth, Scattered versus context-sensitive rewriting, *Acta Informatica*, 27 (1989), 81–95.
7. S. Greibach, J. Hopcroft, Scattered context grammars, *J. Comput. Syst. Sci.*, 3 (1969), 233–247.
8. R. Helm, K. Marriott, M. Odersky, Building visual language parsers, *Proc. ACM Conf. Human Factors in Computing*, 1991.

9. M. Kudlek, C. Martin-Vide, G. Păun, Toward FMT (Formal Macroset Theory), *Pre-Proceedings of the Workshop on Multiset Processing (WMP'2000)* (C.S. Calude, M.J. Dinneen, G. Păun, eds.), CDMTCS–140, 2000, 149–158.

10. M. Latteux, Cônes rationnels commutatifs, *J. of Computer and System Sciences*, 18 (1979), 307–333.

11. K. Marriott, Constraint multiset grammars, *Proc. IEEE Symposium on Visual Languages* (A. L. Ambler, T. Kimura, eds.), IEEE Comp. Soc. Press, 1994, 118–125.

12. G. Păun, G. Rozenberg, A. Salomaa, *DNA Computing. New Computing Paradigms*, Springer-Verlag, Berlin, 1998.

13. G. Păun, Computing with membranes *J. of Computer and System Sciences*, 61, 1 (2000), 108–143

14. G. Păun, Computing with membranes (P systems): Twenty six research topics, *CDMTCS Report No. 119, Auckland University*, 2000.

15. G. Rozenberg, A. Salomaa, eds., *Handbook of Formal Languages*, 3 volumes, Springer-Verlag, Berlin, 1997.

16. Y. Suzuki, H. Tanaka, Symbolic chemical system based on abstract rewriting system and its behavior pattern, *Artificial Life Robotics*, 1 (1997), 211–219.

17. Y. Suzuki, H. Tanaka, Chemical evolution among artificial proto-cells, *Proc. of Artificial Life VIII Conf.*, MIT Press, 2000.

Parikh Mapping and Iteration

Jürgen Dassow

Otto-von-Guericke-Universität Magdeburg
Fakultät für Informatik
PSF 4120, D–39016 Magdeburg
`dassow@iws.cs.uni-magdeburg.de`

Abstract. The Parikh mapping maps each word over an alphabet with n letters to an n-dimensional vector whose components give the number of occurrences of the letters in the word. We consider the Parikh images of sequences and languages obtained by iterated applications of morphisms (or sets of substitutions). Furthermore we modify the Parikh mapping such that it can be iterated and study the corresponding sequences.

1 Introduction

Let V be an alphabet. By $\#(V)$ and V^* we denote the cardinality of V and the set of all words over V including the empty word λ, respectively. We consider a fixed order of the elements $a_1, a_2, \ldots, a_n$ of V and define the Parikh mapping[1] $p_V : V^* \to \mathbf{N}^n$ by

$$p_V(w) = (\#_{a_1}(w), \#_{a_2}(w), \ldots, \#_{a_n}(w)),$$

where, for a word $w \in V^*$ and a letter $a \in V$, we denote the number of occurrences of a in w by $\#_a(w)$ and $\mathbf{N}$ denotes the set of non-negative integers. If V is clear from the context, we write p instead of p_V.

This mapping has also a meaning in multiset theory. Instead of considering a word w, i.e., an ordered sequence over V, we regard the unordered version, i.e., the multiset consisting of $\#_a(w)$ elements a.

In this paper we consider combinations of the Parikh mapping and the well-known concept of iteration. There are at least two possibilities for this combination.

1. The first one is to regard an iteration process on a word leading to a sequence of words and to look for the Parikh images of these words. The simplest iteration process is the iterated application of a morphism. We mention that this idea has – besides its mathematical attraction – also some biological interest (for details we refer to [8]). We shall investigate this approach in Section 2.
2. The second one consists in an iteration of the Parikh mapping itself. In order to ensure this we have to consider the vector $p_V(w)$ as a word. This restricts

---

[1] Named after R. J. PARIKH

C.S. Calude et al. (Eds.): Multiset Processing, LNCS 2235, pp. 85–101, 2001.
© Springer-Verlag Berlin Heidelberg 2001

the alphabet; the elements of V have to be digits in order to represent the numbers of occurrences and the comma and the brackets. However, since the number of occurrences of the comma equals $n - 1$ and each bracket occurs exactly once, we modify the Parikh mapping using numbers as delimiters between the numbers of occurrences. In Section 3 we shall give the detailed definitions and shall give some results on the sequences obtained in this way.

2   Parikh Sequences and Parikh Languages Obtained by Iterating Functions

2.1   D0L Systems

A *D0L system*[2] G is a triple $G = (V, h, w)$ where

- V is an alphabet,
- h is a morphism which maps V^* into V^*, and
- w is a non-empty word over V.

By the iterated application of h to w we obtain an infinite sequence

$$S(G) = (w,\ h(w),\ h^2(w),\ h^3(w),\ \dots)$$

and a language

$$L(G) = \{h^i(w) \mid i \geq 0\}.$$

Here we are interested in the Parikh images of $S(G)$ and $L(G)$, i.e., in the sequence

$$PS(G) = (p(w),\ p(h(w)),\ p(h^2(w)),\ p(h^3(w)),\ \dots)$$

and the set

$$PL(G) = \{p(h^i(w)) \mid i \geq 0\}.$$

To illustrate the concept we consider the following example.

Example 1. Let $G = (V, h, w)$ be the D0L system with $V = \{a, b, c, d\}$, $w = c$ and h given by

$$h(a) = a,\ h(b) = ab,\ h(c) = bcd,\ h(d) = da.$$

Then we obtain

$$h(c) = bcd,\ h^2(c) = h(bcd) = abbcdda, h^3(c) = aababbcddadaa$$

and, in general, for $i \geq 1$,

$$h^i(c) = a^{i-1}ba^{i-2}b \dots aababbcddadaa \dots da^{i-2}da^{i-1}.$$

---

[2] This is the abbreviation of deterministic (D) interactionless (0) Lindenmayer (L) system. We refer to [8] and [11] for a detailed discussion of Lindenmayer systems including their biological motivation.

Therefore, for $i \geq 1$,

$$p(h^i(c)) = \left(2 \cdot \sum_{j=1}^{i-1} j, \; i, \; 1, \; i\right) = (i(i-1), \; i, \; 1, \; i).$$

Hence we get

$$PS(G) = (\; (0,0,1,0), (0,1,1,1), (2,2,1,2), (6,3,1,3), \ldots, (i(i-1), i, 1, i), \ldots \;)$$

and

$$PL(G) = \{(0,0,1,0)\} \cup \{(i(i-1), i, 1, i) \mid i \geq 1\}.$$

Let $G = (V, h, w)$ be a D0L system with $V = \{a_1, a_2, \ldots a_n\}$. For two letters a and b of V and $z \in V^+$, the number of occurrences of a in $h(z)$ generated by one occurrence of b in z is $\#_a(h(b))$. Therefore

$$\#_a(h(z)) = \sum_{b \in V} \#_b(z) \cdot \#_a(h(b)).$$

We now define the growth matrix M_G associated with G as the $n \times n$ matrix $(m_{k,j})$ where $m_{k,j} = \#_{a_k}(h(a_j))$. Then

$$p(h(z)) = p(z)M_G$$

and, for $i \geq 0$, we can prove by induction

$$p(h^i(w)) = p(w) \cdot M_G^i.$$

We now discuss the status of decidability of the following decision problems:

- *Parikh membership problem:* Given a vector $v \in \mathbf{N}^n$ and a D0L system $G = (V, h, w)$ with $\#(V) = n$, decide whether or not $v \in PL(G)$.
- *Parikh finiteness problem:* Given a D0L system G, decide whether or not $PL(G)$ is finite.
- *Parikh sequence equivalence problem:* Given two D0L systems G and G', decide whether or not $PS(G) = PS(G')$.
- *Parikh language equivalence problem:* Given two D0L systems G and G', decide whether or not $PL(G) = PL(G')$.

Theorem 1. *The following problems are decidable for D0L systems:*
  i)   the Parikh membership problem,
  ii) the Parikh finiteness problem,
  iii) the Parikh sequence equivalence problem, and
  iv) the Parikh language equivalence problem.

Proof. We only prove ii) and iii). i) can be shown by analogous considerations. For a proof of iv) we refer to [9]. Moreover, we give proofs based on the Parikh vectors. Using the fact that a set of Parikh vectors is finite if and only if the corresponding set of words is finite, one can give another proof for ii) based on the decidability of the finiteness of the word set.

ii) It is well-known that any infinite set of vectors of $\mathbf{N}^n$ contains at least two vectors x and y such that $x \leq y$ or $y \leq x$, where the order on $\mathbf{N}^n$ is defined componentwise (see [1]).

Now let $G = (V, h, w)$ be a D0L system with $\#(V) = n$. Then we can determine two integers i and j such that $i < j$ and $p(h^i(w))$ and $p(h^j(w))$ are comparable with respect to the partioal order on $\mathbf{N}^n$.

Case 1. $p(h^j(w)) \leq p(h^i(w))$. Then

$$
\begin{aligned}
p(h^{i+1}(w)) - p(h^{j+1}(w)) &= p(h^i(w))M_G - p(h^j(w))M_G \\
&= (p(h^i(w)) - p(h^j(w)))M_G \\
&\geq \underline{0}
\end{aligned}
$$

($\underline{0}$ denotes the zero-vector where all components are 0). This implies $p(h^{j+1}(w)) \leq p(h^{i+1}(w))$ and, by induction, we obtain

$$
p(h^{j+r}(w)) \leq p(h^{i+r}(w)) \text{ for } r \geq 0.
$$

Therefore the set of maximal vectors of $PL(G)$ is contained in the initial part $\{p(w), p(h(w)), p(h^2(w)), \ldots, p(h^j(w))\}$. Thus, $PL(G)$ is finite.

Case 2. $p(h^i(w)) < p(h^j(w))$. Then, as above, we can show that

$$
p(h^{j+r}(w)) \geq p(h^{i+r}(w)) \text{ for } r \geq 0. \tag{1}
$$

Now we consider the sequence

$$
\begin{aligned}
q_0 &= p(h^i(w)) = p(w)M_G^i, \\
q_1 &= p(h^j(w)) = p(h^{i+(j-i)}(w)) = p(w)M_G^{i+(j-i)} = q_0 M_G^{j-i}, \\
q_2 &= p(h^{i+2(j-i)}(w)) = q_0 M_G^{2(j-i)} = q_1 M_G^{j-i}, \\
q_3 &= p(h^{i+3(j-i)}(w)) = q_0 M_G^{3(j-i)} = q_2 M_G^{j-i},
\end{aligned}
$$

etc. By (1),

$$
q_0 \geq q_1 \geq q_2 \geq q_3 \geq \ldots. \tag{2}
$$

Moreover, we set

$$
\begin{aligned}
d_0 &= q_1 - q_0, \\
d_1 &= q_2 - q_1 = q_1 M_G^{j-i} - q_0 M_G^{j-i} = d_0 M_G^{j-i}, \\
d_2 &= q_3 - q_2 = q_2 M_G^{j-i} - q_1 M_G^{j-i} = d_1 M_G^{j-i} = d_0 (M_G^{j-i})^2, \\
d_3 &= q_4 - q_3 = d_2 M_G^{j-i} = d_0 (M_G^{j-i})^3,
\end{aligned}
$$

etc.

We now prove that $PL(G)$ is finite if and only if $d_k = \underline{0}$ for some k.

First let us assume that $PL(G)$ is finite. By $q_l \in PS(G)$ for $l \geq 0$ and (2), there is a number k such that $q_k = q_{k+1}$. This implies $d_k = \underline{0}$.

On the other hand, let $d_k = \underline{0}$ for some k. Then $d_r = d_k(M_G^{j-i})^{r-k} = \underline{0}$ for $r \geq k$. Hence $q_k = q_{k+1} = q_{k+2} = \ldots$ or equivalently

$$q_r = q_k \quad \text{for} \quad r \geq k \tag{3}$$

We now have by definition, known properties of matrix multiplication and finally (3)

$$
\begin{aligned}
p(h^{i+r(j-i)+t}(w)) - p(h^{i+k(j-i)+t}(w)) &= p(w)M_G^{i+r(j-i)+t} - p(w)M_G^{i+k(j-i)+t} \\
&= (p(w)M_G^{i+r(j-i)} - p(w)M_G^{i+k(j-i)})M_G^t \\
&= (p(h^{i+r(j-i)}(w)) - p(h^{i+k(j-i)}(w)))M_G^t \\
&= (q_r - q_k)M_G^t = \underline{0}M_G^t = \underline{0}.
\end{aligned}
$$

This implies

$$p(h^{i+k(j-i)+t}(w)) = p(h^{i+r(j-i)+t}(w)) \text{ for } r \geq k \text{ and } 1 \leq t < (j-i)$$

and

$$PL(G) = \{p(w), p(h(w)), p(h^2(w)), \ldots, p(h^{i+k(j-i)+(j-i-1)})\}.$$

The last statement proves the finiteness of $PL(G)$.

Obviously, the sequence $(d_0, d_1, d_2, \ldots)$ is generated by the D0L system $H = (V, h^{j-i}, v)$ where v is a word with $p(v) = d_0$ (the growth matrix of h^{j-i} is M_G^{j-i}). As above, there are integers i' and j' with $i' < j'$ and comparable vectors $d_{i'}$ and $d_{j'}$. Again, if $d_{j'} \leq d_{i'}$, then

$$\{d_0, d_1, d_2, \ldots\} = \{d_0, d_1, \ldots, d_{j'+(j'-i'-1)}\}$$

and we can easily test whether this set contains the zero-vector $\underline{0}$. If $d_{i'} < d_{j'}$, then we get $d_{i'+r} \leq d_{j'+r}$ for $r \geq 0$, and $d_k = \underline{0}$ for some k implies $d_{k'} = \underline{0}$ for some $k' \leq j' + (j' - i' - 1)$, i.e., we can check whether $d_k = \underline{0}$ holds for some k.

iii) Let $G = (V, h, w)$ and $H = (V, g, v)$ be the two given D0L systems. Further let $n = \#(V)$. We prove that

$$PS(G) = PS(H) \quad \text{if and only if} \quad p(h^i(w)) = p(g^i(v)) \text{ for } 0 \leq i \leq n.$$

Obviously, this criterion is effective and gives an algorithm to decide the Parikh sequence equivalence.

By definition, $PS(G) = PS(H)$ implies $p(h^i(w)) = p(g^i(v))$ for any $i \geq 0$. Therefore $p(h^i(w)) = p(g^i(v))$ holds for $0 \leq i \leq n$.

Let us now assume that $p(h^i(w)) = p(g^i(v))$ for $0 \leq i \leq n$. We determine the smallest r such that the vector $p(h^r(w))$ is linearly dependent on the vectors $p(w), p(h(w)), p(h^2(w)), \ldots, p(h^{r-1}(w))$. Since all Parikh vectors are elements of the n-dimensional vector space $\mathbf{Z}^n$, we have $r \leq n$. We now show that, for $s \geq r$, there are $\alpha_i \in \mathbf{Z}$, $0 \leq i \leq r-1$ such that

$$p(h^s(w)) = \sum_{i=0}^{r-1} \alpha_i p(h^i(w)) = \sum_{i=0}^{r-1} \alpha_i p(g^i(v)) = p(g^s(v))$$

i.e., $p(h^s(w))$ is a linear combination of $p(w), p(h(w)), p(h^2(w)), \ldots, p(h^{r-1}(w))$, $p(h^s(w)) = p(g^s(v))$, and $p(g^s(v))$ can be represented as a linear combination using the same coefficients. We use induction on s. For $s = r$ the statement follows by the choice of r and the supposition. The induction step follows from the following relations:

$$
\begin{aligned}
p(h^{s+1}(w)) &= p(h^s(w))M_G \\
&= \left(\sum_{i=0}^{r-1} \alpha_i p(h^i(w))\right) M_G \quad \text{(by induction assumption)} \\
&= \sum_{i=0}^{r-1} \alpha_i p(h^i(w))M_G = \sum_{i=0}^{r-1} \alpha_i p(h^{i+1}(w)) \\
&= \alpha_{r-1} p(h^r(w)) + \sum_{i=0}^{r-2} \alpha_i p(h^{i+1}(w)) \\
&= \sum_{i=0}^{r-1} \beta_i p(h^i(w)) + \sum_{i=0}^{r-2} \alpha_i p(h^{i+1}(w)) \\
&\qquad \text{(by induction assumption for } r) \\
&= \beta_0 p(w) + \sum_{i=1}^{r-1} (\beta_i + \alpha_{i-1}) p(h^i(w)) \\
&= \beta_0 p(v) + \sum_{i=1}^{r-1} (\beta_i + \alpha_{i-1}) p(g^i(v)) \\
&\qquad \text{(by } p(h^j(w)) = p(g^j(v)) \text{ for } 0 \leq j \leq r) \\
&= \sum_{i=0}^{r-1} \alpha_i p(g^{i+1}(v)) \quad \text{(do the above steps backwards)} \\
&= \sum_{i=0}^{r-1} \alpha_i p(g^i(v))M_H = \left(\sum_{i=0}^{r-1} \alpha_i p(g^i(v))\right) M_H \\
&= p(g^s(v))M_H \quad \text{(by induction assumption)} \\
&= p(g^{s+1}(v))
\end{aligned}
$$

2.2   Some Generalizations

In the preceding subsection we considered D0L systems where a language is generated by iterated applications of a morphism to a word. In order to generalize this concept one can use another mapping or several morphisms or a combination of both these ideas. Thus we get a T0L system which is formally defined as follows.

A T0L system is a triple $G = (V, P, w)$ where V and w are specified as in a D0L system and $P = \{\tau_1, \tau_2, \ldots, \tau_r\}$ is a set of n finite substitutions $\tau_i : V^* \to V^*$, $1 \leq i \leq r$. The language $L(G)$ generated by G is defined as

$$L(G) = \{\tau_{i_1}(\tau_{i_2}(\ldots(\tau_{i_m}(w))\ldots)) \mid m \geq 0, 1 \leq i_j \leq r, 1 \leq j \leq m\}.$$

The associated Parikh language $PL(G)$ is given by

$$PL(G) = \{p(v) \mid v \in L(G)\}.$$

We say that a T0L system $G = (V, \{\tau_1, \tau_2, \ldots, \tau_r\}, w)$ is a
- DT0L system, if τ_i is a morphism for $1 \leq i \leq r$,
- 0L system, if $r = 1$.

Then, a D0L system satisfies $r = 1$ and τ_1 is a morphism which gives precisely our definition above.

A T0L system is called propagating or λ-free, or shortly PT0L system, if $\lambda \notin \tau_i(V)$ for $1 \leq i \leq r$. Adding this restriction to T0L, DT0L, 0L and D0L systems, we obtain PT0L, PDT0L, P0L and PD0L systems, respectively.

Example 2. We consider the PDT0L system $G = (\{a, b\}, \{h_1, h_2\}, a)$ where the morphisms are given by $h_1(a) = a^2$, $h_1(b) = b$, $h_2(a) = b^3$, $h_2(b) = b$. Obviously, iterating h_1 we get all words a^{2^i} with $i \geq 0$. If we apply h_2 then we obtain a word $b^{3 \cdot 2^i}$ and this word is not changed by further application of one of the morphisms. Therefore

$$L(G) = \{a^{2^i} \mid i \geq 0\} \cup \{b^{3 \cdot 2^i} \mid i \geq 0\}$$

and

$$PL(G) = \{(2^i, 0) \mid i \geq 0\} \cup \{(0, 3 \cdot 2^i) \mid i \geq 0\}.$$

Now we show that $PL(G)$ is not the Parikh language of some 0L system. Assume that $PL(G) = PL(H)$ for some 0L system $H = (\{a, b\}, \tau, v)$. Obviously, if $z \in \tau(a)$, then $p(z) = (x, 0)$ or $p(z) = (0, y)$ for some non-negative integers x and y. If $x = 0$ or $y = 0$, then we can generate from a word $a^{2^i} \in L(H)$ (note that $(2^i, 0) \in PL(G) = PL(H)$) the empty word λ. Thus $\lambda \in L(H)$ but $(0, 0) \notin PL(H)$. Analogously, we can show that $p(z') \neq (0, 0)$ for any $z' \in \tau(b)$. Therefore, $v = a$ has to hold. If $p(\tau(a)) \subset \mathbf{N} \times \{0\}$, then we cannot generate any word with a Parikh vector in $\{0\} \times \mathbf{N}$. Hence we have $(x, 0), (0, y) \in \tau(a)$ for some $x \geq 1$ and $y \geq 1$. Thus we can generate the word $a^x b^y$ from $a^2 \in L(H)$ (since $(2, 0) \in PL(G) = PL(H)$). Therefore $p(a^x b^y) = (x, y) \in PL(H)$, but $(x, y) \notin PL(G)$ in contradiction to $PL(H) = PL(G)$.

Example 3. We now consider the 0L system $G' = (\{a\}, \tau, a)$ with $\tau(a) = \{a, a^2\}$. Then it is easy to see that

$$L(G') = \{a^j \mid j \geq 1\} \quad \text{and} \quad PL(G') = \{(j) \mid j \geq 1\}.$$

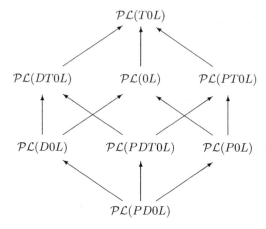

Fig. 1. Hierarchy of the Parikh language families of L systems. $\mathcal{PL}(X) \rightarrow \mathcal{PL}(Y)$ means the proper inclusion $\mathcal{PL}(X) \subset \mathcal{PL}(Y)$, and if two families are not connected, then they are incomparable.

We prove that $PL(G')$ cannot be generated by some DT0L system. Assume that $PL(G') = PL(H')$ for some 0L system $H' = (\{a\}, \{h_1, h_2, \ldots h_n\}, v')$. Since the underlying alphabet only consists of one letter a, any morphism h_i is given by $h_i(a) = a^{k_i}$ for $1 \leq i \leq n$. Without loss of generality, we can assume that $k_i \neq 1$ for $1 \leq i \leq n$ since the identity does not change the words. Moreover, $k_i \neq 0$ for $1 \leq i \leq n$ since we have $(0) \notin PL(H')$ and therefore $\lambda \notin L(H')$. Thus any application of a morphism h_i strongly increases the length. This implies $v' = a$. Then

$$L(H') = \{a^{k_1^{q_1} k_2^{q_2} \ldots k_n^{q_n}} \mid q_j \geq 0, 1 \leq j \leq n\}$$

and

$$PL(H') = \{(k_1^{q_1} k_2^{q_2} \ldots k_n^{q_n}) \mid q_j \geq 0, 1 \leq j \leq n\}.$$

Now let p be a prime number such that $p > k_i$ for $1 \leq i \leq n$. Obviously, $(p) \notin PL(H')$ in contrast to $(p) \in PL(G')$.

For $X \in \{D0L, 0L, DT0L, T0L, PD0L, P0L, PDT0L, PT0L\}$, we denote the family of Parikh languages generated by X systems by $\mathcal{PL}(X)$.

By definition, we obtain $\mathcal{PL}(PX) \subseteq \mathcal{PL}(X)$ for $X \in \{D0L, 0L, DT0L, T0L\}$ and $\mathcal{PL}(YT0L) \subseteq \mathcal{PL}(Y0L)$ for $Y \in \{\lambda, P, D, PD\}$. Moreover, by the Examples 2 and 3 we have the incomparabilty of $\mathcal{PL}(DT0L)$ and $\mathcal{PL}(0L)$. Without proof we summarize the relations between the families of Parikh languages generated by different types of 0L systems. For a complete proof we refer to [2].

Theorem 2. *The diagram of Figure 1 holds.*

With a morphism h we can associate a relation R_h by

$$(w, v) \in R_h \quad \text{if and only if} \quad v = h(w).$$

Further, the iteration of the morphism h corresponds to the reflexive and transitive closure R_h^* of the relation R_h. Thus the language $L(G)$ generated by a D0L system $G = (V, h, w)$ is the set $\{z \mid (w, z) \in R_h^*\}$. Analogously, we can redefine the languages obtained by iteration of a substitution or a set of morphisms or substitutions. This leads to the general concept.

A *language generating device* is a triple $G = (V, R, w)$ where V and w are specified as in a D0L system and R is a binary relation on V^*. The language and the Parikh language generated by G are defined as

$$L(G) = \{z \mid (w, z) \in R^*\} \quad \text{and} \quad PL(G) = \{p(z) \mid z \in L(G)\},$$

respectively.

There is not too much knowledge on Parikh languages of language generating devices. We here only mention some results on context-free grammars which are a special language generating device defined as follows.

A *context-free grammar* is a quadruple $G = (N, T, P, S)$ where N and T are disjoint finite non-empty sets, P is a finite subset of $N \times (N \cup T)^*$ and S is an element of N. We define a relation $\Longrightarrow$ on $(N \cup T)^*$ by $(w, w') \in \Longrightarrow$ if and only if the following conditions are satisfied: $w = uAu'$, $w' = uvu'$ for some words $u, u' \in (N \cup T)^*$ and $(A, v) \in P$. As usual, we write $w \Longrightarrow w'$ instead of $(w, w') \in \Longrightarrow$. Then the language and Parikh language of G are defined as

$$L(G) = \{z \mid z \in T^*, S \Longrightarrow^* z\} \text{ and } PL(G) = \{p(z) \mid z \in L(G)\},$$

respectively.

Example 4. Let

$$G = (\{S, A, B\}, \{a, b, c\}, \{(S, AB), (A, aAb), (A, ab), (B, aBc), (B, ac)\}, S).$$

Since from A we produce pairs of letters a and b and from B pairs of a and c, we get

$$L(G) = \{a^n b^n a^m c^m \mid n \geq 1, m \geq 1\}$$

and

$$PL(G) = \{(n + m, n, m) \mid n \geq 1, m \geq 1\}.$$

In order to characterize the Parikh languages of context-free grammars we need the following notions.

A set $M \subset \mathbf{N}^n$ is called linear, if there are a natural number k and vectors $x_0, x_1, \ldots, x_k$ such that

$$M = \{x_0 + \sum_{i=1}^{k} \alpha_i x_i \mid \alpha_i \in \mathbf{N}, 1 \leq i \leq k\}.$$

M is called semilinear iff it is a finite union of linear sets.

For instance, $PL(G)$ of Example 4 is a linear set since

$$PL(G) = \{(2, 1, 1) + n(1, 1, 0) + m(1, 0, 1) \mid n \geq 0, m \geq 0\}.$$

Due to R.J. PARIKH (see [10]), the following characterization of context-free Parikh languages is known.

Theorem 3. *i) For any context-free grammar G, $PL(G)$ is a semilinear set.*

ii) For any semilinear set M, there is a context-free grammar G such that $M = PL(G)$.

Proof. For a proof of i) we refer to [12]. However, we note that the proof is constructive, i.e., it determines the vectors which describe the linear sets whose union is $PL(G)$.

Let G and G' be two context-free grammars. Then there is a context-free grammar H such that $L(H) = L(G) \cup L(G')$ (see [12]). Moreover,

$$PL(H) = p(L(H)) = p(L(G) \cup L(G')) = p(L(G)) \cup p(L(G')) = PL(G) \cup PL(G').$$

Therefore it is sufficient to prove that, for any linear set M there is a context-free grammar G with $PL(G) = M$.

Let

$$M = \{x_0 + \sum_{i=1}^{k} \alpha_i x_i \mid \alpha_i \in \mathbf{N}, 1 \leq i \leq k\} \subseteq \mathbf{N}^n$$

for some vectors $x_0, x_1, \ldots, x_k$. Further, let T be an alphabet with $\#(T) = n$ and, for $0 \leq i \leq k$, let $w_i \in T^*$ be a word with $p_T(w_i) = x_i$. Then we construct the context-free grammar

$$G = (\{A_i \mid 1 \leq i \leq k\}, T, \{A_k \to w_k, A_k \to w_0\} \cup \bigcup_{i=1}^{k-1} \{A_i \to w_i, A_i \to A_{i+1}\}, A_1).$$

Then we get

$$
\begin{aligned}
A_1 &\Longrightarrow w_1 A_1 \Longrightarrow w_1 w_1 A_1 \Longrightarrow \ldots \Longrightarrow w_1^{n_1} A_1 \Longrightarrow w_1^{n_1} A_2 \\
&\Longrightarrow w_1^{n_1} w_2 A_2 \Longrightarrow w_1^{n_1} w_2 w_2 A_2 \Longrightarrow \ldots \Longrightarrow w_1^{n_1} w_2^{n_2} A_2 \Longrightarrow w_1^{n_1} w_2^{n_2} A_3 \\
&\Longrightarrow w_1^{n_1} w_2^{n_2} w_3 A_3 \Longrightarrow w_1^{n_1} w_2^{n_2} w_3 w_3 A_3 \Longrightarrow \ldots \\
&\Longrightarrow w_1^{n_1} w_2^{n_2} w_3^{n_3} A_3 \Longrightarrow w_1^{n_1} w_2^{n_2} w_3^{n_3} A_4 \\
&\Longrightarrow \ldots \Longrightarrow w_1^{n_1} w_2^{n_2} w_3 n_3 \ldots w_k^{n_k} A_k \\
&\Longrightarrow w_1^{n_1} w_2^{n_2} w_3 n_3 \ldots w_k^{n_k} w_0 = z.
\end{aligned}
$$

Since

$$p_T(z) = x_0 + n_1 x_1 + n_2 x_2 \ldots n_k x_k,$$

it is easy to prove that $M = PL(G)$.

We mention the following decidability properties.

Theorem 4. *The following problems are decidable for context-free grammars:*

i)   the Parikh membership problem,

ii) the Parikh finiteness problem,

iii) the Parikh language equivalence problem.

Proof. i) Given G we first construct the semilinear set M. To check whether or not a given vector belongs to M it is sufficient to check whether or not it belongs to one of the linear sets whose union is M. The last check can be done by solving a system of linear equations.

ii) $PL(G)$ is finite if and only if $L(G)$ is finite. The finiteness of context-free languages is decidable (see [12]).

iii) We have to check the equality of two semilinear sets which can be done algorithmically (see [5]).

For further information on semilinear sets (in connection with languages) we refer, for example, to [5], [6] and [7].

3   Iteration of a Modified Parikh Mapping

In this section we study the iteration of a modified version of the Parikh mapping. The modification has to ensure that the image is contained in V^*, again. Therefore the numbers of occurrences have to be considered as words which can be done by considering q-adic representations of the numbers.

We consider the alphabet

$$V_q = \{0, 1, 2, \ldots, q-1\}.$$

Then the representation $w(m, q)$ of the integer m in the base q, $q \geq 2$, is a word over V_q and we define $\alpha(w(m, q))$ as the string

$$w(\#_0(w(m,q)), q)0w(\#_1(w(m,q)), q)1 \ldots w(\#_{q-1}(w(m,q)), q)(q-1)$$

(we count, in the base q, the number of occurrences of the digits in the word $w(m, q)$, and we used $(q-1)$ instead of $q-1$ in order to ensure that the digit $q-1$ is meant). In the sequel we shall use the notation

$$w(\#_0(w(m,q)), q)\underline{0}w(\#_1(w(m,q)), q)\underline{1} \ldots w(\#_{q-1}(w(m,q)), q)\underline{q-1}$$

where the digits used as delimiters are underlined.

The Parikh mapping of $w(m, q)$ gives

$$p_{V_q}(w(m,q)) = ((\#_0(w(m,q), \#_1(w(m,q)), \ldots, \#_{q-1}(w(m,q))),$$

i.e., essentially, we use the comma instead of the delimiters and the numbers of occurrences instead of the representations of the numbers of occurrences.

The function α can be iterated, and, for given m, p as above, we obtain the sequence $\alpha^j(w(m, q))$, $j = 0, 1, 2, \ldots$, of strings over the alphabet V_q. The set of all these strings will be denoted by $L(m, q)$.

Example 5. We take the base 2 and start with $m = 140$. Therefore,

$$w(m, q) = 10001100.$$

We have 5 occurrences of 0 and 3 occurrences of 1, hence

$$\alpha(w(140,2)) = w(5,2) \underline{0} \; w(3,2) \underline{1} \; = \; 1\,0\,1\,\underline{0}\,1\,1\,\underline{1}.$$

Now,

$$\alpha^2(w(140,2)) = w(2,2) \underline{0} \; w(5,2) \underline{1} \; = \; 1\,0\,\underline{0}\,1\,0\,1\,\underline{1},$$
$$\alpha^3(w(140,2)) = w(3,2) \underline{0} \; w(4,2) \underline{1} \; = \; 1\,1\,\underline{0}\,1\,0\,0\,\underline{1},$$
$$\alpha^4(w(140,2)) = w(3,2) \underline{0} \; w(4,2) \underline{1} \; = \; 1\,1\,\underline{0}\,1\,0\,0\,\underline{1}.$$

Since $\alpha^3(w(140,2)) = \alpha^4(w(140,2))$, we obtain

$$\alpha^i(w(140,2)) = 1\,1\,\underline{0}\,1\,0\,0\,\underline{1} \text{ for } i \geq 3$$

and

$$L(140,2) = \{10001100, 1010111, 1001011, 1101001\}.$$

In our example $L(140,2)$ is a finite language. We shall prove that this holds for all m and q as above.

Theorem 5. *For all positive m and for all $q \geq 2$, the language $L(m,q)$ is finite.*

Proof. In $\alpha(w(m,q))$ some substrings $w(\#_i(w(m,q)),q)$, $0 \leq i \leq q-1$, can be equal to 0 (no symbol i appears in $w(m,q)$), but in $\alpha(w(m,q))$ each digit $0,\,1,...,(p-1)$ appears at least once and similarly in all subsequent strings $\alpha^j(w(m,q))$, $j \geq 2$. Thus, in

$$\alpha^{j+1}(w(m,q)) = w(\#_0(\alpha^j(w(m,q))),q) \underline{0} \; \cdots \; w(\#_{q-1}(\alpha^j(w(m,p))),q) \underline{q-1}$$

all the substrings $w(\#_i(\alpha^j(w(m,p))),q)$, $0 \leq i \leq q-1$, correspond to non-null numbers.

We now compare the length of the strings $\alpha^j(w(m,q))$ and $\alpha^{j+1}(w(m,q))$ for $j \geq 2$. Assume that $\alpha^j(w(m,q))$ contains t_i occurrences of the digit i, $i \in \{0,1,\ldots,(q-1)\}$. Then

$$|\alpha^j(w(m,q))| = \sum_{i=0}^{q-1} t_i,$$

$$|\alpha^{j+1}(w(m,q))| = \sum_{i=0}^{q-1} |w(t_i,q)| + q.$$

Clearly,

$$|w(t_i,q)| = \left\lceil \log_q t_i \right\rceil + 1,$$

hence

$$|\alpha^{j+1}(w(m,q))| = \sum_{i=0}^{q-1} \left\lceil \log_q t_i \right\rceil + 2q. \tag{4}$$

Trivially, $\lceil \log_q t_i \rceil < t_i$, but for $s > q^3$ we have

$$\log_q s < s - 2q$$

(indeed, if $r > 1$ and $s = rp^3$, then $\log_q rq^3 = \log_q r + 3 < r(q^3 - 1) - 2q + r$, because $\log_q r < r$ and $3 < r(q^3 - 1) - 2q$ for all $q \geq 2$, $r > 1$). We now consider $\alpha(w(m, q))$ with the length strictly greater than q^4. At least one of the above considered numbers t_i, $0 \leq i \leq p - 1$, is strictly greater than q^3 (the pigeon principle), hence for such a number t_i we have

$$\lceil \log_q t_i \rceil < t_i - 2q.$$

By (4), this implies

$$|\alpha^{j+1}(w(m, q))| < \sum_{i=0}^{p-1} t_i = |\alpha^j(w(m, q))|.$$

In conclusion, starting from $\alpha(w(m, q))$ of some length greater than q^4, the length of $\alpha^j(w(m, q))$, for $j = 2, 3, \ldots$ decreases strictly, until obtaining a string of length smaller than q^4.

Conversely, having a short string $\alpha(w(m, q))$, the length of $\alpha^j(w(m, q))$, $j = 2, 3, \ldots$ can increase several steps, but each such increase is at most $2q$, the constant term in relation (4). Therefore, the length can increase at most until $q^4 + 2q$, when it must again decrease.

In conclusion, the language $L(m, q)$ contains only strings of length at most $q^4 + 2q$, plus possibly a finite number of strings of greater length, when starting from a string of greater length. Hence this language is finite.

By this proof we know that, for words such that for some i, $1 \leq i \leq q - 1$, the number t_i of occurrences of i in $w(m, q)$ is larger than q^3, the length is decreased by α. We now want to improve this bound.

Let $q \geq 3$, and let us assume that $3q < t_i \leq q^3$ for some i. Then

$$[t_i] \leq 3 \leq q = 3q - 2q < t_i - 2q.$$

By (4), it follows that $|\alpha(w(m, q))| < |w(m, q)|$.

By arguments analogous to those in the proof of Theorem 5, we obtain $3q \cdot q + 2q = 3q^2 + 2q$ as a lower bound for the length of words whose length is decreased by α.

By the finiteness of $L(m, q)$, the problem of the structure of the sequence $\alpha^j(w(m, q))$, $j \geq 2$, arises. Two situations can appear:

- there is a string $\alpha^j(w(m, q))$ with $\alpha^{j+k}(w(m, q)) = \alpha^j(w(m, q))$ for $k \geq 1$
  (we call such a string $\alpha^j(w(m, q))$ *singleton-attractor*)

  or

- there are strings $\alpha^{j+s}(w(m, q))$, $s = 1, 2, \ldots, k$, such that $\alpha^{j+k}(w(m, q)) = \alpha^j(w(m, q))$ and all strings $\alpha^{j+s}(w(m, q))$, $s = 0, 1, 2, \ldots, k - 1$, are pairwise different
  (we say that a *cycle-attractor* appears).

Example 5 shows that $\alpha^4(w(140,2)) = 1101001$ is a singleton-attractor.

Example 6. Let now $q = 3$ and $m = 2189$. Then we obtain

$$z = w(2189,3) = 10000002,$$
$$\alpha(z) = 200\underline{1}1\underline{12},$$
$$\alpha^2(z) = 20\underline{1}01\underline{22},$$
$$\alpha^3(z) = 20\underline{2}110\underline{2},$$
$$\alpha^4(z) = 10\underline{2}110\underline{2},$$
$$\alpha^5(z) = 20\underline{1}01\underline{22} = \alpha^2(z).$$

Therefore we have a cycle-attractor which consists of three elements.

How many singleton-attractors and cycle-attractors exist for a given base p? For a given base p, are there both singleton- and cycle-attractors? Are there cycles of arbitrary large size? We have only partial answers to these questions.

Theorem 6. *For each base $q \geq 2$, there is at least one singleton-attractor.*

Proof. For $q = 2$, by our Example 5, the statement holds.
For $q \geq 3$, we consider the string

$$z = 1\,\underline{0}\,1\,1\,\underline{1}\,2\,\underline{2}\,1\,\underline{3}\,1\,\underline{4} \ldots 1\,p-2\,1\,\underline{p-1}.$$

Obviously, $\alpha(z) = z$.

Theorem 7. *For each base $q \geq 8$, there is at least one cycle-attractor (consisting of two strings).*

Proof. Consider the strings

$$z_1 = 1\,\underline{0}\,(p-3)\,\underline{1}\,4\,2\,1\,\underline{3}\,1\,\underline{4}\,1\,\underline{5} \ldots 1\,p-3\,2\,p-2\,1\,p-1,$$
$$z_2 = 1\,\underline{0}\,(p-2)\,\underline{1}\,2\,\underline{2}\,1\,\underline{3}\,2\,\underline{4}\,1\,\underline{5} \ldots 2\,p-3\,1\,p-2\,1\,p-1.$$

It is easy to see that $\alpha(z_1) = z_2$ and $\alpha(z_2) = z_1$. Hence we have a cycle.

For $q = 8$, by this construction, we obtain the cycle

$$1\,\underline{0}\,5\,\underline{1}\,4\,2\,1\,\underline{3}\,1\,\underline{4}\,1\,\underline{5}\,2\,\underline{6}\,1\,\underline{7},$$
$$1\,\underline{0}\,6\,\underline{1}\,2\,\underline{2}\,1\,\underline{3}\,2\,\underline{4}\,2\,\underline{5}\,1\,\underline{6}\,1\,\underline{7}.$$

By the proof of Theorem 5, we obtain that any attractor contains a word z such that, for $1 \leq i \leq q-1$,

$$\#_i(z) \leq \begin{cases} 8 & \text{for } q = 2 \\ 3q & \text{for } q \geq 3. \end{cases}$$

The following theorem improves this relation.

Theorem 8. *Let $q \geq 6$. Let z be in some attractor. Then*

$$\#_i(z) \leq \begin{cases} 2 & i = 0 \\ q+3 & i = 1 \\ q & i = 2 \\ q-1 & i \geq 3 \end{cases}.$$

For a proof of Theorem 8 we refer to [3]

Theorem 9. *For any $q \geq 4$, any attractor contains a string z with $\#_0(z) = 1$.*

Proof. Since we are interested in elements of attractors we can assume that z is obtained by application of α, i.e., z can be represented in the form

$$z = z^{(0)}\underline{0}z^{(1)}\underline{1}\ldots z^{(p-1)}\underline{p-1}.$$

The number $z^{(i)}$, $0 \leq i \leq q-1$, is called the i-th coefficient of z. Moreover, $\#_k(z)$ is the k-th coefficient of $\alpha(z)$.

By Theorem 8, it is sufficient to show that there is no attractor with $z^{(0)} = 2$ for all elements z of the attractor.

Let us assume that such an attractor A exists and let z be an element of A. Then one coefficient of $\alpha^{-1}(z)$ contains a zero. By Theorem 8, this coefficient has to be q and has to be the coefficient of 1 or 2. By assumption, these relations also hold for $\alpha^{-2}(z)$. Moreover, $q-1$ occurrences of 1 or 2 exist in the coefficients of $\alpha^{-2}(z)$ depending on $(\alpha^{-1}(z))^{(1)} = q$ or $(\alpha^{-1}(z))^{(2)} = q$, respectively. By these facts, $\alpha^{-2}(z)$ is one of the following four elements:

$$2\underline{0}10\underline{1}2\underline{2}2\underline{3}2\underline{4}\ldots 2q - 2\underline{2q-1},$$
$$2\underline{0}2\underline{1}10\underline{2}2\underline{3}2\underline{4}\ldots 2q - 2\underline{2q-1},$$
$$2\underline{0}10\underline{1}1\underline{2}1\underline{3}1\underline{4}\ldots 1q - 2\underline{1q-1},$$
$$2\underline{0}1\underline{1}10\underline{2}1\underline{3}1\underline{4}\ldots 1q - 2\underline{1q-1}.$$

It is easy to compute that in all four cases $(\alpha(z))^{(0)} = 1$.

Theorem 8 and Theorem 9 give the possibility to determine by the computer all single attractors and all cycle-attractors for small values. The result of our computation of single-attractors is given in the table of Figure 2.

Generalizing the single attractors which we found for $q \in \{7, 8, 9\}$ it is easy to see that, for $q \geq 7$, there is the single attractor

$$1\,\underline{0}\,(p-3)\,\underline{1}\,3\,2\,2\,\underline{3}\,1\,4\,\underline{1}\,5\,\ldots\,1\,(p-4)\,2\,(p-3)\,1\,(p-2)\,1\,(p-1)$$

For $q \in \{2, 4, 5\}$, there is no cycle-attractor. For $q = 3$, there is only the cycle-attractor consisting of the elements

$$10\,\underline{0}\,11\,\underline{1}\,1\,2, \quad 2\,\underline{0}\,12\,\underline{1}\,1\,2, \quad 1\,\underline{0}\,10\,\underline{1}\,10\,\underline{2},$$

which we found already in Example 6. For $q = 6$, there is exactly one cycle-attractor, namely

q	single attractors
2	11 0 100 1
3	10 0 10 1 2 2
	1 0 11 1 2 2
	2 0 2 1 10 2
4	1 0 11 1 2 2 1 3
	1 0 2 1 3 2 2 3
	1 0 3 1 1 2 3 3
5	1 0 11 1 2 2 1 3 1 4
	1 0 3 1 2 2 3 3 1 4
6	1 0 11 1 2 2 1 3 1 4 1 5
7	1 0 11 1 2 2 1 3 1 4 1 5 1 6
	1 0 4 1 3 2 2 3 2 4 1 5 1 6
8	1 0 11 1 2 2 1 3 1 4 1 5 1 6 1 7
	1 0 5 1 3 2 2 3 1 4 2 5 1 6 1 7
9	1 0 11 1 2 2 1 3 1 4 1 5 1 6 1 7 1 8
	1 0 6 1 3 2 2 3 1 4 1 5 2 6 1 7 1 8

Fig. 2. List of all single-attractors for $2 \le q \le 9$.

$$1 \, 0 \, 3 \, 1 \, 4 \, 2 \, 1 \, 3 \, 2 \, 4 \, 1 \, 5, \qquad 1 \, 0 \, 4 \, 1 \, 2 \, 2 \, 2 \, 3 \, 2 \, 4 \, 1 \, 5,$$

and for $q = 7$ we found the only cycle-attractor

$$1 \, 0 \, 4 \, 1 \, 4 \, 2 \, 1 \, 3 \, 1 \, 4 \, 2 \, 5 \, 1 \, 6, \; 1 \, 0 \, 5 \, 1 \, 2 \, 2 \, 1 \, 3 \, 3 \, 4 \, 1 \, 5 \, 1 \, 6, \; 1 \, 0 \, 5 \, 1 \, 2 \, 2 \, 2 \, 3 \, 1 \, 4 \, 2 \, 5 \, 1 \, 6.$$

For $p = 8$ and $p = 9$ there exist only the cycle-attractors of Theorem 7.

Results on further modified functions can be found in [4].

References

1. D. KÖNIG, *Theorie der endlichen und unendlichen Graphen*. Chelsea, New York, 1959.
2. J. DASSOW, On Parikh-languages of L systems without interaction. *Rostock. Math. Colloq.* **15** (1980) 103–110.
3. J. DASSOW, S. MARCUS and GH. PĂUN, Iterative reading of numbers and "black holes". *Periodica Math. Hungarica* **27** (1993) 137–152.
4. J. DASSOW, S. MARCUS and GH. PĂUN, Iterative reading of numbers: the ordered case. In: G. ROZENBERG and A. SALOMAA, *Developments in Language Theory*, World Scientific, Singapore, 1994, 157–168.
5. S. GINSBURG, *The Mathematical Theory of Context-Free Languages*. McGrw Hill Book Co., New York, 1966.
6. S. GINSBURG and E.H. SPANIER, Bounded ALGOL-like languages. ???
7. E.M. GURARI and O.H. IBARRA, The complexity of the equivalence problem for counter languages, semilinear sets, and simple programs. In: *Conference Record Tenth Annual ACM Symposium Theory of Computing*, Atlanta, 1979, 142–152.
8. G.T. HERMAN and G. ROZENBERG, *Developmental Systems and Languages*, North-Holland, 1974.

9. M. NIELSEN, On the decidability of some equivalence problems for D0L systems. *Inform. Control* **25** (1974) 166–193.
10. R.J. PARIKH, On context-free languages. *Journal of the ACM* **13** (1966) 570–581.
11. G. ROZENBERG and A. SALOMAA, *The Mathematical Theory of L Systems.* Academic Press, 1980.
12. A. SALOMAA, *Formal Languages.* Academic Press, 1973.

Multiset Constraints and P Systems[*]

Agostino Dovier[1], Carla Piazza[2], and Gianfranco Rossi[3]

[1] Dip. di Informatica, Univ. di Verona
Strada Le Grazie 15, 37134 Verona, Italy
dovier@sci.univr.it
[2] Dip. di Matematica e Informatica, Univ. di Udine
Via Le Scienze 206, 33100 Udine, Italy
piazza@dimi.uniud.it
[3] Dip. di Matematica, Univ. di Parma
Via M. D'Azeglio 85/A, 43100 Parma, Italy
gianfr@prmat.math.unipr.it

Abstract. Multisets are the fundamental data structure of P systems. In this paper we relate P systems with the language and theory for multisets presented in [9.] This allows us, on the one hand, to define and implement P systems using multiset constraints in a constraint logic programming framework, and, on the other hand, to define and implement constraint solving procedures used to test multiset constraint satisfiability in terms of P systems with active membranes. While the former can be exploited to provide a precise formulation of a P system, as well as a working implementation of it, based on a first-order theory, the latter provides a way to obtain a P system for a given problem (in particular, NP problems) starting from a rather natural encoding of its solution in terms of multiset constraints.

1 Introduction

The notion of *multiset* has been recently re-discovered, analyzed, and employed in various areas of Logic and Computer Science. From the point of view of Computer Science, a multiset is nothing but a data structure that can contain elements; differently from sets, the number of occurrences of each element is taken into account and differently from lists, the order of the elements does not matter. This peculiarity makes them suitable for modeling phenomena where resources are generated or consumed. For instance, multisets are used to model linear logic proofs, where tokens are consumed to deduce new ones [24]. Multisets are the fundamental data structure of a number of computational frameworks, such as the Gamma coordination language [3], the Chemical Abstract Machine [4], and P systems modeling *membrane computing* [21]. In all these three approaches, which have been developed with different aims, multisets are viewed as entities

---

[*] Partially supported by MURST project *Certificazione automatica di programmi mediante interpretazione astratta*.

C.S. Calude et al. (Eds.): Multiset Processing, LNCS 2235, pp. 103–121, 2001.

containing data (possibly, other multisets), and programs are, basically, collections of *multiset rewriting rules* that can be applied with an high degree of parallelism.

Efforts have also recently been put forward for introducing multisets as first-class citizens in programming languages. Their high level of abstraction makes it natural to introduce multisets into *declarative programming paradigms* using constraint (logic) programming languages (e.g., [2,9]). In order to allow the user to write declarative executable code, the interpreter of the language should be able to find the solutions of (i.e., solving) the formulae (i.e., the constraints) written by the user to describe the problem at hand. The correctness of such a solver must be proved on the basis of the properties that characterize multisets. Thus, it is important to develop a simple (first-order) theory of multisets that forces the desired behavior of the multiset data structure. If the solver works correctly on a minimal theory, then it will work correctly in any consistent extension of this theory. In [9] a minimal theory (called *Bag*) of multisets is developed parametrically together with minimal theories for lists, compact lists, and sets, and a solution to the multiset unification problem is provided (see also [8]). In [10] the results are extended to more complex constraints and the constraint logic programming language $CLP(\mathcal{BAG})$, able to deal with this kind of constraints, is presented.

This paper aims at establishing a connection between P systems and the constraint language of [9] based on multiset processing as the common denominator of the two proposals. The approach adopted is, as usual, to show how to define one formalism in terms of the other one and *vice versa*. Precisely, first we show how to define P systems using the language and the theory for multisets presented in [9]. This provides a precise formulation of P systems inside a first-order theory. Moreover, it allows us to define a translation from P systems into $CLP(\mathcal{BAG})$ programs that can be used as a simple, executable—though sequential—implementation of P systems.

In defining this mapping we will exploit the flexible multiset processing facilities offered by our multiset constraint solver, in particular *multiset unification*. This kind of "semantic" unification can be used not only as a means to compare two multisets, but also as an easy way to arbitrarily select a subset of elements satisfying some property from a given multiset. The possibility offered by our constraint language to deal with also partially specified multisets turns out to be essential to allow a concise definition of P systems inside our logic framework.

On the opposite side, we show how to implement part of the constraint solver of $CLP(\mathcal{BAG})$—namely, the fundamental part dealing with equality constraints (hence, multiset unification)—using P systems. The extension of this implementation to the whole constraint solver of $CLP(\mathcal{BAG})$ is trivial. Thus, once we have a description of the problem expressed as a multiset constraint over $\mathcal{L}_{Bag}$ we also have immediately a corresponding implementation as a P system. This constitutes an interesting viable alternative to obtain P systems algorithms for all those problems (typically, NP-complete problems) which admit a rather "natural" formulation based on multiset constraints.

The paper is organized as follows. First we review the above-mentioned multiset language, the axiomatic theory Bag, the constraints allowed in that language, and the constraint logic programming language $CLP(\mathcal{BAG})$ (Section 2). Then, in Section 3 we show how to use the multiset constraint language to define P systems and in Section 4 we use the $CLP(\mathcal{BAG})$ language to provide a simple implementation of P systems. In Section 5 we show, instead, how both the standard unification and the multiset unification algorithms can be encoded using P systems. Finally, some conclusions are drawn in Section 6.

2   Multiset Theory, Constraints, and Language

2.1   The Theory Bag

The first-order language $\mathcal{L}_{Bag}$ we consider here is based on a signature $\Sigma = \langle \mathcal{F}_{Bag}, \Pi_{Bag} \rangle$ and a denumerable set $\mathcal{V}$ of logical variables. $\mathcal{F}_{Bag}$ is a set of constant and function symbols containing the constant symbol $\mathtt{nil}$, the binary multiset constructor $\{\!\!|\,\cdot\,|\,\cdot\,|\!\!\}$, plus possibly other (free) constant and function symbols. $\Pi_{Bag} = \{\,\text{``}{=}\text{''}, \text{``}{\in}\text{''}\,\}$ is the set of predicate symbols.

The distinguishing properties of multisets can be precisely stated by the first-order axiomatic theory Bag, defined parametrically in [9]. Bag is a *hybrid theory*: the objects it deals with are built out of interpreted as well as uninterpreted symbols. In particular, multisets may contain uninterpreted Herbrand terms as well as other multisets. A multiset can be built starting from any ground uninterpreted Herbrand term—called the *kernel* of the data structure—and then adding to this term the other elements that compose the multiset. We refer to this kind of data structures as *colored* hybrid multisets.

We use simple syntactic conventions and notations for terms denoting multisets. In particular, the multiset $\{\!\!|\, s_1 \,|\, \{\!\!|\, s_2 \,|\, \cdots \,\{\!\!|\, s_n \,|\, t \,|\!\!\} \cdots |\!\!\} |\!\!\}$ will be denoted by $\{\!\!|\, s_1, \ldots, s_n \,|\, t \,|\!\!\}$ or simply by $\{\!\!|\, s_1, \ldots, s_n \,|\!\!\}$ when t is $\mathtt{nil}$ (i.e., the empty multiset). We will use capital letters for variables, small letters for constant and function symbols.

Example 1 ($\mathcal{L}_{Bag}$ terms).

1. $\{\!\!|\, a, b, a \,|\!\!\}$ (i.e., $\{\!\!|\, a \,|\, \{\!\!|\, b \,|\, \{\!\!|\, a \,|\, \mathtt{nil} \,|\!\!\} |\!\!\} |\!\!\}$) is a term in $\mathcal{L}_{Bag}$ which, intuitively, denotes the multiset containing two occurrences of the element a and one occurrence of the element b;
2. $\{\!\!|\, a \,|\, X \,|\!\!\}$ is a term in $\mathcal{L}_{Bag}$ which, intuitively, denotes any multiset containing at least one occurrence of the element a (in other words, it denotes a *partially specified multiset*);
3. $\{\!\!|\, \{\!\!|\, a \,|\!\!\}, \{\!\!|\, b \,|\!\!\} |\!\!\}$ is a $\mathcal{L}_{Bag}$ term which, intuitively, denotes a multiset containing two nested singleton multisets.
4. $\{\!\!|\, a, b \,|\, c \,|\!\!\}$, where a, b, c are terms built from constant symbols is a multiset containing (one occurrence of) a and b and colored by the kernel c. We also say that $\mathsf{ker}(\{\!\!|\, a, b \,|\, c \,|\!\!\}) = c$.

$$
\begin{array}{llll}
(K) & \forall x\,\bar{y} & (x \notin f(y_1,\dots,y_n)) & f \in \mathcal{F}_{Bag},\, f \not\equiv \{\!\!\{\,\cdot\,|\,\cdot\,\}\!\!\} \\[4pt]
(W) & \forall x\,y\,v & (x \in \{\!\!\{\,y\,|\,v\,\}\!\!\} \leftrightarrow x \in v \vee x = y) & \\[4pt]
(F_1') & \forall \bar{x}\bar{y} & \left(\begin{array}{l} f(x_1,\dots,x_n) = f(y_1,\dots,y_n) \\ \to x_1 = y_1 \wedge \cdots \wedge x_n = y_n \end{array}\right) & f \in \mathcal{F}_{Bag},\, f \not\equiv \{\!\!\{\,\cdot\,|\,\cdot\,\}\!\!\} \\[8pt]
(F_2) & \forall \bar{x}\bar{y} & (f(x_1,\dots,x_m) \neq g(y_1,\dots,y_m)) & f,g \in \mathcal{F}_{Bag},\, f \not\equiv g \\[4pt]
(F_3) & \forall x & (x \neq t[x]) & \\
\end{array}
$$

$$\textit{where } t[x] \textit{ denotes a term, having } x \textit{ as proper subterm}$$

$$
\begin{array}{lll}
(F_3^m) & \forall \bar{x}\bar{y}z & \left(\begin{array}{l} \{\!\!\{\,x_1,\dots,x_m\,|\,z\,\}\!\!\} = \{\!\!\{\,y_1,\dots,y_n\,|\,z\,\}\!\!\} \\ \to \{\!\!\{\,x_1,\dots,x_m\,\}\!\!\} = \{\!\!\{\,y_1,\dots,y_n\,\}\!\!\} \end{array}\right) \quad m,n > 0 \\[10pt]
(E_k^m) & \forall \bar{y}\bar{v} & \left(\begin{array}{l} \{\!\!\{\,y_1\,|\,v_1\,\}\!\!\} = \{\!\!\{\,y_2\,|\,v_2\,\}\!\!\} \leftrightarrow \\ (y_1 = y_2 \wedge v_1 = v_2) \vee \\ \exists z\,(v_1 = \{\!\!\{\,y_2\,|\,z\,\}\!\!\} \wedge v_2 = \{\!\!\{\,y_1\,|\,z\,\}\!\!\}) \end{array}\right) \\
\end{array}
$$

Fig. 1. The theory Bag

The axiomatic theory Bag is shown in Fig. 1. For a detailed discussion, see [9].

We focus here only on the meaning of axiom (E_k^m). The behavior of the multiset constructor symbol $\{\!\!\{\,\cdot\,|\,\cdot\,\}\!\!\}$ is regulated by the following *equational axiom*

$$(E_p^m) \qquad \forall xyz\,\{\!\!\{\,x,y\,|\,z\,\}\!\!\} = \{\!\!\{\,y,x\,|\,z\,\}\!\!\}$$

which, intuitively, states that the order of elements in a multiset is immaterial (*permutativity property*). This axiom forces syntactically different terms to possibly represent the same multiset. However, axiom (E_p^m) lacks in a criterion for establishing disequalities between multisets, even if it is considered together with $KWF_1'F_2F_3$. Therefore, axiom (E_k^m) is introduced in [9] to model (E_p^m) and characterizing the *multiset extensionality* property: *Two (hybrid) multisets are equal if and only if they have the same number of occurrences of each element, regardless of their order.*

2.2 The Constraints

A *multiset constraint* is a conjunction of atomic formulae or negation of atomic formulae in $\mathcal{L}_{Bag}$, that is, literals of the form $t_1 = t_2, t_1 \neq t_2, t_1 \in t_2, t_1 \notin t_2$, where t_i is a first-order term in $\mathcal{L}_{Bag}$.

Example 2 ($\mathcal{L}_{Bag}$ constraints).

1. $X \in \{\!\!\{\,a,b\,|\,Y\,\}\!\!\} \wedge X \notin Y \wedge X \neq a$;
2. $\{\!\!\{\,\{\!\!\{\,h,h,o\,\}\!\!\}, \{\!\!\{\,o,o\,\}\!\!\}\,\}\!\!\} = \{\!\!\{\,W,O\,\}\!\!\} \wedge Y_1 \in W \wedge Y_2 \in W \wedge Y_1 \neq Y_2$ (remember that capital letters denote variables, while small letters represent constant symbols);
3. $\{\!\!\{\,e,e,e,e,e\,\}\!\!\} = \{\!\!\{\,X\,|\,R\,\}\!\!\} \wedge X \notin \{\!\!\{\,e\,|\,S\,\}\!\!\}$.

Let E_{Bag} be the equational theory consisting of the equational axiom (E_p^m). The *standard model* $\mathcal{BAG}$ for the theory Bag is defined in the following way:

– Let $T(\mathcal{F}_{Bag})$ be the set of first-order ground terms built from symbols in $\mathcal{F}_{Bag}$ (i.e., the ordinary Herbrand Universe).
– The *domain* of the model is the quotient $T(\mathcal{F}_{Bag})/\equiv_{E_{Bag}}$ of $T(\mathcal{F}_{Bag})$ over the smallest congruence relation $\equiv_{E_{Bag}}$ induced by the equational theory E_{Bag} on $T(\mathcal{F}_{Bag})$.
– The *interpretation* of a term t is its equivalence class $[\,t\,]$.
– $=$ is interpreted as the identity on the domain $T(\mathcal{F}_{Bag})/\equiv_{E_{Bag}}$.
– The interpretation of membership is the following: $[\,t\,] \in [\,s\,]$ is **true** if and only if there is a term in $[\,s\,]$ of the form $\{\!\{\,t\,|\,r\,\}\!\}$ for some term r.

This model is a very important one. As a matter of fact in [10] we prove that if C is a constraint, its satisfiability (unsatisfiability) in $\mathcal{BAG}$ implies that it will be satisfiable (resp., unsatisfiable) in all the models of Bag (*correspondence property*).

The notion of satisfiability of a formula φ in a model $\mathcal{A}$ is based on the notion of valuation. A *valuation* σ is a function from a subset of $\mathcal{V}$ to the domain A of $\mathcal{A}$, extended to terms and formulas as usual. σ is said to be a *successful valuation* of a formula φ if $\sigma(\varphi) = $ **true**. Thanks to the correspondence property here we can safely restrict ourselves to consider the satisfiability in the model $\mathcal{BAG}$.

Example 3 (Valuations).

1. Consider the constraints of Example 2. The first two are satisfiable. In particular, any successful valuation for constraint 1 needs to map X to b (actually to the equivalence class $[\,b\,]$ containing b. We admit this slight abuse of notation), while any successful valuation for constraint 2 needs to map W to $\{\!\{\,h, h, o\,\}\!\}$ (water) and O to $\{\!\{\,o, o\,\}\!\}$ (oxygen). The third constraint, instead, is unsatisfiable.
2. Consider the constraint $X = \{\!\{\,Y, Z\,\}\!\} \wedge a \neq Y \wedge \mathtt{nil} \notin Y \wedge b \neq Z$. A successful valuation for it is $X \mapsto \{\!\{\,a, b\,\}\!\}, Y \mapsto b, Z \mapsto a$.

The technique used in [10] to check satisfiability of multiset constraints is based on a constraint solving procedure which is able to reduce non-deterministically any given constraint C to a simplified special form—called the solved form—if and only if C is satisfiable (otherwise, C is reduced to **false**).

A constraint C is in *solved form* if all its literals are in one of the following forms:

– $X = t$ and X does not occur neither in t nor elsewhere in C
– $X \neq t$ and X does not occur in t
– $t \notin X$ and X does not occur in t.

For instance, the constraint of Example 3.2 is in solved form. If C is a $\mathcal{L}_{Bag}$ constraint in solved form, then C is satisfiable in $\mathcal{BAG}$. A solved form constraint therefore can be seen as a finite representation of the possible infinite set of its solutions. From a solved form constraint C it is possible to assign terms to the variables occurring in it in order to obtain a solution. Moreover, all the solutions of C can be obtained in this way.

Solved form constraints can be obtained via suitable rewriting procedures. In Section 5 we will show how to implement some of them into P systems.

Example 4 (3-SAT). Consider the instance of 3-SAT:

$$(X_1 \vee \neg X_2 \vee X_3) \wedge (\neg X_1 \vee \neg X_2 \vee X_3) \wedge (X_1 \vee X_2 \vee \neg X_3)$$

It can be translated into the following $\mathcal{L}_{Bag}$ constraint:

$$\{\!\!\{\,\{\!\!\{X_1, Y_1\}\!\!\}, \{\!\!\{X_2, Y_2\}\!\!\}, \{\!\!\{X_3, Y_3\}\!\!\}, $$
$$\{\!\!\{X_1, Y_2, X_3\}\!\!\}, \{\!\!\{Y_1, Y_2, X_3\}\!\!\}, \{\!\!\{X_1, X_2, Y_3\}\!\!\}\,\}\!\!\} $$
$$= \{\!\!\{\,\{\!\!\{0,1\}\!\!\}, \{\!\!\{0,1\}\!\!\}, \{\!\!\{0,1\}\!\!\}, \ \{\!\!\{1 \mid R_1\}\!\!\}, \{\!\!\{1 \mid R_2\}\!\!\}, \{\!\!\{1 \mid R_3\}\!\!\}\,\}\!\!\}$$

where the variables Y_i take the place of $\neg X_i$. Each successful valuation for this constraint is also a solution for the given 3-SAT problem and, vice versa, from each solution for the SAT problem it is immediate to assign values to the variables Y_i and R_i in order to obtain a successful valuation for the constraint. One such solution is, for instance:

$$X_1 = 1 \wedge X_2 = 1 \wedge X_3 = 1$$

(and $Y_1 = 0, Y_2 = 0, Y_3 = 0, R_1 = \{\!\!\{0,1\}\!\!\}, R_2 = \{\!\!\{0,0\}\!\!\}, R_3 = \{\!\!\{0,1\}\!\!\}$).

2.3 The Language $CLP(\mathcal{BAG})$

The availability of a constraint solver which is able to prove satisfiability of multiset constraints in a selected structure makes it immediate to define a CLP language dealing with multiset constraints of the kind we have considered so far. This language is obtained as an instance of the general CLP scheme [16,17] by instantiating it over the language, the theory, and the structure of multisets presented in the previous sections.

The resulting language—called $CLP(\mathcal{BAG})$—can be viewed as a natural extension of traditional logic programming languages, e.g., PROLOG. The syntactic form of a $CLP(\mathcal{BAG})$ program, as well as its operational, algebraic, and logical semantics are those of the general CLP scheme [16,17] suitably instantiated on the specific constraint domain of multisets. This is very similar to what is described in greater detail in [12] for the case of sets. A working prototype of the CLP language which embeds both sets and multisets is available at [23].

A $CLP(\mathcal{BAG})$ *program* P is a finite set of rules of the form

$$A : -c, \bar{B}.$$

where A is an atomic formula (the *head* of the rule), c is a $\mathcal{L}_{Bag}$-constraint and $\bar{B}$ is a (possibly empty) conjunction of atomic formulae. A *goal* is a rule with empty head. As usual, we assume that all the variables in a clause are universally quantified in front of the clause itself.

Example 5 (CLP(BAG) program). The following is a $CLP(\mathcal{BAG})$ program that defines a predicate once(X,M) which is true if X occurs exactly once in the multiset M:

$$\text{once}(\text{X}, \{\!\!\{X \mid R\}\!\!\}) : -$$
$$\text{X} \notin \text{R}.$$

A sample successful goal for this program is $: - \text{once}(\text{b}, \{\!\!\{a, b, a\}\!\!\})$.

Note that clause selection in a $CLP(\mathcal{BAG})$ program is performed using multiset unification, that is two multisets are compared disregarding the order of their elements. Thus, for instance, unifying $\{\!\{X \,|\, R\}\!\}$ and $\{\!\{a, b, a\}\!\}$ will non-deterministically bind X to all possible elements of the second multiset and R to the corresponding multisets composed of the remaining elements (that is, $X = a \wedge R = \{\!\{b, a\}\!\}$, $X = b \wedge R = \{\!\{a, a\}\!\}$, $X = a \wedge R = \{\!\{a, b\}\!\}$).

$CLP(\mathcal{BAG})$ defines a simple, yet very expressive, framework that allows us to program using multisets in a very flexible way.

3   Defining P Systems in $\mathcal{L}_{Bag}$

A P system is a computational model based on transitions governed by multiset rewriting rules. P systems have been defined to mimic the evolution of biological systems (see, e.g., [21]). Their definition formalizes the intuitive idea that biological reactions can be seen as computations. In this section we show how P systems can be naturally defined using the first-order language $\mathcal{L}_{Bag}$ and the theory Bag.

3.1   Membrane Structures

The language MS of membrane structures is defined in [21] to be the set of strings over the alphabet $\{$ "[", "]" $\}$ generated by the *infinite* set of rules:

- $[\,] \in MS$
- for $n \geq 1$, if $s_1, \ldots, s_n \in MS$, then $[s_1, \ldots, s_n] \in MS$.

A *membrane* is characterized by a pair of open and closed parentheses $[,]$. The *degree* of a membrane structure is the number of membranes in it. An equivalence relation $\sim$ is defined to govern membrane equivalence and an *ad-hoc* notion of *depth* is given. The outer membrane is called the *skin*.

Moving to $\mathcal{L}_{Bag}$, this is equivalent to say that MS is the set $T(\Sigma)$ of first-order ground terms over the signature $\Sigma = \{\texttt{nil}, \{\!\{\cdot \,|\, \cdot\}\!\}\}$. Such set is generated by the *two* rules:

- $\texttt{nil} \in MS$
- if $s, t \in MS$, then $\{\!\{s \,|\, t\}\!\} \in MS$.

Membranes are therefore identified with multisets. $\sim$ is exactly the equivalence relation $\equiv_{E_{Bag}}$ induced on $T(\Sigma)$ by the (E_p^m) permutativity axiom (c.f. Section 2).

To compute the *degree* of a term s it is sufficient to count the number of occurrences of the constant symbol $\texttt{nil}$ in s. This assertion is justified by the following observations:

- if $\texttt{nil}$ occurs as *element* of a multiset (i.e., as first argument of a term of the form $\{\!\{\_ \,|\, \_\}\!\}$), such as in $\{\!\{s_1, \ldots, s_n, \texttt{nil} \,|\, r\}\!\}$ then it must be counted as a pair of open and immediately closed brackets.

– Otherwise, if `nil` occurs as the tail of a multiset (i.e., as the second argument of a term of the form $\{\!\!\{\,\_\,|\,\_\,\}\!\!\}$), such as in $\{\!\!\{\,s_1,\ldots,s_n\,|\,\mathtt{nil}\,\}\!\!\}$ then it exactly identifies one non-empty multiset. We recall that the multiset $\{\!\!\{\,s_1,\ldots,s_n\,\}\!\!\}$ is represented by the term $\{\!\!\{\,s_1\,|\,\{\!\!\{\,s_2\,|\cdots\{\!\!\{\,s_n\,|\,\mathtt{nil}\,\}\!\!\}\cdots\}\!\!\}\,\}\!\!\}$.

The notion of *depth* is the well-established notion of *rank* of a (multi)set:

$$rank(s) = \begin{cases} 0 & \text{if } s = \mathtt{nil} \\ 1 + \max\{rank(t) : t \in s\} & \text{otherwise} \end{cases}$$

A compact way to state that a multiset $\{\!\!\{\,s_1,\ldots,s_m\,\}\!\!\}$ is included in a multiset $\{\!\!\{\,t_1,\ldots,t_n\,\}\!\!\}$ is that of imposing the constraint (R is a logic variable):

$$\{\!\!\{\,t_1,\ldots,t_n\,\}\!\!\} = \{\!\!\{\,s_1,\ldots,s_m\,|\,R\,\}\!\!\}.$$

3.2   Super-cells

A *super-cell* is a membrane structure s in which each membrane (i.e., each multiset) can contain elements coming from a (possibly denumerable) universe U of other objects (i.e., non-membrane objects) [21]. Each membrane is uniquely identified by a natural number (its *label*).

An *active super-cell* is a super-cell which may contain more than one occurrence of membranes with the same label. Active super-cells are introduced in [22] to model the notion of membrane division.

To naturally represent labeling of membranes in $\mathcal{L}_{Bag}$ we will adopt the notion of *kernel*.[1] We fix the signature Σ to be defined as

$$\Sigma = \{\!\!\{\,\cdot\,|\,\cdot\,\}\!\!\} \cup \mathbb{N}_{\mathtt{nil}} \cup U$$

where $\mathbb{N}_{\mathtt{nil}} = \{\mathtt{nil} = \mathtt{nil}_1, \mathtt{nil}_2, \mathtt{nil}_3, \ldots\}$ is the denumerable set of constant symbols (to be used as multiset kernels), and U is a set of constant symbols (the *objects*) disjoint from $\mathbb{N}_{\mathtt{nil}}$.

An *active super-cell* is a ground Σ-term such that for all subterms of the form $\{\!\!\{\,s_1,\ldots,s_m\,|\,k\,\}\!\!\}$, where $m > 0$ and k is not of the form $\{\!\!\{\,\_\,|\,\_\,\}\!\!\}$, it holds that $k \in \mathbb{N}_{\mathtt{nil}}$ (in other words, symbols from U cannot be used as kernels). A *super-cell* is an active super-cell such that each $\mathtt{nil}_i$ occurs at most once.

Figure 2 shows a super-cell of degree 5 and the corresponding representation as an $\mathcal{L}_{Bag}$ term.

As for membrane structures, *equivalence* between super-cells is multiset equivalence $\equiv_{E_{Bag}}$. The notion of degree, instead, must be refined: the *degree* of an active super-cell s is the number of occurrences of constant symbols of $\mathbb{N}_{\mathtt{nil}}$ in it. Hereafter, we usually refer to super-cells where there is exactly one occurrence of each constant $\mathtt{nil}_1,\ldots,\mathtt{nil}_{degree(s)}$.

---

[1]   As explained in Section 2, if there is at least one constant symbol $c \in \Sigma$, then it is possible to write a term t of the form $\{\!\!\{\,s_1,\ldots,s_n\,|\,c\,\}\!\!\}$. t identifies a multiset with the n elements $s_1,\ldots,s_n$ (not necessarily all distinct) and *colored* (labeled) by the constant c. We also say that c is the kernel of t, briefly $ker(t) = c$.

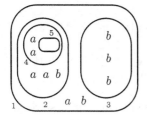

$$\{\, a,$$
$$b,$$
$$\{\, a, a, b, \{\, a, a, \mathtt{nil}_5 \,|\, \mathtt{nil}_4 \,\} \,|\, \mathtt{nil}_2 \,\},$$
$$\{\, b, b, b \,|\, \mathtt{nil}_3 \,\} \qquad\qquad\qquad |\, \mathtt{nil}_1 \,\}$$

Fig. 2. A super-cell of degree 5 and the corresponding term

3.3 Transition P Systems

We give here the notion of P system using the language $\mathcal{L}_{Bag}$. A *transition P system of degree n* can be defined as a tuple

$$T_n = \langle U, \mu, (R_1, \rho_1), \dots, (R_n, \rho_n), i_\mathcal{O} \rangle$$

where U is a universe of objects, μ is a super-cell of degree n, $i_\mathcal{O}$ is the output membrane, and, for all $i = 1, \dots, n$, R_i is a finite set of evolution rules (to be described below) and ρ_i is a partial order relation over R_i.

A *computation* of T_n is a sequence of super-cells $\mu = \mu_0, \mu_1, \dots, \mu_m$ where μ_{j+1} is obtained from μ_j by applying one or more evolution rules.

Observe that the n membranes occurring in μ are uniquely identified by their integer label i. Referring to the multiset representation, this means that the multisets in μ are distinguished each other by their kernels, $\mathtt{nil}_i, \dots, \mathtt{nil}_n$. Hereafter, we will denote these multisets by $m_1, \dots, m_n$.

An *evolution rule* is a pair of the form $u \to v$ where u is a string of elements of U and v is either of the form v' or $v'\delta$ where v' is a string over

$$(V \times \{\mathsf{here}, \mathsf{out}\}) \cup (V \times \{\mathsf{in}_1, \dots, \mathsf{in}_n\}).$$

and V is an alphabet. A rule in R_i applies to the multiset m_i. Applying a rule causes some effects to occur either on m_i or on the multiset possibly containing m_i according to the form of the right-hand side of the rule. More precisely, the semantics of evolution rules, expressed in terms of the multiset representation introduced so far, can be described as follows. Let us consider first rules of the form $u \to v'$, that is without δ. Consider a rule in R_i with the following general form:

$$u_1 \dots u_h \mapsto (v_1, \mathsf{here}) \cdots (v_{k_1}, \mathsf{here})(w_1, \mathsf{out}) \cdots (w_{k_2}, \mathsf{out})(z_1, \mathsf{in}_{x_1}) \cdots (z_{k_3}, \mathsf{in}_{x_{k_3}})$$

The rule can be applied to the multiset m_i if m_i contains (in any order) all the objects $u_1, \dots, u_h$. In the language of multiset constraints, this fact can be expressed by requiring that the constraint

$$m_i = \{\, u_1, \dots, u_h \,|\, M \,\} \wedge \mathsf{ker}(M) = \mathtt{nil}_i$$

holds, for some multiset M (assume ker is the function returning the kernel of a ground multiset). Applying this rule yields the following effects:

$a.$ **local effect:**

replace m_i in μ by $\{v_1, \ldots, v_{k_1} \,|\, M\}$ (if $k_1 = 0$ replace m_i by M).

$b.$ **exit effect:**

if $m_i \in m_j$ then replace m_j in μ by $\{w_1, \ldots, w_{k_2} \,|\, m_j\}$; if m_i is the skin then simply remove the elements $w_1, \ldots, w_{k_2}$ from it.

$c.$ **adjacency effect:**

for $j = 1, \ldots, k_3$, if $m_i \in m_{x_j}$ or $m_{x_j} \in m_i$ (i.e. the two membranes are adjacent), then add z_j to m_{x_j}. If for some $j = 1, \ldots, k_3$ the adjacency condition is not fulfilled, the application of the rule is <u>not</u> allowed.

Consider now the scheme of a δ rule (*dissolving rule*):

$$u_1 \ldots u_h \mapsto (v_1, \text{here}) \cdots (v_{k_1}, \text{here})(w_1, \text{out}) \cdots (w_{k_2}, \text{out})(z_1, \text{in}_{x_1}) \cdots (z_{k_3}, \text{in}_{x_{k_3}})\delta$$

After having obtained the local, exit, and adjacency effects as in the rule of the first kind, the δ causes the further effect of destroying m_i and carrying out its elements (unless m_i is the *skin* that cannot be destroyed). Thus, if $m_j = \{m_i \,|\, m_j'\}$ (i.e. $m_i \in m_j$) and $m_i = \{c_1, \ldots, c_\ell \,|\, \text{nil}_i\}$ then replace m_j by $\{c_1, \ldots, c_\ell \,|\, m_j'\}$.

3.4 P Systems with Active Membranes

In [22] new rules are allowed to occur in P systems. In particular we are interested here in the possibility of creating a *copy* of a membrane. Thus we consider the following simplified version of the *division rule* [22]

$$u_1 \ldots u_k \to ((v_1 \ldots v_h)(w_1 \ldots w_j))$$

The semantics of this rule, expressed in terms of the multiset representation of super-cells introduced in this section, can be described as follows. The rule can be applied—like other rules—to the multiset m_i if m_i contains $u_1, \ldots, u_k$, that is if the constraint

$$m_i = \{u_1, \ldots, u_h \,|\, M\} \wedge \text{ker}(M) = \text{nil}_i$$

holds, for some multiset M. Applying this rule yields the following effects:

– remove elements $u_1, \ldots, u_k$ from m_i; since $m_i = \{u_1, \ldots, u_k \,|\, M\}$, M represents what remains of the multiset m_i;

– make two copies of M (both having the same kernel nil_i) and add all the elements $v_1, \ldots, v_h$ to the first copy and $w_1, \ldots, w_j$ to the second copy; that is, in terms of our multiset representation, replace m_i in μ with the two new multisets

$$\{v_1, \ldots, v_h \,|\, M\} \qquad \{w_1, \ldots, w_j \,|\, M\}.$$

Fig. 3 shows an example of the effect of the application of the membrane division rule. Observe that this rule can be applied to more than one membrane, possibly simultaneously.

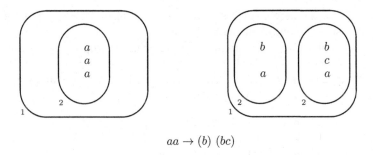

$$aa \rightarrow (b)\,(bc)$$

Fig. 3. Membrane division

4   Implementing P Systems in $CLP(\mathcal{BAG})$

We show that a transition P system can be written in an easy way as a program in a CLP language which supports a decision procedure for multiset constraints. In particular, we show how to translate a P system into a $CLP(\mathcal{BAG})$ program (see Section 2.3).

The main predicate governing the application of the evolution rules is:

```
p_interpreter(Membrane, Membrane) : −
    not rule(Membrane, _).
p_interpreter(MembraneIn, MembraneOut) : −
    rule(MembraneIn, MembraneInt),
    p_interpreter(MembraneInt, MembraneOut).
```

Note that the multiset transformation process terminates as soon as no rule can be applied successfully to the input multiset (first argument of p_interpreter). Each evolution rule of each set R_i of the given P system can be (automatically) translated into a single $CLP(\mathcal{BAG})$ clause, according to the multiset-based representation of membranes and of computations described in Section 3.3. As an example, consider the rule:

$$u_1 \ldots u_h \mapsto (v_1, \text{here}) \cdots (v_{k_1}, \text{here})(w_1, \text{out}) \cdots (w_{k_2}, \text{out}).$$

Assume that m_i is not the outer membrane (in that case the clause must be slightly changed):

```
rule(MembrIn, MembrOut) : −
    choose(MembrIn, nilᵢ, Mᵢ, Mⱼ),
    Mᵢ = {| u₁, …, u_h | M′ᵢ |},
    replace(MembrIn, Mᵢ, {| v₁, …, v_{k₁} | M′ᵢ |}, M′ⱼ),
    replace(MembrIn, Mⱼ, {| w₁, …, w_{k₂} | M′ⱼ |}, MembrOut).
```

M_j is the membrane such that $M_i \in M_j$. Multiset unification in the second sub-goal is used to test applicability of this rule to M_i (find whether $u_1, \ldots, u_h$ belong to

M_i). $v_1, \ldots, v_{k_1}$ are the here elements. The last sub-goal updates the membrane containing M_i using the out elements $w_1, \ldots, w_{k_2}$.

The auxiliary predicates choose and replace can be simply programmed in $CLP(\mathcal{BAG})$ as follows.

choose(Membrane, Kernel, M_i, M_j) chooses a multiset M_i in Membrane with kernel Kernel and the multiset M_j that contains it.

$$\begin{aligned} &\text{choose}(\{\!\!\{\, M_i \mid R \,\}\!\!\}, \text{Kernel}, M_i, \{\!\!\{\, M_i \mid R \,\}\!\!\}) : - \\ &\quad \text{kernel}(M_i, \text{Kernel}). \\ &\text{choose}(\{\!\!\{\, M \mid R \,\}\!\!\}, \text{Kernel}, M_i, M_j) : - \\ &\quad \text{choose}(M, \text{Kernel}, M_i, M_j). \end{aligned}$$

The predicate replace(A, B, C, D) finds one occurrence of B in A and replaces it with C obtaining D.

$$\begin{aligned} &\text{replace}(A, A, C, C). \\ &\text{replace}(\{\!\!\{\, B \mid R \,\}\!\!\}, B, C, \{\!\!\{\, C \mid R \,\}\!\!\}). \\ &\text{replace}(\{\!\!\{\, M \mid R \,\}\!\!\}, B, C, \{\!\!\{\, M' \mid R \,\}\!\!\}) : - \\ &\quad M \neq B, \\ &\quad \text{replace}(M, B, C, M'). \end{aligned}$$

The predicate kernel(M, K) stating that the kernel of the multiset M is K can be programmed by:

$$\text{kernel}(\text{nil}_1, \text{nil}_1).$$
$$\vdots \quad \vdots \quad \vdots$$
$$\text{kernel}(\text{nil}_n, \text{nil}_n).$$
$$\begin{aligned} &\text{kernel}(\{\!\!\{\, A \mid B \,\}\!\!\}, K) : - \\ &\quad \text{kernel}(B, K). \end{aligned}$$

To implement the division rule it is convenient to introduce also a predicate replace2($A, B, C1, C2, D$) whose structure is the same as that of replace but it replaces B in A with both $C1$ and $C2$.

We give here an example running on the interpreter of $CLP(\mathcal{BAG})$ [23] of a P system computing the function n^2.

Example 6. Consider the P system such that $U = \{a, b\}$,

$$\mu = \{\!\!\{ \{\!\!\{\, a, \underbrace{b, \ldots, b}_{n} \mid \text{nil}_2 \,\}\!\!\} \mid \text{nil}_1 \}\!\!\}$$

The set of rules R_1 is empty, while the rules in R_2 are:

$$\begin{aligned} &(1) \; abb \mapsto ((bb)(ab)) \\ &(2) \; ab \mapsto b \\ &(3) \; bb \mapsto b(b, \text{out})(b, \text{out}) \\ &(4) \quad b \mapsto (b, \text{out})\delta \end{aligned}$$

ordered as $(1) > (2) > (3) > (4)$. (As common in P systems, we assume that $u \mapsto (v_1, \text{here})$ can be abbreviated as $u \mapsto v_1$). At the end of each computation,

n^2 occurrences of the object b are present in the outer membrane. All other membranes in it have disappeared. Basically, we have used the well-known fact that $n^2 = \sum_{i=1}^{n}(2i-1)$. We first produce membranes containing $b^n, b^{n-1}, \ldots, b^1$. Then we move out the tokens, in accordance with the mathematical law. The following $CLP(\mathcal{BAG})$ clauses can be used to simulate these rules:

```
%%%(1)
rule(MembrIn, MembrOut) : −
    choose(MembrIn, nil₂, Mᵢ, Mⱼ),
    Mᵢ = { a, b, b | M'ᵢ },
    replace2(MembrIn, Mᵢ, { b, b | M'ᵢ }, { a, b | M'ᵢ }, MembrOut).
%%%(2)
rule(MembrIn, MembrOut) : −
    choose(MembrIn, nil₂, Mᵢ, Mⱼ),
    Mᵢ = { a, b | M'ᵢ },
    replace(MembrIn, Mᵢ, { b | M'ᵢ }, MembrOut).
%%%(3)
rule(MembrIn, MembrOut) : −
    choose(MembrIn, nil₂, Mᵢ, Mⱼ),
    Mᵢ = { b, b | M'ᵢ },
    replace(MembrIn, Mᵢ, { b | M'ᵢ }, MembrInt),
    replace(MembrIn, Mⱼ, { b, b | MembrInt }, MembrOut).
%%%(4)
rule(MembrIn, MembrOut) : −
    choose(MembrIn, nil₂, Mᵢ, Mⱼ),
    Mᵢ = { b | M'ᵢ },
    Mⱼ = { Mᵢ | MembrInt },      δ removes Mᵢ from Mⱼ
    replace(MembrIn, Mⱼ, { b | MembrInt }, MembrOut).
```

The $CLP(\mathcal{BAG})$ program that simulates the given P system, therefore, is composed by the clauses for predicates `p_interpreter` and `rule` along with those for the auxiliary predicates `choose`, `replace`, and `replace2`. A goal for this program can be:

$$: -\text{p\_interpreter}(\{\!\{\, a, b, b, b \,|\, nil_2 \,\}\!\} \,|\, nil_1 \,\}\!\}, \text{MembrOut}),$$

from which we get the computed answer

$$\{\!\{\, b, b, b, b, b, b, b, b, b \,|\, nil_1 \,\}\!\}$$

Note that rule ordering is obtained by exploiting the corresponding $CLP(\mathcal{BAG})$ clause ordering. This solution works correctly for the first successful computation; however, if other alternative computations are attempted through backtracking, clause ordering is no longer sufficient to guarantee rule ordering. To force the desired ordering is always preserved one could add a unique identifier to each rule and slightly modify the definition of predicate `p_interpreter` so as to apply rules according to a list of rule identifiers which explicitly states the rule ordering for each membrane.

5   Implementing Constraint Solvers by P Systems

Given a first-order language $\mathcal{L}$, an interpretation structure $\mathcal{A}$, and the corresponding theory, we show how to implement the algorithm for deciding satisfiability in $\mathcal{A}$ of a $\mathcal{L}$-constraint C using a transition P system derived from C. Basically, the condition for this translation to be fully automatized is that the constraint solving algorithm is given via rewriting rules (a condition often fulfilled in constraint decision procedures). In this section we focus our attention on the solution of equality constraints, that is on the unification problem. We show how to implement both standard and extended unification (namely, multiset unification) algorithms using P systems.

5.1   Standard Unification

We start from the classical first-order unification problem—c.f., e.g., [15,20,1]. Let Σ be a first-order signature and C, $s_1 = t_1 \wedge \cdots \wedge s_n = t_n$, be a constraint composed by conjunctions of equations (i.e., an *equation system*). The unification problem, in its decision version, consists in checking if there is a *substitution* θ (a function from the variables occurring in C to the set of first-order terms) such that for all $i = 1, \ldots, n$, $\theta(s_i)$ and $\theta(t_i)$ are syntactically equal. A survey of the problem with the main results can be found in [1].

Let $T(C)$ be the transition P system

$$\langle U, \mu, (R_1, \rho_1), (R_2, \rho_2), i_{\mathcal{O}} \rangle$$

where:

- the universe U is the *set of all ground equations*, $\{s = t : s, t \in T(\Sigma)\}$
- $\mu = \{\!| \, m_= \, | \, \mathtt{nil}_1 \, |\!\}$
- $m_= = \{\!| \, s_1 = t_1, \ldots, s_n = t_n \, | \, \mathtt{nil}_2 \, |\!\}$
- $i_{\mathcal{O}} = 1$ (i.e. the output membrane is the skin)
- R_1 and R_2 are the sets of evolution rules for multisets with kernels $\mathtt{nil}_1$ and $\mathtt{nil}_2$, respectively, that implement the unification test.

Evolution rules for R_2 are obtained from the non-deterministic unification algorithm Unify [15,20] and are shown in Figure 4. Actually, each rule is a meta-rule consisting of all its infinite possible instances. However, only a *finite* number of instances depending on C is needed, as explained below.

R_1, instead, is assumed to be empty in this case.

Classical results on Herbrand's non-deterministic algorithm ensure termination of the computation of the transition P system independently of the ordering chosen when rules are applied. Moreover, each non-deterministic execution is equivalent (equivalence of m.g.u.'s modulo variants) [1]. Thus, no ordering is needed between rewriting rules to ensure termination, and all the possible non-deterministic computations of the P system can be considered equivalent.

By analyzing the computations of the P system it is easy to prove that the dimension of the terms involved can be bounded by the size of the initial

(r_1)	$\left. \begin{array}{r} X = X \end{array} \right\}$	$\mapsto \varepsilon$ (remove tautologies)
(r_2)	$\left. \begin{array}{r} t = X \\ t \text{ is not a variable} \end{array} \right\}$	$\mapsto X = t$ (i.e., $(X = t, \mathsf{here})$)
(r_3)	$\left. \begin{array}{r} X = t \quad X = s \\ X \text{ does not occur in } t \end{array} \right\}$	$\mapsto X = t \quad t = s$
(r_4)	$\left. \begin{array}{r} X = t \quad Y = X \\ X \text{ does not occur in } t \end{array} \right\}$	$\mapsto X = t \quad Y = t$
(r_5)	$\left. \begin{array}{r} X = t \\ X \not\equiv t \text{ and } X \text{ occurs in } t \end{array} \right\}$	$\mapsto (\mathsf{false}, \mathsf{out})$
(r_6)	$\left. \begin{array}{l} f(s_1, \ldots, s_m) = g(t_1, \ldots, t_n) \\ \hfill f \not\equiv g \end{array} \right\}$	$\mapsto (\mathsf{false}, \mathsf{out})$
(r_7)	$\left. f(s_1, \ldots, s_m) = f(t_1, \ldots, t_m) \right\}$	$\mapsto s_1 = t_1 \quad \cdots \quad s_m = t_m$

Fig. 4. Unification as evolution rules (R_2)

constraint C used to define $T(C)$. More precisely, let *size* and *height* be defined as follows:

$$size(t) = \begin{cases} 0 & \text{if } t \text{ is a variable} \\ 1 + \sum_{i=1}^{n} size(t_i) & \text{if } t = f(t_1, \ldots, t_n) \\ \sum_{i=1}^{m} size(s_i) + size(t_i) & \text{if } t = s_1 = t_1 \wedge \cdots \wedge s_m = t_m \end{cases}$$

$$height(t) = \begin{cases} 0 & \text{if } t \text{ is a variable} \\ 1 + \max_{i=1}^{n} height(t_i) & \text{if } t = f(t_1, \ldots, t_n) \end{cases}$$

Proposition 1. *Let $s = t$ be an equation occurring at some point during the computation of $\mathsf{Unify}(C)$. Then $height(s) \leq size(C)$ and $height(t) \leq size(C)$.*

Proof. Immediate, by classical results. $\qquad\square$

Thus, the number of rules in R_2 can be chosen to be finite. Let S be the set of terms built using the function, constant, and variable symbols occurring in C of height bounded by $size(C)$. R_2 is chosen as the set of instances of the meta-rules of Fig. 4 over the set S.

Proposition 2. *Let $\mu = \mu_0, \ldots, \mu_m$ be a terminating computation of the P system $T(C)$. C does not admit solutions if and only if $\mathsf{false} \in m_1 \in \mu_m$.*

Proof. The evolution of m_2 mimics the execution of standard unification algorithm Unify (that is terminating and correct). A successful termination does not produce false. Otherwise, false is stored in the membrane m_1 and never removed. $\qquad\square$

Observe that the P system obtained in this way, although generated by an instance C, is able to deal with a wide family of equation systems as inputs. Precisely, all input systems with variables, function and constant symbols from those of C, and with initial size bounded by that of C.

5.2   Multiset Unification

When one or more function symbols from Σ have some semantic properties, the unification problem becomes more complex. For instance, in the case of multiset unification (as it happens in Bag) we use a binary function symbol whose semantics is governed by axiom (E_p^m); in this case the unification problem turns out to be NP-complete [9]. We show here how to map the multiset unification algorithm of [9] to a P system with active membranes (using membrane division). This allows one, for instance, to obtain an alternative (polynomial-time) implementation of SAT using P systems, since SAT can be reduced to multiset unification (cf. Example 4).

Let $\Sigma = \{\, \mathtt{nil}, \{\!\cdot\,|\,\cdot\}, \dots \}$ be the signature of $\mathcal{L}_{Bag}$. The transition system $T(C)$ is the same as in Section 5.1, save for the rewriting rules R_1 and R_2. They are chosen in order to simulate the multiset unification algorithm of [9] and are reported in Fig. 5 and Fig. 6. Moreover, some ordering on the rules of R_2 should be imposed to ensure termination in all possible cases as described in [9].

(r_1)	$X = X$ $\}\mapsto \varepsilon$ (remove tautologies)
(r_2)	$\dfrac{t = X}{t \text{ is not a variable}}\Big\}\mapsto X = t$ (i.e., $(X = t, \text{here})$)
(r_3)	$\dfrac{X = t \ \ X = s}{X \text{ does not occur in } t}\Big\}\mapsto X = t \ \ t = s$
(r_4)	$\dfrac{X = t \ \ Y = X}{X \text{ does not occur in } t}\Big\}\mapsto X = t \ \ Y = t$
(r_5)	$\dfrac{X = t}{X \not\equiv t \text{ and } X \text{ occurs in } t}\Big\}\mapsto \delta$
(r_6)	$\dfrac{f(s_1,\dots,s_m) = g(t_1,\dots,t_n)}{f \not\equiv g}\Big\}\mapsto \delta$
(r_7)	$\dfrac{f(s_1,\dots,s_m) = f(t_1,\dots,t_m)}{f \not\equiv \{\!\cdot\,|\,\cdot\}}\Big\}\mapsto s_1 = t_1 \ \cdots \ s_m = t_m$
(r_8)	$\{t_1,\dots,t_m \,|\, N\} = \{s_1,\dots,s_n \,|\, N\} \}\mapsto \{t_1,\dots,t_m\} = \{s_1,\dots,s_n\}$
(r_9)	$\{t \,|\, s\} = \{t' \,|\, s'\} \}\mapsto \begin{array}{l}((t = t' \ \ s = s')\\ (s = \{t' \,|\, N\} \ \ \{t \,|\, N\} = s'))\end{array}$

Fig. 5. Multiset Unification as evolution rules (R_2)

$$\boxed{\boxed{\begin{array}{|c|c|} \hline (r_1) & s = t \\ \hline \end{array} \Big\} \mapsto \varepsilon \ \textit{(remove equations)}}}$$

Fig. 6. Evolution rules for the skin (R_1)

Consider the rules for membranes with label 2 (i.e., multisets with kernel nil_2). Rules r_1–r_4 are the same as for the standard unification case. Rule r_7 is now restricted to non-multiset terms. Rules r_8 and r_9 treat the case of equalities between two multisets. These two rules introduce a slight notational ambiguity. As a matter of fact, the multisets occurring in those rules are not membranes but multiset terms in the first-order language in which the equations are written. This kind of confusion should be avoided by using different function symbols for building multiset terms in equations (e.g., $[\![\ ]\!]$ and $[\![\cdot\,|\,\cdot]\!]$). Rule r_8 deals with the special case of two non-ground multiset terms with the same tail variable. Rule r_9, instead, deals with the general case. It requires the ability to express a (don't know) non-deterministic choice between two possible computations, corresponding to the two alternatives in the right-hand part of axiom (E_k^m). It is implemented by membrane division.

Rules r_5 and r_6 dealing with failures cause now the deletion of the membrane with label 2 in which the condition is fulfilled, putting all its remaining elements (if any) in the outer membrane. Therefore, equalities can be present in membrane 1 as effect of deletion of membranes with label 2. Rule r_1 in R_1 removes all of them. At the end of the execution, membrane 1 is either empty or it contains some membranes with label 2, each one corresponding to a successful non-deterministic branch of the computation.

Proposition 3. *Let $\mu = \mu_0, \ldots, \mu_m$ be a terminating computation of the P system $T(C)$. C is unsatisfiable if and only if m_1 is empty.*

Proof. Immediate from the fact that we mimic the algorithm in [9]. $\square$

As an example consider the equality constraint C that models the instance of 3-SAT of Example 4. The P system $T(C)$ which tests the satisfiability of this constraint can be seen as (yet another) implementation of an algorithm to solve the 3-SAT problem using P systems. Computations of $T(C)$ run in non-deterministic polynomial time. If membrane division is implemented in an effective way, this yields a polynomial time method for solving 3-SAT (see also [22]).

The constraint solver of $CLP(\mathcal{BAG})$ (see [10]) deals not only with equality constraints, but also with disequality, membership, and not-membership constraints. The same technique used for implementing unification as evolution rules can be applied almost unaltered to implement also the rewriting procedures for the other kinds of constraints. Thus the whole constraint solver of $CLP(\mathcal{BAG})$ can be implemented as a P system.

Therefore, any problem which can be easily expressed as a multiset constraint (containing equalities, disequalities, memberships, and negation of memberships) can be automatically implemented using a P system.

6   Conclusions

We have described how to use the language $\mathcal{L}_{Bag}$ to define P systems in a fairly natural way. This gives a precise formulation of P systems inside a first-order theory. Moreover, we have shown that the CLP language $CLP(\mathcal{BAG})$ is particularly well-suited to simulate P systems, thanks to its capabilities of manipulating multisets in a very general and flexible way. Then, showing that the constraint solvers of $CLP(\mathcal{BAG})$ can be implemented by P systems (endowed with the membrane division rule), we have suggested an alternative way to encode, in a rather natural way, NP problems which are easily described using $\mathcal{L}_{Bag}$ constraints into P systems.

As a future work it could be interesting to analyze if using a Prolog system that supports some form of parallelism (e.g., [13]) as the execution environment where to embed our CLP language would provide a way to enhance also our simulation of P systems in the direction of a real parallel execution.

References

1. K. R. Apt. *From Logic Programming to Prolog*. International Series in Computer Science. Prentice Hall, 1997.
2. P. Arenas-Sánchez, F. J. López-Fraguas, M. Rodríguez-Artalejo Embedding Multiset Constraints into a Lazy Functional Logic Language In C. Palamidessi, H. Glaser, K. Meinke, editors, *Principles of Declarative Programming*, LNCS 1490, Springer-Verlag, pp. 429–444, 1998.
3. J. Bânatre and D. Le Métayer. Programming by Multiset Transformation. *Communications of the ACM*, 36(1):98–111. January 1993.
4. G. Berry and G. Boudol. The Chemical Abstract Machine. *Theoretical Computer Science*, vol. 96 (1992) 217–248.
5. J.-P. Bodeveix, C. Percebois, S. Majoul. An Object-Oriented Coordination Model based on Multiset Rewriting. In Proc. of *Ninth International Conference on Parallel and Distributed Computing Systems*. Dijon, France, 1996.
6. I. Cervesato, N. Durgin, P. Lincoln, J. Mitchell, and A. Scedrov. Relating Strands and Multiset Rewriting for Security Protocol Analysis In P. Syverson, ed., *13th IEEE Computer Security Foundations Workshop—CSFW'00*, pp. 35–51, 2000.
7. P. Ciancarini, D. Fogli, and M. Gaspari. A Logic Language Based on GAMMA-like Multiset Rewriting. In R. Dyckhoff, H. Herre, P. Schroeder-Heister eds., *Extensions of Logic Programming*, 5th International Workshop, LNCS 1050, 1996, pp. 83–101.
8. E. Dantsin and A. Voronkov. A Nondeterministic Polynomial-Time Unification Algorithm for Bags, Sets and Trees. In W. Thomas ed., *Foundations of Software Science and Computation Structure*, LNCS Vol. 1578, pages 180–196, 1999.
9. A. Dovier, A. Policriti, and G. Rossi. A uniform axiomatic view of lists, multisets, and sets, and the relevant unification algorithms. *Fundamenta Informaticae*, 36(2/3):201–234, 1998.
10. A. Dovier, C. Piazza, and G. Rossi. A uniform approach to constraint-solving for lists, multisets, compact lists, and sets. Technical Report, Dipartimento di Matematica, Università di Parma, no. 235, July 2000 (available at http://prmat.math.unipr.it/~gianfr/PAPERS/RR_PR235.ps).

11. A. Dovier, E. G. Omodeo, E. Pontelli, and G. Rossi. {log}: A Language for Programming in Logic with Finite Sets. *Journal of Logic Programming*, 28(1):1–44, 1996.
12. A. Dovier, C. Piazza, E. Pontelli, and G. Rossi. Sets and constraint logic programming. *ACM Transaction on Programming Language and Systems* (TOPLAS), 22(5) 2000, pp. 861–931.
13. G. Gupta and E. Pontelli. Optimization Schemas for Parallel Implementation of Nondeterministic Languages. *Int. Parallel Proc. Symposium*, IEEE, pp. 428–435, 1997.
14. C. Hankin, D. Le Métayer, and D. Sands. A Parallel Programming Style and Its Algebra of Programs. In *Proc. Conf. on Parallel Architecture and Languages Europe (PARLE 93)*, vol. 694 of LNCS, 367-378, Springer-Verlag, Berlin, 1993.
15. J. Herbrand. Recherches sur la theorie de la demonstration. Master's thesis, Université de Paris, 1930. Also in *Ecrits logiques de Jacques Herbrand*, PUF, Paris, 1968.
16. J. Jaffar and M. J. Maher. Constraint Logic Programming: A Survey. *Journal of Logic Programming*, 19–20:503–581, 1994.
17. J. Jaffar, M. J. Maher, K. Marriott, and P. J. Stuckey. The Semantics of Constraint Logic Programs. *Journal of Logic Programming 37*(1–3), 1–46, 1998.
18. K. Marriott, B. Meyer, and K. B. Wittenburg. A survey of visual language specification and recognition. In K. Marriott and B. Meyer, editors, *Visual Language Theory* pages 5–85, Springer, 1998.
19. A. Mal'cev. Axiomatizable Classes of Locally Free Algebras of Various Types. In *The Metamathematics of Algebraic Systems*, Collected Papers, Ch. 23. North Holland, 1971.
20. A. Martelli and U. Montanari. An efficient unification algorithm. *ACM Transactions on Programming Languages and Systems 4* (1982), 258–282.
21. G. Păun. Computing with Membranes. Journal of Computer and System Science, 61(1):108–143, 2000.
22. G. Păun. Attacking NP Complete Problems. Journal of Automata, Languages and Combinatorics, 6(1):75–90, 2001.
23. G. Rossi. The Languages $CLP(\mathcal{SET})$ and $CLP(\mathcal{BAG})$. User Manuals and Running Interpreters. In http://prmat.math.unipr.it/~gianfr/setlog.Home.html.
24. A. Tzouvaras. The Linear Logic of Multisets. *Logic Journal of the IGPL*, Vol. 6, No. 6, pp. 901–916, 1998.

Toward a Formal Macroset Theory

Manfred Kudlek[1], Carlos Martín-Vide[2], and Gheorghe Păun[3],*

[1] Fachbereich Informatik, Universität Hamburg,
Vogt-Kölln-Str. 30, D-22527 Hamburg, Germany
kudlek@informatik.uni-hamburg.de
[2] Research Group on Mathematical Linguistics, Rovira i Virgili University,
Pl. Imperial Tàrraco 1, 43005 Tarragona, Spain
cmv@astor.urv.es
[3] Institute of Mathematics of the Romanian Academy,
PO Box 1-764, 70700 Bucureşti, Romania
gpaun@imar.ro

Abstract. A *macroset* is a (finite or infinite) set of multisets over a finite alphabet. We introduce a Chomsky-like hierarchy of multiset rewriting devices which, therefore, generate macrosets. Some results are proved about the power of these devices and some open problems are formulated. We also present an algebraic characterization of some of the macroset families as least fixed point solutions of algebraic systems of equations.

1 Introduction

In the last years, the idea of multiset processing has appeared more and more frequently in various domains: nondeterministic programming [1], the chemical abstract machine [6], DNA computing [28], membrane computing (P systems) [25], [26] (see a current bibliography of the area in [27]). Multisets also appear in logic [4], linguistics [7], [14], artificial life [30], etc. The thesis [32] is entirely devoted to mathematically formalizing multisets. Several papers have considered fuzzy variants of multisets ([24], [21], [35]), while [12] and [5] deal with *pomsets* (partially ordered multisets).

A multiset over a given set of *objects* corresponds to a string over the alphabet naming those objects. To a language it corresponds a sets of multisets, called here a *macroset*.

Observing that P system theory is devoted to a generative approach to multiset processing in a distributed way, in [27] one formulates the problem of producing a non-distributed theory of multiset generating, like Chomsky grammar theory. We systematically address here this question, by considering a natural translation of notions related to sequential rewriting of strings to rewriting of multisets. A Chomsky-like hierarchy of *multiset grammars* is obtained, but with some relationships between classes of macrosets which are different from those among the corresponding families of languages.

---

* Research supported by the Direcció General de Recerca, Generalitat de Catalunya (PIV).

C.S. Calude et al. (Eds.): Multiset Processing, LNCS 2235, pp. 123–133, 2001.
© Springer-Verlag Berlin Heidelberg 2001

In this framework, an important question could be solved: is erasing important in the case of non-context-free macroset grammars? This corresponds to the problem whether or not the Parikh sets of string context-free matrix languages generated with and without using erasing rules are the same, which, in turn, is related to he problem still open in the regulated rewriting area whether or not erasing rules increase the power of string matrix grammars (without appearance checking).

It should be noted that in [30], [31] one already considers an abstract rewriting multiset system (ARMS, for short), which corresponds to our arbitrary multiset grammars, without introducing particular (Chomsky-like) variants; the ARMS are mainly used as a formal framework for simulating the development of populations in artificial life-like questions. Moreover, languages of *multisets* (or *bags*), also called *commutative* languages, and language families (trios, semi AFL's, AFL's) of commutative languages, have been considered in [8], [18], [19], [20], [16]. None of these papers proposes a Chomsky-like hierarchy of multiset processing grammars.

2 Multiset Grammars

All formal language theory prerequisites we use can be found in [29].

Let V be a finite alphabet; its elements are called both *symbols* and *objects*. The set of strings over V is denoted by V^*; the empty string is denoted by λ, the length of $x \in V^*$ is denoted by $|x|$, and the number of occurrences of a symbol $a \in V$ in a string $x \in V^*$ is denoted by $|x|_a$.

A multiset over an alphabet V is a mapping $\mu : V \longrightarrow \mathbf{N}$. We denote by $V^\#$ the set of all multisets over V. We can represent the multiset μ as a vector, $\mu(V) = (\mu(a_1), \ldots, \mu(a_n))$, for $V = \{a_1, \ldots, a_n\}$. Note that the ordering of the objects in V is relevant. The multiset μ can also be represented by any permutation of the string $w_\mu = a_1^{\mu(a_1)} \ldots a_n^{\mu(a_n)}$. Clearly, for each permutation w' of w_μ we have $\Psi_V(w') = \mu(V)$, where Ψ_V is the Parikh mapping associated with V. Sometimes we shall also write w_μ for a multiset μ.

Conversely, with each string $x \in V^*$ we can associate the multiset $\mu_x : V \longrightarrow \mathbf{N}$ defined by $\mu_x(a) = |x|_a$, for each $a \in V$.

For a multiset μ we denote by $|\mu| = \sum_{a \in V} \mu(a) = |w_\mu|$ its *weight* or *norm*; for $U \subseteq V$, we also denote $|\mu|_U = \sum_{a \in U} \mu(a) = |w_\mu|_U$.

A (finite or infinite) set of multisets over an alphabet V is called a *macroset*. Thus, $V^\#$ is the *universal macroset* over V (it corresponds to the "universal language" V^* over V). A macroset $M \subseteq V^\#$, for $V = \{a_1, \ldots, a_n\}$, can be naturally represented by the set of vectors $\{(\mu(V)) \mid \mu \in M\}$.

For two multisets μ_1, μ_2 over the same V we define the *inclusion* $\mu_1 \subseteq \mu_2$, the *sum* $\mu_1 + \mu_2$, the *union* $\mu_1 \cup \mu_2$, the *intersection* $\mu_1 \cap \mu_2$, and, for the case $\mu_1 \subseteq \mu_2$ only, the *difference* $\mu_1 - \mu_2$ in the following ways:

$$\mu_1 \subseteq \mu_2 \text{ iff } \mu_1(a) \leq \mu_2(a), \text{ for all } a \in V;$$
$$(\mu_1 + \mu_2)(a) = \mu_1(a) + \mu_2(a), \text{ for each } a \in V;$$

$$(\mu_1 \cup \mu_2)(a) = \max(\mu_1(a), \mu_2(a)), \text{ for each } a \in V;$$
$$(\mu_1 \cap \mu_2)(a) = \min(\mu_1(a), \mu_2(a)), \text{ for each } a \in V;$$
$$(\mu_1 - \mu_2)(a) = \mu_1(a) - \mu_2(a), \text{ for each } a \in V.$$

A usual set $U \subseteq V$ is a multiset μ_U with $\mu_U(a) = 1$ iff $a \in U$. The empty multiset is denoted by $\emptyset$ and it corresponds to the empty string λ.

The correspondence among strings and multisets (hence among languages and macrosets) suggests considering the following macrosets generating devices.

A *multiset grammar* is a construct $G = (N, T, A, P)$, where N, T are disjoint alphabets, the *nonterminal* and the *terminal* one, respectively (we denote their union by V), $A \subseteq V^{\#}$ is a finite macroset over V (its elements are called *axioms*, and P is a finite set of *multiset rewriting rules* (in short, *rules*) of the form $\mu_1 \to \mu_2$, where μ_1, μ_2 are multisets over V and $|\mu_1|_N \geq 1$. (Taking advantage of the string representation of multisets, we will usually write strings instead of multisets in A and in rules.)

For two multisets α_1, α_2 over V, we write $\alpha_1 \Longrightarrow_r \alpha_2$ for some $r : \mu_1 \to \mu_2 \in P$ if $\mu_1 \subseteq \alpha_1$ and $\alpha_2 = (\alpha_1 - \mu_1) + \mu_2$. If r is understood, then we write $\Longrightarrow$ instead of $\Longrightarrow_r$. We denote by $\Longrightarrow^*$ the reflexive and transitive closure of the relation $\Longrightarrow$. The macroset *generated* by G is defined by

$$M(G) = \{\beta \in T^{\#} \mid \alpha \Longrightarrow^* \beta, \text{ for some } \alpha \in A\}.$$

A natural Chomsky-like classification of such grammars is the following one:

1. Grammars G as above are said to be *arbitrary*.
2. If $|\mu_1| \leq |\mu_2|$ for all rules $\mu_1 \to \mu_2$ in P, then G is said to be *monotone*.
3. If $|\mu_1| = 1$ for all rules $\mu_1 \to \mu_2$ in P, then G is said to be *context-free*.
4. If $|\mu_1| = 1$ and $|\mu_2|_N \leq 1$ for all rules $\mu_1 \to \mu_2$ in P, then G is said to be *linear*.
5. If $|\mu_1| = 1, |\mu_2| \leq 2$, and $|\mu_2|_N \leq 1$ for all rules $\mu_1 \to \mu_2$ in P, then G is said to be *regular*.

   We also consider the following subclass of linear grammars, which corresponds to no Chomsky class:
6. If G is a linear multiset grammar such that for each rule $\mu_1 \to \mu_2$ in P such that $\mu_2(a_i) > 0, \mu_2(a_j) > 0$ for some $1 \leq i < j \leq n$, imply $\mu_2(a_k) > 0$ for all $i \leq k \leq j$, then G is said to be *local*. (We always increase the number of copies of objects which are adjacent in the ordering of V.)

We denote by **mARB, mMON, mCF, mLIN, mREG, mLOC** the families of macrosets generated by arbitrary, monotone, context-free, linear, regular, and local multiset grammars, respectively. By **FIN, REG, LIN, CF, CS, RE** we denote the families of finite, regular, linear, context-free, context-sensitive, and recursively enumerable languages, respectively. For a family **F** of languages we denote by **PsF** the family of Parikh sets of vectors associated with languages in **F**. The family of all semilinear languages is denoted by **SLin**.

We also consider here *matrix grammars*, with and without appearance checking, both for the string and the multiset case. The definition for multisets is a direct extension of the definition for strings. Always we consider only matrix grammars with context-free rules. We denote by **MAT** the family of languages generated by matrix grammars without λ rules and without appearance checking; if erasing rules are used, then the superscript λ is added, if appearance checking features are used, then the subscript **ac** is added. For the multiset case we add the letter **m** in front of these notations, thus obtaining the families **mMAT**, **mMAT**$^\lambda$, **mMAT**$_{\mathbf{ac}}$, and **mMAT**$^\lambda_{\mathbf{ac}}$, respectively.

3   Generative Power

We start by proving a normal form result which says that instead of a finite set of axioms, we may consider one axiom only, in the form of a single nonterminal symbol (like in Chomsky grammars).

Lemma 1. *For each multiset grammar* $G = (N, T, A, P)$, *there is an equivalent multiset grammar* $G' = (N', T, A', P')$, *where* A' *contains only one multiset* S *over* $N' \cup T$, *such that* $|S| = 1$. *Moreover,* G' *is of the same type as* G.

Proof. For arbitrary, monotone, and context-free grammars the assertion is obvious: take one more nonterminal symbol, S, add it to N and add all rules $S \to w$, for $w \in A$, to P. We obtain an equivalent grammar, which is monotone (context-free) if G was monotone (context-free).

For the linear case we proceed as follows. Assume that $A = \{w_1, \ldots, w_n\}$, with $w_i \in (N \cup T)^{\#}$. Let $N(w_i), T(w_i)$ be the multisets of nonterminals and of terminals, respectively, which appear in $w_i, 1 \le i \le n$, and let $t = \max\{|N(w_i)| \mid 1 \le i \le n\}$. Then, consider

$$N' = \{[w] \in N^* \mid 1 \le |w| \le t\} \cup \{S\},$$
$$A' = \{S\},$$
$$P' = \{S \to [N(w_i)]T(w_i) \mid 1 \le i \le n\}$$
$$\cup \{[w] \to [w']x \mid [w], [w'] \in N', w' = (w - X) + Y, \text{ for a rule}$$
$$X \to Yx \in P, X, Y \in N, x \in T^*\}$$
$$\cup \{[X] \to x \mid \text{for } X \to x \in P, X \in N, x \in T^*\}.$$

The equality $M(G) = M(G')$ is obvious and also obvious is the fact that G' is linear, regular, local if G is linear, regular, local, respectively.   □

Because most of the relationships between the above mentioned families of macrosets are easy to be established, we directly give a synthesis result (also including a relation which will be proved later):

Theorem 1. (Macroset Hierarchy Theorem) *The relations in the diagram from Figure 1 hold; the arrows denote inclusions of the lower families into the upper families; all these inclusions are proper, with the exception of the arrow marked with a question mark, for which the properness is open.*

Proof. In view of Lemma 1, we can always assume that the set of axioms contains only one axiom, consisting of a single nonterminal.

The equalities **mREG** = **PsREG**, **mLIN** = **PsLIN** and **mCF** = **PsCF** are obvious; because **PsREG** = **PsLIN** = **PsCF** = **SLin**, we get the equality of all these families (see also Section 4).

The equality **PsREG** = **SLin** follows from the definition of semilinear sets as finite unions of linear sets : $M = \bigcup_{i=1}^{n} M_i$ with $M_i = \{c_i + \sum_{j=1}^{m(i)} k_{ij} x_{ij} \mid k_{ij} \in \mathbf{N}\}$ which can be generated by a regular multiset grammar $S \to S_i$, $S_i \to A_i + c_i$, $A_i \to A_i + x_{ij}$, $A_i \to x_{ij}$, and the fact that **PsCF** $\subseteq$ **SLin** ([11], Theorem 5.2.1).

It is obvious that **mFIN** = **PsFIN**, and that **mLOC** contains infinite macrosets.

Consider the macroset $M = \{(n, 1, n) \mid n \geq 1\}$, for $V = \{a_1, a_2, a_3\}$. It is easy to see that we cannot generate this macroset by a local multiset grammar (all vectors $(n, 1, n), n \geq 1$, belong to M, hence we have to increase the number of a_1 and a_3 occurrences in a synchronized way, but we cannot do this in a local way), but we have $M \in$ **mLIN**. This shows that **mLOC** $\subset$ **mLIN** is a proper inclusion.

Given a monotone multiset grammar $G = (N, T, A, P)$, we can construct a matrix multiset grammar in the following way. For each $X \in N \cup T$ consider the new symbols X', X''. For each rule $X_1 X_2 \ldots X_k \to Y_1 Y_2 \ldots Y_r$ in P, with $k \leq r$ (we use the string representation of rules), consider the matrix

$$(X_1' \to Y_1'', \ldots, X_{k-1}' \to Y_{k-1}'', X_k' \to Y_k' Y_{k+1}' \ldots Y_r', Y_1'' \to Y_1', \ldots, Y_{k-1}'' \to Y_{k-1}').$$

(The use of double primed symbols prevents the rewriting of an object introduced by a previous rule of the same matrix; this is allowed in matrix grammars but not in monotone grammars.) The matrix grammar with these matrices, with all primed symbols as nonterminals, and with the axioms obtained from the axioms of A by priming their objects generate the same macroset as G. The technical details are left to the reader. Thus, **mMON** $\subseteq$ **mMAT**.

Now, the equality **mMAT** = **PsMAT** is obvious. (A normal form as that in Lemma 1 we have started with also holds for matrix grammars. See also [8].) From a string matrix grammar $G = (N, T, S, R)$ in the binary normal form we can immediately pass to a multiset matrix grammar G' (for $(X \to Y, A \to x) \in R$ consider the multiset rewriting rule $XA \to Yx$, etc) such that $M(G') = \Psi_T(L(G))$. Consequently, **mMAT** $\subseteq$ **mMON**, and therefore **mMON**=**mMAT** = **PsMAT**.

The inclusion **mCF** $\subset$ **mMON** is proper because **MAT** contains non-semilinear languages. The simplest example, from Petri net theory, is $ac \to add, bd \to bc, a \to b, b \to ae, a \to e$ with axiom ac, generating the non-semilinear multiset language $L = \{\mu \in \{c, e\}^{\#} \mid \mu(c) \leq 2^{\mu(e)}\}$.

The proper inclusion **mMON** $\subset$ **mARB** is a consequence from the facts **mMON** $\subseteq$ **PSPACE**, that the uniform word problem is **EXPSPACE**-hard for **mARB** [15], and that **PSPACE** $\subset$ **EXPSPACE**.

The equality **mMAT**$_{\mathbf{ac}}$ = **PsMAT**$_{\mathbf{ac}}$ is obvious. **PsMAT** $\subset$ **PsMAT**$_{\mathbf{ac}}$ is a proper inclusion, because the one-letter languages in **MAT**$^\lambda$ are regular, while **MAT**$_{\mathbf{ac}}$ contains the language $\{a^{2^n} \mid n \geq 1\}$. Therefore, both the inclusions **mMON** $\subset$ **mMAT**$_{\mathbf{ac}}$ and **mMAT**$^\lambda$ $\subset$ **mMAT**$^\lambda_{\mathbf{ac}}$ are proper.

The equalities $\mathbf{mARB} = \mathbf{mMAT}^\lambda = \mathbf{PsMAT}^\lambda$ follow in the same way as $\mathbf{mMON} = \mathbf{mMAT} = \mathbf{PsMAT}$.

The equality $\mathbf{mMAT}^\lambda_{\mathrm{ac}} = \mathbf{PsMAT}^\lambda_{\mathrm{ac}}$ is obvious, $\mathbf{PsRE} = \mathbf{PsMAT}^\lambda_{\mathrm{ac}}$ follows from $\mathbf{RE} = \mathbf{MAT}^\lambda_{\mathrm{ac}}$. Because there are one-letter languages in $\mathbf{RE}$ which are not in $\mathbf{CS}$, the inclusion $\mathbf{PsCS} \subset \mathbf{mMAT}^\lambda_{\mathrm{ac}}$ is proper.

The inclusion $\mathbf{PsMAT}_{\mathrm{ac}} \subseteq \mathbf{PsCS}$ follows from $\mathbf{MAT}_{\mathrm{ac}} \subset \mathbf{CS}$.

$\mathbf{mETOL} = \mathbf{PsETOL}$ is obvious since derivations are context-independent. The proper inclusion $\mathbf{mCF} \subset \mathbf{mETOL}$ follows from the fact that the non-semilinear language $L = \{a^{2^n} \mid n \geq 1\}$ is in $\mathbf{mETOL}$. $\mathbf{mETOL} \subseteq \mathbf{mMAT}_{\mathrm{ac}}$ is a consequence of $\mathbf{ETOL} \subset \mathbf{MAT}_{\mathrm{ac}}$.   □

Observe the unexpected relations $\mathbf{mMON} = \mathbf{mMAT}$, $\mathbf{mARB} = \mathbf{mMAT}^\lambda$, $\mathbf{mMON} \subset \mathbf{mMAT}^\lambda \cap \mathbf{mMAT}_{\mathrm{ac}}$, $\mathbf{mARB} \subset \mathbf{mMAT}^\lambda_{\mathrm{ac}}$, $\mathbf{mMON} \subset \mathbf{PsCS}$, which correspond to relations from the language case which are different, unknown, or even opposite (this last case appears for $\mathbf{mMON} \subset \mathbf{mMAT}_{\mathrm{ac}}$ versus $\mathbf{MAT}_{\mathrm{ac}} \subset \mathbf{CS}$).

Quite interesting is also the fact that $\mathbf{mARB} \subset \mathbf{PsRE}$ is a strict inclusion. This means that arbitrary multiset rewriting rules are not sufficient in order to get a characterization of the power of Turing machines as multiset processing devices. Regulated rewriting of multisets (following the model of regulated rewriting in formal language theory, [9]) is thus necessary and, at least for matrix grammars with appearance checking, we get a characterization of $\mathbf{PsRE}$.

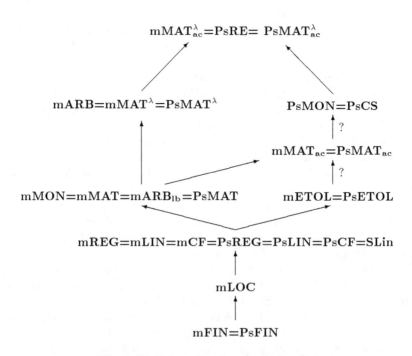

Fig. 1. The hierarchy of macroset families

Note also that **PsRE** is of the power of Turing machines since any word u can be encoded as a multiset $\mu(u)$, even as $\mu(u) \in \mathbf{N}$ such that any $L \in \mathbf{RE}$ can be represented by a multiset language $\mu(L) \in \mathbf{PsRE}$.

Many such characterizations can be found in the distributed framework of P systems; the previous observation that **mARB** $\subset$ **PsRE** can also be seen as a further motivation of membrane computing.

The corresponding hierarchy of families of word languages is shown in Figure 2 [10] where also the family **ETOL** has been added.

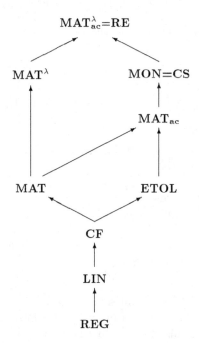

Fig. 2. The hierarchy of word language families

In the diagram in Figure 1 there appears a family which was not defined yet, **mARB**$_{\text{lb}}$. This is the family of macrosets which can be generated by arbitrary multiset grammars with a *linearly bounded workspace*.

For $G = (N, T, A, P)$, let $\sigma : \alpha \Longrightarrow^* \beta$ be a derivation with $\alpha \in A$. The workspace of σ, denoted by $WS(\sigma)$, is the maximal weight of multisets appearing in σ. Then, for $\beta \in T^{\#}$ we put

$$WS_G(\beta) = \inf\{WS(\sigma) \mid \sigma : \alpha \Longrightarrow^* \beta, \alpha \in A\}.$$

We say that G has a linearly bounded workspace if there is a constant k such that for each non-empty $\beta \in M(G)$ we have $WS(\beta) \leq k|\beta|$.

The family of all macrosets generated by grammars with a linearly bounded workspace is denoted by **mARB**$_{\text{lb}}$.

The following result corresponds to the workspace theorem in formal language theory; the proof for multisets is simpler than in the case of languages.

Theorem 2. (Workspace Theorem) $\mathbf{mMON} = \mathbf{mARB_{lb}}$.

Proof. We have only to prove the inclusion $\mathbf{mARB_{lb}} \subseteq \mathbf{mMON}$.

Consider an arbitrary multiset grammar $G = (N, T, A, P)$ and a constant k such that $WS_G(\beta) \le k|\beta|$ for all non-empty $\beta \in M(G)$. We construct the monotone multiset grammar $G' = (N', T, S, P')$ with

$$N' = \{\langle u \rangle \mid u \in (N \cup T \cup \{b\})^\#, |u| = k\} \cup \{S\},$$

where S and b are new symbols, and the following rules (we denote $V = N \cup T \cup \{b\}$):

1. $S \to \langle v_1 \rangle \ldots \langle v_q \rangle \langle v_{q+1} b^t \rangle S$,
   $S \to \langle v_1 \rangle \ldots \langle v_q \rangle \langle v_{q+1} b^t \rangle$
   for $v_1 \ldots v_q v_{q+1} \in A, |v_i| = k, 1 \le i \le q, q \ge 0$, and $|v_{q+1} b^t| = k, t \ge 0$,
2. $S \to \langle b^k \rangle S$,
3. $\langle u \rangle \langle v \rangle \to \langle u' \rangle \langle v' \rangle$
   for all $u, v, u', v' \in V^\#, |u| = |v| = |u'| = |v'| = k, u + v = u' + v'$,
4. $\langle u_1 \rangle \ldots \langle u_r \rangle \to \langle v_1 \rangle \ldots \langle v_r \rangle$,
   for each $\langle u_i \rangle, \langle v_i \rangle \in N', 1 \le i \le r, r \ge 1$, such that $u_1 + u_2 + \ldots + u_r \Longrightarrow v_1 + v_2 + \ldots + v_r$ by a rule $u \to v \in P$,
5. $\langle u_1 \rangle \ldots \langle u_r \rangle \to \langle v_1 \rangle \ldots \langle v_r \rangle \langle v_{r+1} \rangle \ldots \langle v_t \rangle$,
   for $\langle u_i \rangle, \langle v_j \rangle \in N', 1 \le i \le r, 1 \le j \le t, r \ge 1, t > r$, such that $u_1 + u_2 + \ldots + u_r \Longrightarrow v_1 + v_2 + \ldots + v_r + v_{r+1} + \ldots + v_t$ by a rule $u \to v \in P$ and $|v_i|_b = 0$
   for all $1 \le i \le t - 1, v_t \in V^\#, |v_t| = k$,
6. $\langle u \rangle \to u$, for all $u \in (T \cup \{b\})^\#$ with $|u|_T \ge 1$.

The nonterminals of G' are multisets of weight k, maybe with multiple occurrences of the object b. By rules of type 1 we introduce the axioms of G, splitted among nonterminals of G'; further nonterminals of the form $\langle b^k \rangle$ can be introduced, if needed in the derivation, by rules of type 2. By rules of type 3, the contents of multisets in the nonterminal symbols of G' can be freely interchanged. By rules of types 4 and 5 we can simulate the rules from P, by making use of the "free space"; made available by the occurrences of b in the nonterminals of G'. When no nonterminal with respect to G is present in a nonterminal $\langle u \rangle$, then by rules of type 6 we introduce the terminal objects of u, providing that at least one exist; the occurrences of b are ignored. Thus, one can easily see that we get $M(G) = M(G')$.   □

4   An Algebraic Characterization

Another way to characterize the families $\mathbf{mREG}$, $\mathbf{mLIN}$, $\mathbf{mCF}$, $\mathbf{PsREG}$, $\mathbf{PsLIN}$, and $\mathbf{PsCF}$ is to define macrosets as least fixed points (solutions) of corresponding systems of equations on the ω-complete semiring with vector addition as underlying operation (for the theory of semirings, see [13], [17]). This can be done in a quite general manner for other operations, giving rational (corresponding to regular), linear, and algebraic (corresponding to context-free) languages/macrosets. If the underlying operation is commutative all these families of languages/macrosets coincide and this is the case for vector addition on $\mathbf{N}^n$.

Let $\mathcal{M}$ be a monoid with binary operation $\circ$ and unit element $\mathbf{1}$, or with a binary operation $\circ : \mathcal{M} \times \mathcal{M} \to \mathcal{P}(\mathcal{M})$ and unit element $\mathbf{1}$, i.e., $\mathbf{1} \circ \alpha = \alpha \circ \mathbf{1} = \{\alpha\}$.

Extend $\circ$ to an associative operation $\circ : \mathcal{P}(\mathcal{M}) \times \mathcal{P}(\mathcal{M}) \to \mathcal{P}(\mathcal{M})$, distributive with union $\cup$ ($A \circ (B \cup C) = (A \circ B) \cup (A \circ C)$ and $(A \cup B) \circ C = (A \circ B) \cup (B \circ C)$), with unit element $\{\mathbf{1}\}$ ($\{\mathbf{1}\} \circ A = A \circ \{\mathbf{1}\} = A$), and zero element $\emptyset$ ($\emptyset \circ A = A \circ \emptyset = \emptyset$).

Then $\mathcal{S} = (\mathcal{P}(\mathcal{M}), \cup, \circ, \emptyset, \{\mathbf{1}\})$ is an ω-complete semiring, i.e., if $A_i \subseteq A_{i+1}$ for $0 \leq i$ then $B \circ \bigcup_{i \geq 0} A_i = \bigcup_{i \geq 0}(B \circ A_i)$ and $(\bigcup_{i \geq 0} A_i) \circ B = \bigcup_{i \geq 0}(A_i \circ B)$.

Define also $A^{(0)} = \{\mathbf{1}\}$, $A^{(1)} = A, A^{(k+1)} = A \circ A^{(k)}$, $A^\circ = \bigcup_{k \geq 0} A^{(k)}$.

Let $\mathcal{X} = \{X_1, \ldots, X_n\}$ be a set of variables such that $\mathcal{X} \cap \mathcal{M} = \emptyset$.

A *monomial* over $\mathcal{S}$ with variables in $\mathcal{X}$ is a finite string of the form: $A_1 \circ A_2 \circ \ldots \circ A_k$, where $A_i \in \mathcal{X}$ or $A_i \subseteq \mathcal{M}, |A_i| < \infty, i = 1, \ldots, k$. Without loss of generality, $A_i = \{\alpha_i\}$ with $\alpha_i \in \mathcal{M}$ suffices. The α_{ij} (or $\{\alpha_{ij}\}$) will be called *constants*. A *polynomial* $p(\underline{X})$ over $\mathcal{S}$ is a finite union of monomials where $\underline{X} = (X_1, \cdots, X_n)$.

A *system of equations* over $\mathcal{S}$ is a finite set of equations

$$\mathcal{E} := \{X_i = p_i(\underline{X}) \mid i = 1, \ldots, n\},$$

where $p_i(\underline{X})$ are polynomials. This will also be denoted by $\underline{X} = \underline{p}(\underline{X})$.

The *solution* of $\mathcal{E}$ is a n-tuple $\underline{L} = (L_1, \ldots, L_n)$ of sets over $\mathcal{M}$, with $L_i = p_i(L_1, \ldots, L_n)$ and the n-tuple is minimal with this property, i.e., if $\underline{L}' = (L'_1, \ldots, L'_n)$ is another n-tuple satisfying $\mathcal{E}$, then $\underline{L} \leq \underline{L}'$ (where the order is defined componentwise with respect to inclusion, i.e., $\underline{A} = (A_1, \cdots, A_n) \leq (B_1, \cdots, B_n) = \underline{B}$ if and only if for all $1 \leq i \leq n$ we have $A_i \subseteq B_i$).

From the theory of semirings it follows that any system of equations over $\mathcal{S}$ has a unique solution, and this is the least fixed point starting with

$$\underline{X}^{(0)} = (X_1^{(0)}, \cdots, X_n^{(0)}) = (\emptyset, \cdots, \emptyset) = \underline{\emptyset}, \text{ and } \underline{X}^{t+1} = \underline{p}(\underline{X}^{(t)}).$$

Then the following fact holds: $\underline{X}^{(t)} \leq \underline{X}^{(t+1)}$ for $0 \leq t$.

This is seen by induction and the property of the polynomial with respect to inclusion, as $\underline{\emptyset} \leq \underline{X}^{(1)}$ and $\underline{X}^{(t+1)} = \underline{p}(\underline{X}^{(t)}) \leq \underline{p}(\underline{X}^{(t+1)}) = \underline{X}^{(t+2)}$.

A general system of equations is called *algebraic*; it is *linear* if all monomials are of the form $A \circ X \circ B$ or A, and *rational* if they are of the form $X \circ A$ or A, with $A \subseteq M$ and $B \subseteq M$. Corresponding families of macrosets (solutions of such systems of equations) are denoted by $\mathbf{ALG}(\circ)$, $\mathbf{LIN}(\circ)$, and $\mathbf{RAT}(\circ)$. In the case $\circ$ is commutative then all families are identical: $\mathbf{ALG}(\circ) = \mathbf{LIN}(\circ) = \mathbf{RAT}(\circ)$.

Interpreting the variables of the system of equations as non-terminals, then also rational (regular), linear and algebraic (context-free) grammars can be defined. They generate the same languages as those defined by least fixed points.

Since the operation of vector addition is commutative, it follows that rational, linear and algebraic languages coincide. In this way, we have again obtained the corresponding equalities from Theorem 1.

5 Conclusion

Of course, many other problems remain to be investigated; the whole program of formal language theory can/must be repeated for macrosets. For instance, we can define pure, distributed (like in grammar system area), or parallel (like in Lindenmayer and in P system area) multiset grammars; also other regulating mechanisms than the matrix one can be considered. We leave to the interested reader the task to make the further step toward Formal Macroset Theory.

References

1. J.P. Banâtre, A. Coutant, D. Le Metayer, A parallel machine for multiset transformation and its programming style, *Future Generation Computer Systems*, 4 (1988), 133–144.
2. J.P. Banâtre, D. Le Metayer, The Gamma model and its discipline of programming, *Science of Computer Programming*, 15 (1990), 55–77.
3. J. P. Banâtre, D. Le Metayer, Gamma and the chemical reaction model: ten years after, *Coordination Programming: Mechanisms, Models and Semantics*, Imperial College Press, 1996.
4. H.P. Barendregt, *The Lambda Calculus; Its Syntax and Semantics*, North-Holland, Amsterdam, 1984.
5. T. Basten, Parsing partially ordered multisets, *Intern. J. Found. Computer Sci.*, 8, 4 (1997), 379–407.
6. G. Berry, G. Boudol, The chemical abstract machine, *Theoretical Computer Sci.*, 96 (1992), 217–248.
7. A. Colmerauer, Equations and inequalities on finite and infinite trees, *Proc. of Second Intern. Conf. on Fifth Generation Computer Systems*, Tokyo, 1984, 85–99.
8. S. Crespi-Reghizzi, D. Mandrioli, Commutative Grammars, *Calcolo*, 13, 2 (1976), 173–189.
9. J. Dassow, Gh. Păun, *Regulated Rewriting in Formal Language Theory*, Springer-Verlag, Berlin, 1989.
10. J. Dassow, Gh Păun, A. Salomaa, *Grammars with Controlled Derivations*, in: Handbook of Formal Languages, vol. 2, p 101–154, Springer-Verlag, 1997.
11. S. Ginsburg, *The Mathematical Theory of Context-free Languages*, McGraw Hill, 1966.
12. J.L. Gischer, The equational theory of pomsets, *Theoretical Computer Sci.*, 61, 2-3 (1988), 199–224.
13. J.S. Golan, *The Theory of Semirings with Application in Mathematics and Theoretical Computer Science*, Longman Scientific and Technical, 1992.
14. S. Gorn, Explicit definitions and linguistic dominoes, in *Systems and Computer Science* (J. Hart, S. Takasu, eds.), Univ. of Toronto Press, Toronto, 1967, 77–115.
15. D.T. Hunyh, Commutative grammars: The complexity of uniform word problem, *Inform. Control*, 57 (1983), 21–39.

16. J. Kortelainen, Properties of trios and AFLs with bounded or commutative generators, *Mathematics, University of Oulu*, No. 53, 1980.

17. W. Kuich, A. Salomaa, *Semirings, Automata, Languages*, EATCS Monographs on Theoretical Computer Science 5, Springer-Verlag, Berlin,

18. M. Latteux, Cônes rationnels commutativement clos, *R.A.I.R.O.*, 11, 1 (1977), 29–51.

19. M. Latteux, Langages commutatifs, *Publications de Laboratoire de Calcul de l'Université des Sciences et Techniques de Lille*, May 1978.

20. M. Latteux, Cônes rationnels commutatifs, *Journal of Computer System Sciences*, 18 (1979), 307–333.

21. S. Miyamoto, Fuzzy multisets and application to rough approximation of fuzzy sets, *Techn. Rep. Inst. of Inform. Sciences and Electronics*, Univ. of Tsukuba, 1SE-TR-96-136, June 1996, 1–10.

22. S. Miyamoto, Basic operations on fuzzy multisets, *J. of Japan Soc. of Fuzzy Theory and Systems*, 8, 4 (1996).

23. S. Miyamoto, Fuzzy multisets with infinite collections of memberships, *Proc. 7th Intern. Fuzzy Systems Ass. World. Congress* (IFSA'97), June 1997, Prague, vol. I, 61–66.

24. B. Li, Fuzzy bags and applications, *Fuzzy Sets and Systems*, 34 (1990), 61–71.

25. Gh. Păun, Computing with membranes, *Journal of Computer and System Sciences*, 61, 1 (2000), 108–143

26. Gh. Păun, Computing with membranes. An introduction, *Bulletin of the EATCS*, 67 (Febr. 1999), 139–152.

27. Gh. Păun, Computing with membranes (P systems): Twenty six research topics, *Auckland University, CDMTCS Report* No 119, 2000 (www.cs.auckland.ac.nz/CDMTCS).

28. Gh. Păun, G. Rozenberg, A. Salomaa, *DNA Computing. New Computing Paradigms*, Springer-Verlag, Heidelberg, 1998.

29. G. Rozenberg, A. Salomaa, eds., *Handbook of Formal Languages*, 3 volumes, Springer-Verlag, Berlin, 1997.

30. Y. Suzuki, H. Tanaka, Symbolic chemical system based on abstract rewriting system and its behavior pattern, *Artificial Life Robotics*, 1 (1997), 211–219.

31. Y. Suzuki, H. Tanaka, Chemical evolution among artificial proto-cells, *Proc. of Artificial Life VIII Conf.*, MIT Press, 2000

32. A. Syropoulos, A note on multi-sets, basic pairs and the chemical abstract machine, manuscript, 2000.

33. A. Syropoulos, Fuzzy sets and fuzzy multisets as Chu spaces, manuscript, 2000.

34. A. Syropoulos, *Multisets and Chu Spaces*, PhD Thesis, Univ. of Xanti, Greece, in preparation.

35. R. R. Yager, On the theory of bags, *Intern. J. General Systems*, 13 (1986), 23–37.

Normal Forms of Grammars, Finite Automata, Abstract Families, and Closure Properties of Multiset Languages

Manfred Kudlek[1] and Victor Mitrana[2]

[1] Fachbereich Informatik, Universität Hamburg
kudlek@informatik.uni-hamburg.de
[2] University of Bucharest, Faculty of Mathematics
Str. Academiei 14, 70109 Bucharest, Romania
mitrana@funinf.math.unibuc.ro

Abstract. We investigate closure properties of multiset languages, defined by multiset grammars. To this aim, this abstract families of multiset languages are considered, as well as several normal forms for multiset grammars. Furthermore, a new definition of deterministic finite multiset automata is proposed.

1 Introduction

Recently [4] commutative grammars, i.e., multiset grammars, have been reinvestigated for the description of new computational models inspired from biology and biochemistry [5].

In this paper we investigate closure properties of such multiset languages under multiset operations corresponding to those for word languages.

In Section 3 we present various normal forms for multiset grammars following the lines for word grammars, and a characterization of regular multiset languages by an expression of iteration height 1. This differs entirely from the corresponding star height of regular word languages.

In Section 4 we also introduce a variant of a deterministic finite multiset automaton [1] which is equivalent to the nondeterministic version.

To show some closure properties, especially under inverse homomorphism and iteration, abstract families of multiset languages are considered in Section 5.

In Section 6 we present all results obtained so far. Still open are closure under set difference and complement for monotone and arbitrary multiset languages, as well as under **0**-free substitution and homomorphism for monotone multiset languages. Interesting is also the closure under other multiset operations.

2 Multisets and Semilinear Sets

Definition 2.1. (*Multisets*) Let $V = \{a_1, \cdots, a_n\}$ be an alphabet. A *multiset* over V will either be denoted by $x = \langle \mu_x(a_1) \cdot a_1, \cdots, \mu_x(a_n) \cdot a_n \rangle$ where $\mu_x(a_i)$

C.S. Calude et al. (Eds.): Multiset Processing, LNCS 2235, pp. 135–146, 2001.

is the multiplicity of a_i, or as a vector $x = (\mu_x(a_1), \cdots, \mu_x(a_n)) \in \mathbb{N}^n$. Another notation sometimes used is $a_1^{\mu_x(a_1)} \cdots a_n^{\mu(a_n)}$ with $a_i^0 = \lambda$ (the empty word). Let the set of multisets over V be denoted by $V^\oplus$. The neutral element is denoted by $\mathbf{0}$.

If x is a multiset define $\sigma(x) = \sum_{i=1}^n \mu_x(a_i)$ as its *norm* or *length* or *weight*. Write $\xi \in x$ if $\mu_x(\xi) > 0$.

For two multisets $x = \langle \mu_x(a_i) \cdot a_i \mid i \geq 0 \rangle$ and $y = \langle \mu_y(a_i) \cdot a_i \mid i \geq 0 \rangle$ define $x \subseteq y$ iff $\forall i \geq 0 : \mu_x(a_i) \leq \mu_y(a_i)$. Analogously, define $z = x + y$ by $\mu_z(a_i) = \mu_x(a_i) + \mu_y(a_i)$ for $i \geq 0$, and by $z = x - y$ by $\mu_z(a_i) = max\{0, \mu_x(a_i) - \mu_y(a_i)\}$.

Define also $z = x \cap y$ and $z = x \cup y$ by $\mu_z(a_i) = min\{\mu_x(a_i), \mu_y(a_i)\}$ and $\mu_z(a_i) = max\{\mu_x(a_i), \mu_y(a_i)\}$, respectively.

The next definitions follow [2].

Definition 2.2. (*Semilinear sets*) A *linear set* M over $\mathbb{N}^n$ is defined by a multiset c (*constant*) and finite set $P \subseteq \mathbb{N}^n = \{p_1, \cdots, p_m\}$ of multisets (*periods*) as follows:

$M = \{c + \sum_j^m k_j \cdot p_j \mid k_j \in \mathbb{N}\}$. Note that $M = \{c\} + P^\oplus$.

A *semilinear set* M over $\mathbb{N}^n$ is a finite union of linear sets, with sets of periods $P_i = \{p_{i1}, \cdots, p_{im(i)}\}$:

$M = \bigcup_{i=1}^s M_i$ with $M_i = \{c_i + \sum_{j=1}^{m(i)} k_{ij} \cdot p_{ij} \mid k_{ij} \in \mathbb{N}\}$.

Therefore, $M = \bigcup_{i=1}^s (\{c_i\} + P_i^\oplus)$.

Denote by **sLin** the family of all semilinear sets.

3   Multiset Grammars

Definition 3.1. A *multiset grammar* is a quadruple $G = (N, T, A, P)$, where N, T are disjoint alphabets, the *nonterminal* and the *terminal* one, respectively, with $V = N \cup T$. $A \subseteq V^\oplus$ is a finite set over V (its elements are called *axioms*), and P is a finite set of *multiset rewriting rules* (in short, *productions*) of the form $\mu_1 \to \mu_2$, where μ_1, μ_2 are multisets over V and $|\mu_1|_N \geq 1$.

For two multisets α_1, α_2 over V, write $\alpha_1 \xrightarrow{r} \alpha_2$ for some $r : \mu_1 \to \mu_2 \in P$ if $\mu_1 \subseteq \alpha_1$ and $\alpha_2 = (\alpha_1 - \mu_1) + \mu_2$. If r is understood, then write $\longrightarrow$ instead of $\xrightarrow{r}$. Denote by $\xrightarrow{*}$ the reflexive and transitive closure of the relation $\longrightarrow$. The multiset language *generated* by G is defined by

$$M(G) = \{\beta \in T^\oplus \mid \alpha \xrightarrow{*} \beta, \text{ for some } \alpha \in A\}.$$

A natural Chomsky-like classification of such grammars is the following one:

1. Grammars G as above are said to be *arbitrary*.
2. If $|\mu_1| \leq |\mu_2|$ for all rules $\mu_1 \to \mu_2$ in P, then G is said to be *monotone*.
3. If $|\mu_1| = 1$ for all productions $\mu_1 \to \mu_2$ in P, then G is said to be *context-free*.
4. If $|\mu_1| = 1$ and $|\mu_2|_N \leq 1$ for all productions $\mu_1 \to \mu_2$ in P, then G is said to be *linear*.
5. If $|\mu_1| = 1, |\mu_2| \leq 2$, and $|\mu_2|_N \leq 1$ for all productions $\mu_1 \to \mu_2$ in P, then G is said to be *regular*.

Denote by **mARB, mMON, mCF, mLIN, mREG** the families of macro-sets generated by arbitrary, monotone, context-free, linear, and regular multiset grammars, respectively.

Lemma 3.1. *To each multiset grammar G there exists an equivalent one G' with $A' = \{S\}$ and $S \in N$. If $\mathbf{0} \in M(G)$ for a monotone grammar then $S \to \mathbf{0} \in P'$ and S does not appear in any right hand side μ_2 of P'.*

Proof. Let $G = (N, T, A, P)$ and $A = \{\alpha_1, \cdots, \alpha_m\}$, and $S \notin N$. Define $G = (N \cup \{S\}, T, \{S\}, P')$ by $P' = P \cup \{S \to \alpha_j \mid 1 \le j \le m\}$. If $\mathbf{0} \in M(G)$ for monotone G, let also $S \to \mathbf{0} \in P'$. Then $M(G') = M(G)$. □

Lemma 3.2. (*Separated Form*) *To each context-free, monotone or arbitrary multiset grammar there exists an equivalent one in separated form, i.e., all productions are of the forms $\mu_1 \to \mu_2$ with $\mu_1 \in N^\oplus$ and either $\mu_2 \in N^\oplus$ or $\mu_2 \in T$ or $\mu_2 = \mathbf{0}$.*

Proof. Let $G = (N, T, \{S\}, P)$. Let G be already in form of Lemma 3.1. Define $G' = (N \cup \bar{T}, T, \{S\}, P')$, and the homomorphism h by $h(a) = \bar{a}$ for $a \in T$, $h(A) = A$ for $A \in N$. Let $P' = \{h(\mu_1) \to h(\mu_2) \mid \mu_1 \to \mu_2 \in P\} \cup \{\bar{a} \to a \mid a \in T\}$. Then $M(G') = M(G)$. □

Lemma 3.3. *The multiset grammars can be transformed into equivalent multiset grammars with productions in normal forms.*

Proof. Since $+$ is commutative, **mREG=mLIN=mCF**.

Regular multiset grammars can be transformed into equivalent ones with productions of the forms $A \to B + a$, $A \to a$, and $S \to \mathbf{0}$ in which case S does not appear on any right hand side of a production.

Let $G = (N, T, \{S\}, P)$ a regular multiset grammar. If $B \to \mathbf{0} \in P$ with $B \neq S$ then construct a new regular multiset grammar $G' = (N, T, \{S\}, P')$ with productions $P' = (P - \{B \to \mathbf{0}\}) \cup \{A \to a \mid A \to B + a \in P\}$. Then $M(G') = M(G)$. Repeat this procedure. Since $|\{B \to \mathbf{0} \in P\}| < \infty$ this procedure terminates, either without any erasing production or with the only erasing production $S \to \mathbf{0}$.

For a monotone multiset grammar add new symbols
$$\{(R, A) \mid A \in N\} \cup \{(C_{ri}, A) \mid r \in P, 1 \le i \le m(r), A \in N\}$$
$$\cup \{(C_{ri}, A_{rk}) \mid r \in P, 1 \le i, k \le m(r), A \in N\}$$
$$\cup \{G_{ri}, A) \mid r \in P, 1 \le i \le n(r)\} \cup \{G_{ri}, A_{rk}) \mid r \in P, 1 \le i, k \le n(r)\},$$
and if $r : u \to v \in P$ with $u = A_{r1} + \cdots + A_{rm(r)}$, $v = B_{r1} + \cdots + B_{rn(r)}$, $m(r) \le n(r)$ is a production replace it by new productions
$$(R, A) \to (C_{r1}, A),$$
$$(C_{ri}, A) + A_i \to (C_{r,i+1}, A) + A_{ri} \ (1 \le i < m(r)),$$
$$(C_{rm(r)}, A) + A_{rm(r)} \to (G_{r1}, A) + A_{rm(r)},$$
$$(C_{rk}, A_k) \to (C_{r,k+1}, Akj) \ (k \neq m(r) \ ),$$
$$(C_{ri}, A_{rk}) + A_i \to (C_{r,i+1}, A_{rk}) + A_{ri} \ (k < i < m(r)),$$
$$(C_{rm(r)}, A_{rk}) + A_{m(r)} \to (G_{r1}, A_{rk}) + A_{rm(r)}$$
$$(C_{rm(r)}, A_{m(r)}) \to (G_{r1}, A_{m(r)}),$$
$$(G_{rj}, A) + A_{rj} \to (G_{r,j+1}, A) + B_j \ (1 \le j \le m(r)),$$

$(G_{rj}, A_{rk}) + A_{rj} \to (G_{r,j+1}, A_{rk}) + B_j \ (1 \le j < k)$,
$(G_{rk}, A_{rk}) \to (G_{r,k+1}, B_k)$,
$(G_{rj}, A) \to (G_{r,j+1}, A) + B_j \ (m(r) < j < n(r))$,
$(G_{rn(r)}, A) \to (R, A) + B_{n(r)}$,
$(R, A) \to A$,
$(R, A) + B \to (R, B) + A$ for $A, B \in N$.

Note that this also works for $m(r) = 1$.

In the next step, if $A \to B$ is a production, replace on all right hand sides of productions A by A or B, and remove $A \to B$.

Then all productions are of the forms $A + B \to C + D$, $A \to B + C$, or $A \to a$, and the generated multiset languages are identical.

For an arbitrary multiset grammar add new symbols
$\{R, \Lambda\} \cup \{C_{ri} \mid r \in P, 1 \le i \le m(r)\} \cup \{E_{ri} \mid r \in P, 1 \le i \le m(r)\}$
$\qquad \cup \{G_{rj} \mid r \in P, 1 \le j \le n(r)\}$

and if $r : u \to v \in P$, with $u = A_1 + \cdots + A_{m(r)}$, $v = B_1 + \cdots + B_{n(r)}$ is a production, replace it by new productions

$R \to C_{r1}$,
$C_{ri} + A_i \to C_{r,i+1} + A_{ri} \ (1 \le i < m(r))$,
$C_{rm(r)} + A_{m(r)} \to E_{r1} + A_{rm(r)}$,
$E_{ri} + A_{ri} \to E_{r,i+1} + \Lambda \ (i \le i < m(r))$,
$E_{rm(r)} + A_{rm(r)} \to G_{r1} + \Lambda$,
$G_{rj} \to G_{r,j+1} + B_j \ (1 \le j < n(r))$,
$G_{rn(r)} \to R + B_{n(r)}$,
$\Lambda \to \mathbf{0}$, $R \to \mathbf{0}$.

Note that this also works for $m(r) = 1$.

In the next step, if $A \to B$ is a production, replace on all right hand sides of productions A by A or B, and remove $A \to B$.

Then all productions are of the forms $A + B \to C + D$, $A \to B + C$, $A \to a$, or $A \to \mathbf{0}$, respectively, and the generated multiset languages are identical.   □

Lemma 3.4. **mREG** = **sLin**.

Proof. a) **mREG** $\subseteq$ **sLin** is Parikh's theorem [2], Theorem 5.2.1 since **mREG** = **PsCF**.

b) Let $M = \bigcup_{i=1}^{s} M_i \in \mathbf{sLin}$ with $M_i = \{c_i + \sum_{j=1}^{m(i)} k_{ij} \cdot p_{ij} \mid k_{ij} \in I\!N\}$.

Construct the regular multiset grammar $G = (N, T, S, P)$ with P consisting of the productions $S \to S_i$, $S_i \to A_i + c_i$, $A_i \to A_i + p_{ij}$, $A_i \to \mathbf{0}$ where $1 \le i \le s$ and $1 \le j \le m(i)$.

Then $M = M(G)$, and therefore **sLin** $\subseteq$ **mREG**.   □

Lemma 3.5. (*Normal form of* **mREG**) *Any* $M \in \mathbf{mREG}$ *can be characterized in the normal form* $M = \bigcup_{i=1}^{s}(\{c_i\} + P_i^{\oplus})$ *with multisets* c_i *and finite sets of multisets* P_i.

Thus the $\oplus$-*height of any* $M \in \mathbf{mREG}$ *is at most 1.*

Proof. This follows immediately from Lemma 3.4 and the remark for the definition of semilinear sets.   □

4   Finite Multiset Automata

Definition 4.1. (Finite Multiset Automaton) A *Finite Multiset Automaton (FMA)* is a quintuple $A = (V, Q, q_0, Q_F, \delta)$ with a finite alphabet V, a finite set of states Q, an initial state q_0, a set of terminal states $Q_F \subseteq Q$, and a finite set of instructions $\delta \subseteq Q \times V^\oplus \times Q$.

The automaton works in such a way that if $\mu \in V^\oplus$, $\xi \in V^\oplus$ with $\xi > \mathbf{0}$, $(q, \xi, q') \in \delta$, $\mu - \xi \geq \mathbf{0}$, then $\mu \in V^\oplus$ is reduced to $\mu' = \mu - \xi$.

A accepts a multiset μ if μ is reduced in finitely many steps to $\mathbf{0}$, and A is in a terminal state. The multiset language $M(A)$, accepted by a FMA A consits of all accepted multisets.

Note that due to the commutativity of the basic operation $+$ for multisets the application of instructions (q, a, q') and (q, b, q'') is non-deterministic.

But a FMA is called *deterministic (DFMA)* if δ is a function $\delta : Q \times V \to Q$. Otherwise it is called *non-deterministic*.

As for classical finite automata a number of theorems can be shown to bring a FMA to a simple form.

Lemma 4.1. *To any FMA $A = (V, Q, q_0, Q_F, \delta)$ there exists an equivalent FMA $A' = (V, Q', q_0, Q_F, \delta')$ with $\delta' \subseteq Q' \times (V \cup \{\mathbf{0}\}) \times Q'$.*

Proof. As for classical FA on words replace $(p, \xi, q) \in \delta$ with $\xi = \sum_{j=0}^{k} a_j$ ($k \geq 1$, $a_j \in V$) by a series $p, a_0, q_1)$, (q_j, a_j, q_{j+1}) $(1 \leq j < k)$, (q_k, a_k, q), all in δ', and add $\{q_j \mid 1 \leq j \leq k\}$ to Q. Then $M(A') = M(A)$.   □

Lemma 4.2. (*Removal of* $\mathbf{0}$) *To any FMA $A = (V, Q, q_0, Q_F, \delta)$ an equivalent FMA $A' = (V, Q, q_0, Q'_F, \delta')$ can be constructed, with $\delta' \subseteq Q \times V \times Q$.*

Proof. Construct a relation $\lambda - 0$ as follows:
$M_0 = \{(p, q) \in Q \times Q \mid (p, \mathbf{0}, q) \in \delta\} \cup \{(q, q)\ q \in Q\}$,
$M_{i+1} = M_i \cup \{(p, q) \in Q \times Q \mid \exists q \in M_i : (p, \mathbf{0}, q) \in \delta\}$.
Since $|Q \times Q| < \infty$ there exists a k such that $M_k + j = M_k$ for $j \geq 0$. Define $\lambda_0 = M_k$. Actually, $\lambda_0 = M_0^*$.

Construct δ' by $(q, q, q') \in \delta'$ iff there exist $p, p' \in Q$ with $(q, p) \in \lambda_0$, $(p', q') \in \lambda_0$, $(p, a, p') \in \delta$.

If $(q_0, q_f) \in \lambda_0$ with $q_f \in Q_F$ let $q_0 \in Q'_F$.

Then $M(A') = M(A)$.   □

Lemma 4.3. (Deterministic FMA) *To any FMA $A = (V, Q, q_0, Q_F, \delta)$ there exists an equivalent complete DFMA $A' = (V, Q', q'_0, Q'_F, \delta')$.*

Proof. As for FA on words, using Lemma 4.2, construct the powerset automaton with $Q' = \mathcal{P}(Q)$ by $\delta'(Q_1, a) = \{q_2 \in Q \mid \exists q_1 \in Q_1 : (q_1, a, q_2) \in \delta\}$, $q'_0 = \{q_0\}$, $Q'_F = \{Q_i \in \mathcal{P}(Q) \mid Q_i \cap Q_F \neq \emptyset\}$.

Let Q' consist of those $Q_1 \in \mathcal{P}(Q)$ which are reachable from q'_0, joined with $\{\emptyset\}$. If there is no Q_2 with $\delta'(Q_1, b) = Q_2$ define $\delta'(Q_1, b) = \emptyset$, as well as $\delta'(\emptyset, a) = \emptyset$ for all $a \in V$.

Then A' is a complete DFMA with $M(A') = M(A)$.   □

Lemma 4.4. *The family of multiset languages accepted by FMA is identical to* **mREG**.

Proof. a) Let $A = (V, Q, q_0, Q_F, \delta)$ be a FMA. By Lemma 4.3 it can be assumed that A is a complete DMFA. Construct a regular multiset grammar $G = (N, T, \{S'\}, P)$ as follows:

Let $N = Q \cup \{S'\}$ and $T = V$. For $\delta(q, a) = q'$ let $q \rightarrow q' + a \in P$ $(q \neq q_0)$, and if $\delta(q_0, a) = q'$ let $S' \rightarrow q' + a \in P$. If $q_0 \in Q_F$ let also $S' \rightarrow \mathbf{0} \in P$.

Then $M(G) = M(A)$.

b) Let $G = (N, T, S, P)$ be a regular multiset grammar in normal form. Construct a FMA $A = (V, Q, q_0, Q_F, \delta)$ as follows:

$Q = N \cup \{q_f\}$ and $V = T$. For $A \rightarrow B + a \in P$ let $\delta(A, a) = B$, and for $A \rightarrow a \in P$ let $\delta(A, a) = q_f$. If $S \rightarrow \mathbf{0} \in P$ let $S \in Q_F$.

Then $M(A) = M(G)$. □

5   Abstract Families of Multiset Languages

In this sections some theorems on abstract families of multiset languages are presented. This is done in a similar way as for abstract families of word languages.

Definition 5.1. A homomorphism $h : V_1^\oplus \rightarrow V_2^\oplus$ is a mapping with properties $h(\mathbf{0}) = \mathbf{0}$ and $h(\mu_1 + \mu_2) = h(\mu_1) + h(\mu_2)$.

A homomorphism is called **0**-free if $h(a) \neq \mathbf{0}$ for all $a \in V_1$.

Definition 5.2. A substitution $\sigma : V_1^\oplus \rightarrow 2^{V_2^\oplus}$ is a mapping with properties $\sigma(\mathbf{0}) = \{\mathbf{0}\}$ and $\sigma(\mu_1 + \mu_2) = \sigma(\mu_1) + \sigma(\mu_2)$.

A substitution is called **0**-free if $\mathbf{0} \notin \sigma(a)$ for all $a \in V_1$.

Definition 5.3. Let $M \subseteq T^\oplus$ be a multiset language and $h : T \rightarrow T_1^\oplus$ a homomorphism with $\forall \mu \in M : |\mu| \leq k|h(\mu)|$. Then h is called k-*linear erasing* with respect to M.

A family $\mathcal{M}$ of multiset languages is closed under *linear erasing* if $h(M) \in \mathcal{M}$ for all $M \in \mathcal{M}$ and k-linear erasing homomorphisms h with respect to M.

The proofs of the next theorems follow those for abstract families of (word) languages in [Sal].

Definition 5.4. Let $M \subseteq (T + \{\mathbf{0}, c, \cdots, c^{k-1}\})^\oplus$ with $c \notin T$, and $h : T \cup \{c\} \rightarrow T^\oplus$ a homomorphism defined by $h(c) = \mathbf{0}$ and $h(a) = a$ for $a \in T$.

Then h is called k-*restricted* on M. Obviously, $\forall \mu \in M : |\mu| \leq k|h(\mu)|$. Thus h is also k-linear erasing on M.

A family $\mathcal{M}$ of multiset languages is closed under *restricted* homomorphism if $h(M) \in \mathcal{M}$ for all $M \in \mathcal{M}$ and k-restricted homomorphisms h on M.

Theorem 5.1. *If a family $\mathcal{M}$ of multiset languages is closed under* **0**-*free substitutions, restricted homomorphisms, union with regular multiset languages, and intersection with regular multiset languages, then $\mathcal{M}$ is closed under inverse homomorphisms.*

Proof. Let $M \subseteq T^\oplus$, $T_1 = \{a_1, \cdots, a_n\}$, and $h : T \to T^\oplus$ be a homomorphism with $h(a_i) = \alpha_i \in T^\oplus$ $(1 \leq i \leq n)$.

a) Assume $\mathbf{0} \notin M \in \mathcal{M}$. Let $k = max\{|\alpha_i| \mid 1 \leq i \leq n\} + 1$. Define the set $\overline{T} = \{\overline{a_1}, \cdots, \overline{a_n}\}$, and the regular substitution σ by $\sigma(a) = \{a\} + \overline{T}^\oplus$.

Let $M_1 = \{\overline{a_i} + \alpha_1 \mid 1 \leq i \leq n\}$, a finite set, and $M_2 = \sigma(M) \cap M_1^\oplus$. Then, by the closure properties, $M \in \mathcal{M}$.

Define a $\mathbf{0}$-free homomorphism $h_1 : \overline{T} \cup T \to T \cup \{c\}$ by $h_1(\overline{a}) = a$, $h_1(a) = c$ for $a \in T$. Then $h_1(M_2) \subseteq (T \cup \{c\})^\oplus$ and $h_1(M_2) \in \mathcal{M}$.

Define another homomorphism $h_2 : T \cup \{c\} \to T$ by $h_2(c) = \mathbf{0}$ and $h_2(a) = a$ for $a \in T$. h_2 is k-restricted on $h_1(M_2)$ since with $\beta = \sum_{j=1}^m (\overline{a_{i(j)}} + \alpha_{i(j)})$, follows $h_1(\beta) = h_1(\sum_{j=1}^m (\overline{a_{i(j)}} + \alpha_{i(j)})) = \sum_{j=1}^m (\overline{a_{i(j)}} + c^{|\alpha_{i(j)}|})$, and $h_2(h_1(\beta)) = \sum_{j=1}^m a_{i(j)}$. Therefore $|h_1(\beta)| \leq k \cdot m = k \cdot |h(h_1(\beta))|$.

Furthermore, $h_2(h_1(M_2)) = h^{-1}(M)$, since
$$\beta \in h^{-1}(M) \implies h(\beta) = \sum_{j=1}^m \alpha_{i(j)} \implies \sum_{j=1}^m (\overline{a_{i(j)}} + \alpha_{i(j)}) \in M_2$$
$$\implies \sum_{j=1}^m a_{i(j)} \in h_2(h_1(M)).$$
Thus $h^{-1}(M) \in \mathcal{M}$.

b) $\mathbf{0} \in M$.

Then $M - \{\mathbf{0}\} = M \cap (T + T^\oplus) \in \mathcal{M}$, and therefore $h^{-1}(M - \{\mathbf{0}\}) \in \mathcal{M}$. If h is $\mathbf{0}$-free then $h^{-1}(\mathbf{0}) = \{\mathbf{0}\}$, a regular multiset language. Otherwise, $h^{-1}(\mathbf{0}) = T_2^\oplus$ for some $T_2 \subseteq T$, also a regular mulriset language. Now,
$$h^{-1}(M) = h^{-1}(M - \{\mathbf{0}\}) \cup h^{-1}(\mathbf{0}). \text{ Hence } h^{-1}(M) \in \mathcal{M}. \qquad \square$$

Theorem 5.2. *If a family $\mathcal{M}$ of multiset languages contains* **mREG**, *is closed under intersection with regular multiset languages and substitutions ($\mathbf{0}$-free substitutions), then $\mathcal{M}$ is closed under iteration $^\oplus$.*

Proof. $\mathcal{M}$ contains $\{c\}^\oplus$. If $M \in \mathcal{M}$, then with the substitution ($\mathbf{0}$-free substitution) $\sigma(c) = M$ also $M^\oplus \in \mathcal{M}$.

Note that actually closure under intersection with regular multiset languages is not needed. $\qquad \square$

6   Closure Properties

Lemma 6.1. mREG, MON *and* **ARB** *are closed under union.*

Proof. Let $G_i = (N_i, T_i, S_i, P_i)$ with $i = 1, 2$ be two multiset grammars in separated form, generating $M(G_i)$.

Construct $G = (N_1 \cup \overline{N_2} \cup \{S\}, T_1 \cup T_2, S, P_1 \cup \overline{V_2} \cup \{S \to S_1, S \to \overline{S_2}\}$ where $\overline{P_2}$ denotes the set of productions P_2 with every nonterminal x from N_2 changed into $\overline{x}$. Then G is also in separated form and of the same type as G_i.

Obviously, $M(G) = M(G_1) \cup M(G_2)$. $\qquad \square$

Lemma 6.2. mREG, MON *and* **ARB** *are closed under the operation $+$.*

Proof. Let $G_i = (N_i, T_i, S_i, P_i)$ with $i = 1, 2$ be two multiset grammars in separated form, generating $M(G_i)$.

Construct $G = (N_1 \cup \overline{N_2} \cup \{S\}, T_1 \cup T_2, S, P_1 \cup \overline{V_2} \cup \{S \rightarrow S_1 + \overline{S_2}\}$ with $\overline{P_2}$ as above. Then G is also in separated form and of the same type as G_i.

Obviously, $M(G) = M(G_1) + M(G_2)$. □

Lemma 6.3. **mREG**, **MON** *and* **ARB** *are closed under intersection.*

Proof. Let $G_i = (N_i, T_i, S_i, P_i)$ with $i = 1, 2$ be two multiset grammars in separated and normal form, generating $M(G_i)$.

For multiset languages from **mCF** this is [2], Theorem 5.6.1, since **mCF**= **mREG**.

For multiset languages from **mMON** construct a new grammar $G=(N,T,S,P)$ with $N = (V_1 \cup \{\bullet\}) \times (V_2 \cup \{\bullet\})$, $T = T_1 \cap T_2$, $S = \binom{S_1}{S_2}$, and P as follows:

If $A \rightarrow B + C \in P_1$ then put $\binom{A}{E} \rightarrow \binom{B}{E} + \binom{C}{\bullet}$, $\binom{A}{E} + \binom{\bullet}{F} \rightarrow \binom{B}{E} + \binom{C}{F}$, into P, with $E, F \in V_2 \cup \{\bullet\}$.

If $A \rightarrow B + C \in P_2$ then put $\binom{E}{A} \rightarrow \binom{E}{B} + \binom{\bullet}{C}$, $\binom{E}{A} + \binom{F}{\bullet} \rightarrow \binom{E}{B} + \binom{F}{C}$ into P, with $E, F \in V_1 \cup \{\bullet\}$.

If $A + B \rightarrow C + D \in P_1$ then put $\binom{A}{E} + \binom{B}{F} \rightarrow \binom{C}{E} + \binom{D}{F}$ into P, with $E, F \in V_2 \cup \{\bullet\}$.

If $A + B \rightarrow C + D \in P_2$ then put $\binom{E}{A} + \binom{F}{B} \rightarrow \binom{E}{C} + \binom{F}{D}$ into P, with $E, F \in V_1 \cup \{\bullet\}$.

If $A \rightarrow a \in P_1$ then put $\binom{A}{E} \rightarrow \binom{a}{E}$ into P, with $E \in V_2 \cup \{\bullet\}$.

If $A \rightarrow a \in P_2$ then put $\binom{E}{A} \rightarrow \binom{E}{a}$ into P, with $E \in V_1 \cup \{\bullet\}$.

Put also into P the productions $\binom{A}{E} + \binom{B}{F} \rightarrow \binom{A}{F} + \binom{B}{E}$, with $A, B \in V_1 \cup \{\bullet\}$, $E, F \in V_2 \cup \{\bullet\}$, $S \rightarrow \binom{S_1}{S_2}$, $\binom{\bullet}{\bullet} \rightarrow \bullet$, and $\binom{a}{a} \rightarrow a$, with $a \in T$.

In any derivation of some $\mu \in T^\oplus$ at most $|\mu| - 1$ $\bullet$'s are generated in the upper part, as well in the lower part, namely only by productions $A \rightarrow B + C$.

Consider the generated multiset language $M(G) \subseteq (T \cup \{\bullet\})^\oplus$. By definition $M(G) \in$ **mMON**. The homomorphism $h : T \cup \{\bullet\} \rightarrow T$ defined by $h(\bullet) = \mathbf{0}$ and $h(a) = a$ for $a \in T$ is a linear erasing homomorphism since $|\mu| \leq 2|h(\mu)|$.

By Theorem 6.1, $h(M(G)) \in$ **mMON**, and trivially $h(M(G)) = M(G_1) \cap M(G_2)$.

For languages from **mARB** construct
$$G = (\overline{V_1} \cup \widehat{V_2} \cup \{S\}, T_1 \cap T_2, S, \overline{P_1} \cup \widehat{P_2} \cup \{S \rightarrow \overline{S_1} + \widehat{S_2}\} \cup \{\overline{x} + \widehat{x} \rightarrow x \mid x \in T_1 \cap T_2\}).$$
Then $M(G) = M(G_1) \cap M(G_2)$. □

Corollary 6.1. **mREG**, **MON** *and* **ARB** *are closed under intersection with regular multiset languages.*

Lemma 6.4. **mREG** *and* **ARB** *are closed under substitution* σ, **mMON** *is closed under* **0**-*free substitutions.*

Proof. Let $G = (N, T, S, P)$ be a multiset grammars in separated form, generating $M(G)$.

Let $\sigma(a)$ be generated by a multiset grammar $G_a = (N_a, T_a, S_a, P_a)$ of the same type as G, in separated form, with $N_a \cap N_b = \emptyset$ for $a \neq b$ and $N \cap N_a = \emptyset$ for $a \in T$.

Construct a new multiset grammar $G' = (N', T', S, P')$ with $N' = N \cup \bigcup_{a \in T} N_a$, $T' = \bigcup_{a \in T} T_a$, and $P' = (P - \{A \to a \mid a \in T\}) \cup \{A \to S_a \mid a \in T\} \cup \bigcup_{a \in T} P_a$.

Then, in all cases, $M(G') = \sigma(M(G))$.   □

Corollary 6.2. **mREG** and **ARB** are closed under homomorphisms, **mMON** is closed under **0**-free homomorphisms.

Theorem 6.1. **mMON** is closed under linear erasing, and therefore also under k-restricted homomorphisms.

Proof. Let $M \in$ **mMON**, h a k-linear erasing with respect to M with $h : T \to T_1^\oplus$. Assume $k \geq 1$, since otherwise only $M = \emptyset$ or $M = \{0\}$ are possible. Let $h(M) \subseteq T_1^\oplus$. Trivially, $h(M) \in$ **mARB** since **mMON** is closed under homomorphisms. Let $G = (V, T, S, P)$ be a with $M(G) = M$ in separated form. Then the grammar $G_1 = (V, T_1, S, P_1)$ with $P_1 = (P - \{A \to a \mid a \in T\}) \cup \{A \to h(a) \mid a \in T\}$ generates $h(M)$

Now, let $\mu_1 \in h(M)$. Then there exists a $\mu \in M$ with $h(\mu) = \mu_1$. Since h is k-linear erasing on M, $|\mu| \leq k \cdot |h(\mu)| = k \cdot |\mu_1|$, and therefore it follows that $WS(\mu_1) = max\{WS(\mu), |\mu_1|)\} = max\{|\mu|, |\mu_1|\} \leq k \cdot |\mu_1|$.

Hence, by the workspace theorem, $h(M) \in$ **mMON**.   □

Lemma 6.5. **mREG**, **mMON**, and **mARB** are closed under inverse homomorphisms.

Proof. This follows from Lemmata 6.1, 6.4, Corollary 6.1, 6.2, and Theorems 6.1, 5.1.   □

Lemma 6.6. **mREG**, **mMON**, and **mARB** are closed under iteration $^\oplus$.

Proof. This follows from Corollary 6.1, Lemma 6.4, and Theorem 5.2.   □

Lemma 6.7. **mREG** is closed under set difference, i.e., if $M_1, M_2 \in$ **mREG** then $M_1 - M_2 \in$ **mREG**.

Proof. This is [2], Theorem 5.6.2.   □

Corollary 6.3. **mREG** is closed under complement, i.e., if $M \in$ **mREG** then $T^\oplus - M \in$ **mREG**.

Proof. Since trivially $T^\oplus \in$ **mREG** this is a consequence from Lemma 6.7.
    □

For two multiset languages let $M_1 \wedge M_2 = \{\mu_1 \cap \mu_2 \mid \mu_1 \in M_1, \mu_2 \in M_2\}$ and $M_1 \vee M_2 = \{\mu_1 \cup \mu_2 \mid \mu_1 \in M_1, \mu_2 \in M_2\}$.

Lemma 6.8. (*Closure properties of* **PsCS** = **PsMON**) **PsCS** is closed under union, $+$, iteration $^\oplus$, intersection, complement, set difference, **0**-free homomorphisms, and substitutiona, inverse homomorphisms, and $\vee$, but not under homomorphisms and substitutions.

Proof. Let $\psi(u)$ denote the Parikh vector of a word u. Since **CS** is closed under $\cup$, $\cdot$, *, and $\psi(A \cup B) = \psi(A) \cup \psi(B)$, $\psi(A \cdot B) = \psi(A) + \psi(B)$, $\psi(A^*) = \psi(A)^\oplus$, **PsCS** is closed under $\cup$, $+$, and $^\oplus$.

PsCS is closed under $\cap$, since to any $L \in \mathbf{CS}$ another language $L' \in \mathbf{CS}$ with $\psi(L') = \psi(L)$ can be constructed by adding productions $a_i a_j \to a_j a_i$ to the productions of the grammar for L. Then $L'' = L' \cap \{a_1\}^* \cdots \{a_n\}^* \in \mathbf{CS}$ and all $v \in L'$ have the form $v = a_i^{k_i} \cdots a_n^{k_n}$ where $T = \{a_1, \cdots, a_n\}$. Thus to each $u \in L$ there exists a unique $\phi(u) \in L''$ with $\psi(u) = \psi(\phi(u))$ being a representation of $\psi(u)$. By this one gets a representation of $\psi(L)$ as $\phi(L) \in \mathbf{CS}$. Let $L_1, L_2 \in \mathbf{CS}$. Then $\phi(L_1), \phi(L_2) \in \mathbf{CS}$ are representations of $\psi(L_1), \psi(L_2)$. Since $\phi(L_1) \cap \phi(L_2) \in \mathbf{CS}$ it follows immediately that $\psi(L_1) \cap \psi(L_2) \in \mathbf{PsCS}$.

PsCS is closed under complement. Now $L \in \mathbf{CS}$ implies $\pi(L) \in \mathbf{CS}$ where $\pi(L) = \{u \in T^* \mid \exists v \in L : \psi(v) = \psi(u)\}$, i.e., $\pi(L)$ consists of all permutations of words in L. Trivially, $\psi(L) = \psi(\pi(L))$. Since $\mathbf{CS}$ is closed under complement, $T^* - \pi(L) \in \mathbf{CS}$, and therefore $T^{\oplus} - \psi(L) = \psi(T^* - \pi(L)) \in \mathbf{PsCS}$.

Because of $M_1 - M_2 = M_1 \cap (T^{\oplus} - M_2)$ it follows that **PsCS** is also closed under set difference.

Let $\sigma(a_i) = X_i \in \mathbf{PsCS}$ be a 0-free multiset substitution. Then there exist $Y_i \in \mathbf{CS}$ with $\psi(Y_i) = X_i$. Define a λ-free word substitution $\tau(a_i) = Y_i \in \mathbf{CS}$. Then $\psi(\tau(a_i)) = \psi(Y_i) = X_i = \sigma(a_i)$. Therefore, with $u = x_1 \cdots x_m$ follows

$$\psi(\tau(u)) = \psi(\tau(x_1) \cdots \tau(x_m)) = \sum_{j=1}^{m} \psi(\tau(x_j)) = \sum_{j=1}^{m} \sigma(x_j) = \sigma(\sum_{j=1}^{m} x_j) = \sigma(\psi(u)) \, .$$

With $M = \psi(L)$ and since $\tau(L) \in \mathbf{CS}$ it follows that $\sigma(M) = \sigma(\psi(L)) = \psi(\tau(L))$.

By Theorem 5.1, **PsCS** is also closed under inverse homomorphisms.

PsCS is closed under $\vee$. Let $M_1, M_2 \in \mathbf{PsCS}$ with $M_j = \psi(L_j)$, $L_j \in \mathbf{CS}$, $L_j \subseteq T^*$, and $T = \{a_1, \cdots, a_n\}$. Then also $\phi(L_j) \in \mathbf{CS}$ with $\phi(L_j)$ defined as above.

Assume $\mathbf{0} \notin M_j$. Let $g(a_i) = b_i$, $h(a_i) = c_i$, and $T_1 = g(T)$, $T_2 = h(T)$. Then $L' = ag(L_1)bh(L_2)c \in \mathbf{CS}$ with $\{a, b, c\} \cap (T \cup T_1 \cup T_2 = \emptyset$. L' consists of words $ab_1^{r_1} \cdots b_n^{r_n} bc_1^{s_1} \cdots c_n^{s_n} c$.

Extend a LBA accepting L' by additional instructions checking successively for each i whether $r_i \geq s_i$ or $r_i < s_i$. In the first case it changes all b_i into d_i and all c_i into e_i, in the second case all b_i into e_i and all c_i into d_i. The resulting LBA accepts a language $L'' \subseteq \{a\} T_3^+ \{b\} T_3^+ \{c\}$ where $T_3 = \{d_1, \cdots, d_n\} \cup \{e_1, \cdots, e_n\}$.

Define a homomorphism f by $f(d_i) = a_i$, $f(e_i) = f(a) = f(b) = f(c) = \lambda$. f is k-limited erasing on L'' since $|u| \leq 2 \cdot |f(u)| + 3 \leq 5 \cdot |f(u)|$. Then $f(L'') \in \mathbf{CS}$, and $\psi(f(L'')) = \psi(L_1) \vee \psi(L_2)$.

If $\mathbf{0} \in M_1$ or $\mathbf{0} \in M_2$ then $M_1 \vee M_2 = (M_1' \vee M_2') \cup \{\mathbf{0}\} \in \mathbf{PsCS}$ where $\mathbf{0} \notin M_1' \cap M_2'$.

PsCS is not closed under homomorphisms and substitutions.

Consider $M = \psi(L) \in \mathbf{PsRE}$ with $L \in \mathbf{RE}$, $L \in T^*$, $M \notin \mathbf{PsCS}$. Then there is a effectively constructable $L' \in \mathbf{CS}$ with $L \subseteq T_1^*$, $T_1 = T \cup \{\Lambda\}$, and a homomorphism h with $h(x) = x$ for $x \in T$, $h(\Lambda) = \lambda$, and $L = h(L')$.

Consider $\phi(L') \in \mathbf{CS}$ with ϕ as above such that $\psi(\phi(L')) = \psi(L') \in \mathbf{PsCS}$. Then $\psi(h(\phi(L'))) = \psi(L) = h(\psi(L')) = M$. But then closure of **PsCS** under homomorphism would imply $M \in \mathbf{PsCS}$, a contradiction.

Therefore **PsCS** is not closed under homomorphism and substitution.   □

Lemma 6.9. (*Closure properties of* **PsRE**) **PsRE** *is closed under union,* $+$, *iteration* $^\oplus$, *intersection, homomorphisms, substitutions, inverse homomorphisms,* $\vee$, $\wedge$, *but not under complement and set difference.*

Proof. Let $\psi(u)$ denote the Parikh vector of a word u. Since **RE** is closed under $\cup$, $\cdot$, *, and $\psi(A \cup B) = \psi(A) \cup \psi(B)$, $\psi(A \cdot B) = \psi(A) + \psi(B)$, $\psi(A^*) = \psi(A)^\oplus$, **PsRE** is closed under $\cup$, $+$, and $^\oplus$.

PsRE is closed under $\cap$, since to any $L \in \mathbf{RE}$ another language $L' \in \mathbf{RE}$ with $\psi(L') = \psi(L)$ can be constructed by adding productions $a_i a_j \to a_j a_i$ to the productions of the grammar for L. Then $L'' = L' \cap \{a_1\}^* \cdots \{a_n\}^* \in \mathbf{RE}$ and all $v \in L'$ have the form $v = a_i^{k_i} \cdots a_n^{k_n}$ where $T = \{a_1, \cdots, a_n\}$. Thus to each $u \in L$ there exists a unique $\phi(u) \in L''$ with $\psi(u) = \psi(\phi(u))$ being a representation of $\psi(u)$. By this one gets a representation of $\psi(L)$ as $\phi(L) \in \mathbf{RE}$. Let $L_1, L_2 \in \mathbf{RE}$. Then $\phi(L_1), \phi(L_2) \in \mathbf{RE}$ are representations of $\psi(L_1), \psi(L_2)$. Since $\phi(L_1) \cap \phi(L_2) \in \mathbf{RE}$ it follows immediately that $\psi(L_1) \cap \psi(L_2) \in \mathbf{PsRE}$.

PsRE is of the power of Turing machines since any word u can be encoded as a multiset $\mu(u)$, even as $\mu(u) \in \mathbf{N}$ such that any $L \in \mathbf{RE}$ can be represented by a multiset language $\mu(L) \in \mathbf{PsRE}$. Therefore **PsRE** is not closed under set difference $-$ and complement $^-$.

Let $\sigma(a_i) = X_i \in \mathbf{PsRE}$ be a multiset substitution. Then there exist multisets $Y_i \in \mathbf{RE}$ with $\psi(Y_i) = X_i$. Define a word substitution $\tau(a_i) = Y_i \in \mathbf{RE}$. This gives $\psi(\tau(a_i)) = \psi(Y_i) = X_i = \sigma(a_i)$. With $u = x_1 \cdots x_m$ follows

$$\psi(\tau(u)) = \psi(\tau(x_1) \cdots \tau(x_m)) = \sum_{j=1}^{m} \psi(\tau(x_j)) = \sum_{j=1}^{m} \sigma(x_j) = \sigma(\sum_{j=1}^{m} x_j = \sigma(\psi(u)).$$

With $M = \psi(L)$ and since $\tau(L) \in \mathbf{RE}$ it follows that $\sigma(M) = \sigma(\psi(L)) = \psi(\tau(L))$.

Thus **PsRE** is also closed under homomorphisms.

By Theorem 5.1 **PsRE** is also closed under inverse homomorphisms.

PsRE is closed under $\vee$. Let $M_1, M_2 \in \mathbf{PsCS}$ with $M_j = \psi(L_j)$, $L_j \in \mathbf{RE}$, $L_j \subseteq T^*$, and $T = \{a_1, \cdots, a_n\}$. Then also $\phi(L_j) \in \mathbf{CS}$ with $\phi(L_j)$ defined as above.

Let $g(a_i) = b_i$, $h(a_i) = c_i$, and $T_1 = g(T)$, $T_2 = h(T)$. With new symbols $\{a, b, c\} \cap (T \cup T_1 \cup T_2) = \emptyset$, it follows $L' = ag(L_1)bh(L_2)c \in \mathbf{RE}$. L' consists of words $ab_1^{r_1} \cdots b_n^{r_n} bc_1^{s_1} \cdots c_n^{s_n} c$.

Extend a TM accepting L' by additional instructions checking successively for each i whether $r_i \geq s_i$ or $r_i < s_i$. In the first case it changes all b_i into d_i and all c_i into e_i, in the second case all b_i into e_i and all c_i into d_i. The resulting TM accepts a language $L'' \subseteq \{a\} T_3^+ \{b\} T_3^+ \{c\}$ where $T_3 = \{d_1, \cdots, d_n\} \cup \{e_1, \cdots, e_n\}$.

Define a homomorphism f by $f(d_i) = a_i$, $f(e_i) = f(a) = f(b) = f(c) = \lambda$. Note that the result can also be λ. Then $f(L'') \in \mathbf{RE}$, and $\psi(f(L'')) = \psi(L_1) \vee \psi(L_2)$.

By a similar argument **PsRE** is also closed under $\wedge$. $\square$

Lemma 6.10. *To each* $M \in \mathbf{mARB}$ *there effectively exist a* $M' \in \mathbf{mMON}$ *and a homomorphism* h *such that* $M = h(M')$.

	mCF	mMON	mARB	PsCS	PsRE
$\cup$	Y	Y	Y	Y	Y
$\cap$	Y	Y	Y	Y	Y
$-$	Y			Y	N
$+$	Y	Y	Y	Y	Y
$\oplus$	Y	Y	Y	Y	Y
h_0	Y	Y	Y	Y	Y
h	Y	N	Y	N	Y
h^{-1}	Y	Y	Y	Y	Y
σ_0	Y	Y	Y	Y	Y
σ	Y	N	Y	N	Y
$\cap mR$	Y	Y	Y	Y	Y
$-$	Y			Y	N
$\wedge$					Y
$\vee$				Y	Y

Proof. Consider a multiset grammar $G = (N, T, \{S\}, P)$ with generated language $M(G) = M \in \mathbf{mARB}$. Replace any production $A \to \mathbf{0} \in P$ by $A \to \Lambda$ with $\Lambda \notin N \cup T$, generating a multiset language $M' \in \mathbf{mMON}$. Defining the homomorphism h by $h(x) = x$ for $x \in T$, $h(\Lambda) = \lambda$, it follows that $M = h(M')$.   □

Lemma 6.11. mMON *is not closed under homomorphisms and substitutions.*

Proof. This follows immediately from Lemma 6.11.   □

The results so far obtained for closure properties of multiset languages are given in the following table where also the families **PsCS** and **PsRE** of multisets are added.

Theorem 6.2. *The previous table shows the known closure properties of families of multiset languages.*

The places not filled are open problems.

References

1. E. Csuhaj-Varjú, C. Martín Vide, V. Mitrana, *Multiset automata,* in this volume.
2. S. Ginsburg, *The Mathematical Theory of Context-free Languages,* McGraw Hill, 1966.
3. M. Kudlek, *Rational, Linear and Algebraic Languages of Multisets,* Pre-Proceedings of the Workshop on Multiset Processing (WMP'2000), C. S. Calude, M. J. Dinneen, G. Păun, eds., CDMTCS-140, 2000, 138–148.
4. M. Kudlek, C. Martín Vide, Gh. Păun, *Toward FMT (Formal Macroset Theory),* Pre-Proceedings of the Workshop on Multiset Processing (WMP'2000), C. S. Calude, M. J. Dinneen, G. Păun, eds., CDMTCS-140, 2000, 149–158.
5. Gh. Păun, Computing with Membranes. An Introduction, *Bulletin of EATCS,* 67 (1999), 139–152.
6. A. Salomaa, *Formal Languages,* Academic Press, 1973.

On Multisets in Database Systems

Gianfranco Lamperti, Michele Melchiori, and Marina Zanella

Dipartimento di Elettronica per l'Automazione
Università di Brescia, Via Branze 38, 25123 Brecia, Italy
(lamperti|melchior|zanella)@ing.unibs.it

Abstract. Database systems cope with the management of large groups of persistent data in a shared, reliable, effective, and efficient way. Within a database, a multiset (or bag) is a collection of elements of the same type that may contain duplicates. There exists a tight coupling between databases and multisets. First, a large variety of data models explicitly support multiset constructors. Second, commercial relational database systems, even if founded on a formal data model which is set-oriented in nature, allows for the multiset-oriented manipulation of tables. Third, multiset processing in databases may be dictated by efficiency reasons, as the cost of duplicate removal may turn out to be prohibitive. Finally, even in a pure set-oriented conceptual framework, multiset processing may turn out to be appropriate for optimization of query evaluation. The mismatch between the relational model and standardized relational query languages has led researchers to provide a foundation to the manipulation of multisets. Other research has focused on extending the relational model by relaxing the first normal form assumption, giving rise to the notion of a nested relation and to a corresponding nested relational algebra. These two research streams have been integrated within the concept of a complex relation, where different types of constructors other than relation coexist, such as multiset and list. Several other database research areas cope with multiset processing, including view maintenance, data warehousing, and web information discovery.

1   Introduction

There exists a tight coupling between databases and multisets. A database is a collection of persistent information that is managed by a software package called *database management system* (DBMS) or, simply, *database system*. A database may be very large in nature. Consequently, a DBMS is required to manage information not only in main memory but, more specifically, in mass memory. Usually, a database is *shared*, that is, different applications and users are permitted to access the same database concurrently. This way, both redundancy and inconsistency of data can be avoided. A DBMS provides several other capabilities, including support for *recovery* and *privacy* of data.

A database is defined according to a *data model*, which provides means of data organization based on certain design patterns. A data model supports *structuring constructs* that are similar to the type constructors of general-purpose programming languages, like record, array, set, and file.

C.S. Calude et al. (Eds.): Multiset Processing, LNCS 2235, pp. 147–215, 2001.

Throughout the history of database technology, several data models have been defined, including the *hierarchical data model*, the *network data model*, the *relational data model*, and the *object-oriented data model*. All of them are called *logical* data models, to highlight that the relevant data structures reflect a specific organization, such as a tree, a graph, a table, or an object. Usually, a database is first described based on a *conceptual* data model, which is independent of the specific organization of data, such as the *entity-relationship data model*. Such a conceptual model is then mapped to a logical data model.

A database consists of a *schema* and an *instance*. The database schema defines the structure of data, which is expected to be essentially invariant with time. The database instance is the actual value of data, which is instead time-varying. The database changes its *state* when data are modified either by deleting or updating existing information, or by inserting new information.

The content of a database can be defined, accessed, and manipulated by means of a *database language*, which usually provides both *data definition* and *data manipulation* constructs. By 'data definition' we mean specifying the database schema. By 'data manipulation' we mean either querying or changing the database instance. For the DBMS to be usable, it is expected to retrieve and modify data efficiently, whatever the data model.

To simplify user interaction with the system, the same information can be organized at different *abstraction levels*. Specifically, the *physical level* describes how the data are stored in persistent memory in terms of complex low-level data structures. The *logical level* describes what data are incorporated within the database and what relationships exist among such data. Finally, the *view level* describes what part of the database can be seen by each class of users, as several views can be provided for the same database.

For the different abstraction levels to work as whole, appropriate mappings from the view level to the logical level, and from the logical level to the physical level, are required. Considering a relational DBMS, a view is mapped to one or several relations (tables) of the logical level, while a relation is mapped to one or several low-level data structures of the physical level[1].

Drawing now our attention to why multisets and databases are related to each other, we observe that, first, several DBMS data models support multi-set constructors [15, 47, 28, 45, 36, 44, 23, 12]. Consequently, database query languages are affected by the use of duplicates.

Second, there exist practical multiset-oriented query languages that have been implemented based on previous formalization of some set-oriented languages. Considering relational database technology, relational DBMSs have been developed upon the mathematical notion of a relation and the formalization of set-oriented query languages, namely *relational algebra* and *relational calculus*. However, relational DBMSs, almost invariably extend the nature of relations to multisets and support query languages with additional expressive power.

---

[1] The physical level incorporates in general a variety of additional data structures that are invisible at the logical level, such as, for example, the index of a relation, which allows for fast query evaluation.

Finally, dealing with multisets rather than with sets is generally bound to speed up query evaluation, as there is no need for duplicate elimination, an operation which can be extremely costly in large databases. Note that, in this case, multiset processing may be confined to optimization issues within the query evaluation, even in set-oriented queries, where duplicates are removed only when the set-oriented semantics of a query would be otherwise violated.

The mismatch between the relational model and the standardized relational query languages has led researchers to provide a foundation to the manipulation of multisets. Other research has focused on extending the relational model by relaxing the first normal form assumption (which requires flat tables), giving rise to the notion of a nested relation and to a corresponding nested relational algebra. These two research streams have been integrated within the concept of a complex relation, which is polymorphic in nature, as different types of constructors other than relation coexist, such as multiset and list.

Several other database research areas cope with multiset processing, including view maintenance, data warehousing, and web information discovery.

The remainder of the paper is organized as follows. Section 2 introduces the relational model in a pure set-oriented framework. Section 3 discusses how multisets can be manipulated in relational database systems. Section 4 presents the nested relational model, where sets can be nested within each other to form nested relations. Section 5 provides some hints on how the nested relational model can cope with multisets, thereby yielding the notion of a complex relation. A discussion on related work is provided in Sect. 6. Conclusions are drawn in Sect 7.

2   Relational Model

Most database systems currently on the market are based on the relational paradigm, which was proposed by Codd in 1970 [16], with the main purpose of overcoming the typical flaws of the database systems of that period, among which was the inability to provide data independence, that is, the separation between the logical view of the model from its physical implementation.

However, the success of the relational approach was quite slow. Its high level of abstraction prevented for several years the implementation of efficient structures, as they were significantly different from those used in that epoch, typically, relevant to hierarchical and reticular data models. In fact, even though the first prototypes of relational systems have been developed since the early 1970s, the first relational systems appeared on the market in 1981, and became significant only in the middle 1980s.

The relational data model makes use of a single structure to organize data: the mathematical concept of n-ary relation. Beside being easy to formalize, the relational model has a simple representation: a relation is a table, and the database is viewed as a collection of tables. Another advantage of the relational model with respect to previous data models is that it responds to the requirement of data independence.

Intuitively, a table, which has a unique name in the database, contains one or more columns: each column has a heading, called attribute name, and a corresponding set of possible values, called the domain (for example, integers, strings, etc.). Each row is an ordered n-tuple of values $\langle v_1, v_2, \ldots, v_n \rangle$, each belonging to the domain of the corresponding attribute: each v_i is in the domain of column $i \in [1 .. n]$, where n is the number of columns. A table is therefore an unordered collection of distinct tuples.

Formally, let $\mathbb{U}$ be the set of attribute names, $\mathbb{U} = \{A_1, \ldots, A_n\}$, and $\mathbb{D}$ the set of domains, $\mathbb{D} = \{D_1, \ldots, D_n\}$. Each domain D_i contains only atomic values. We assume the existence of a function $\mathrm{Dom} : \mathbb{U} \mapsto \mathbb{D}$, which associates the appropriate domain with each attribute name. Then, a *tuple* over a set of attributes $X \subseteq \mathbb{U}$ is a function t that associates with each attribute $A_i \in X$ a value of the domain $\mathrm{Dom}(A_i)$; this value is denoted by $t_{(A_i)}$. We shall also write $t_{(Y)}$ with $Y \subseteq X$ to denote the restriction of the function t to the attributes in Y.

A *relation schema* of a relational database has the form $R = (A_1, \ldots A_n)$, where R is the relation name and A_i are distinct attribute names. A *database schema*, denoted by Δ, is a set of relation schemas with distinct relation names.

A *relation instance* (or simply, a *relation*) defined on a relation schema $R = (A_1, \ldots, A_n)$, is a finite set r of tuples over $X = \{A_1, \ldots, A_n\}$.

A *database instance* δ on a database schema Δ is a set of relations $\{r_1, \ldots, r_m\}$ where each r_i is defined on precisely one R_i in Δ.

With respect to the tabular representation of relation, the above definitions lead to the following observable properties:

1. The values within each column are homogeneous, and they all belong to the same domain. The domain corresponds to the attribute of the column.
2. There are no identical rows: a relation is a set, and therefore it does not contain duplicated elements.
3. The order of columns is irrelevant, since they are identified by their name.
4. The order of rows is immaterial, since they are identified by their content.

The relational data model is *value-oriented*, as all the information are represented by means of values: the identity of a tuple in a relation is based only on the values it determines for the attributes, and the relationship among tuples in different relations is based only on attribute values. This is particularly important since not all aspects of a relation might be known at a given time: in this case the relation could contain values which are not specified. The domains of attributes can be extended by including a special *null* value, that represents the absence of information. The theory of null values is especially important for the relational model, where all the information is value-based.

Moreover, part of the content of a relation is inherent to constraints on the values that some or all of its attributes can assume. Two important classes of such constraints are expressed by the notions of a *key* and *functional dependency*.

A subset K of the attributes of a relation r is a *key* of r when the following properties hold:

Table 1. Relation Courses(course, year, teacher)

course	year	teacher
Algebra	1	Sheila
Calculus	1	Gregory
Compilers	5	Jerri
Computers	2	Patricia
Databases	3	Carol
Geometry	2	Martin
Operating systems	4	Angelique
Programming languages	3	Jerri
Robotics	5	Richard
Software engineering	4	Angelique

1. *Unique identification:* r does not contain two distinct tuples t_1, t_2 which agree on all the attributes in K, that is:

$$(\forall t_1)(\forall t_2)(t_{1(K)} \neq t_{2(K)}) \; ; \tag{1}$$

2. *Minimality:* no proper subset of K enjoys the unique identification property.

A set K fulfilling the unique identification property is called a *superkey*, since it is the superset of a key. The set of all the attributes in a relation is always a key (since all the tuples are distinct), so each relation has at least one key.

Given two sets of attributes X and Y of a relation r, we say that Y *functionally depends* on X in r, denoted by $X \rightarrow Y$, if and only if for every pair of tuples $t_1 \in r$, $t_2 \in r$, if $t_{1(X)} = t_{2(X)}$, then $t_{1(Y)} = t_{2(Y)}$.

Intuitively, there is a functional dependency (FD) when the value of one or more attributes in a relation determines the value of another group of attributes. The concept of a key of a relation can be rephrased in terms of FD: a set K of attributes of r is a key if the following properties hold:

1. $K \rightarrow N_K$, where N_K is the set of attributes of r that do not belong to K;
2. No proper subset of K meets the same property.

Example 1. Shown in Tables 1 and 2 are relations Courses and Teachers, respectively. Specifically, the schema of Courses consists of three attributes, namely:

1. course, defined on the domain of course names, which identifies the course, in other words, course is a key for Courses;
2. year, defined on the range $[1 .. 5]$, which identifies the curriculum year in which the course is taught;
3. teacher, defined on the domain of teacher names, which identifies the name of the relevant teacher.

The instance of Courses includes ten tuples. On the other hand, the schema of Teachers involves attributes name, age, and department, where name is a key.

Table 2. Relation Teachers(name, age, department)

name	age	department
Angelique	42	Computer science
Carol	38	Computer science
Gregory	63	Mathematics
Jerri	54	Computer science
Martin	42	Mathematics
Patricia	36	Electronics
Richard	45	Electronics
Sheila	40	Mathematics

Note that, in order to avoid dangling teachers, the domain of name in Teachers is expected to include the domain of teacher in Courses. Therefore attributes teacher and name allow us to join each tuple of Courses with one tuple of Teachers appropriately. Thus, to find out the department of the teacher of a given course, we match the appropriate tuple in Courses with the one in Teachers which agree on attributes teacher and name, respectively.   □

2.1   Relational Algebra

Once a relational database schema has been instantiated, it is possible to retrieve and modify the stored information by means of appropriate languages. Formally, updating the database amounts to changing its state. Denoting with $\mathbb{S}$ the domain of states of a database schema Δ, the update language describes functions Φ between the domain of states, namely $\Phi : \mathbb{S} \mapsto \mathbb{S}$. On the other hand, querying a database amounts to creating new relations, therefore query languages describe functions from $\mathbb{S}$ to the set of relations over possible schemas. Relations yielded by queries are temporary, that is, they disappear automatically from the database after they have been displayed to the user as a side effect of the query evaluation. Thus, we distinguish *base relations* from *temporary relations*, where the former are those relations which instantiate the database schema, while the latter correspond to query results. However, temporary relations may be transformed into base (persistent) relations by special-purpose create statements.

A great deal of effort in the design of relational database systems has been devoted to query languages, among which is *relational algebra*. Relational algebra consists of expressions applied to relations. Intuitively, a relational-algebra expression follows the pattern of arithmetic expressions, where variables are replaced by relation names, while arithmetic operators, such as $+$, $-$, $*$, and $/$, are replaced by relational operators. The application of a relational operator to one or two relations yields a new relation, exactly as the application of an arithmetic operator gives rise to a new number. Likewise, relational operators may be applied to relational expression, exactly as arithmetic operators can be applied to arithmetic expressions, for example, $(x-y)*(z+w)$. Note that relational algebra is in fact an algebra because the application of relational operators on relations

Table 3. Relation `Priorities(course, prerequisite)`

course	prerequisite
Compilers	Software engineering
Computers	Calculus
Geometry	Algebra
Operating Systems	Programming languages
Programming languages	Computers
Robotics	Geometry
Robotics	Programming languages
Software engineering	Programming languages
Software engineering	Databases

produces relations. In other words, relational expressions are closed with respect to the domain of possible relations.

A *query* is an expression of relational algebra. Since relations are essentially sets, relational algebra deals with the manipulations of sets. However, it is possible to extend the semantics of relational operators to multisets, as outlined in Sect. 3. As a matter of fact, concrete relational query languages based somehow on relational algebra, such as SQL (see Sect. 3.4), allows the manipulation of relations in a multiset-oriented way, where, for efficiency reasons, duplicate tuples are not removed unless explicitly specified.

Relational-algebra operators include *projection, selection, product, join, renaming, union, intersection,* and *difference.* Furthermore, for modularity reasons, we introduce the *assignment* statement.

Projection. The projection operator is unary: it takes a relation and produces a new relation containing only a subset of its columns. Let r be a relation defined over the schema R containing the set X of attributes, and let $Y \subseteq X$. The *projection* of r onto Y, denoted by $\pi_Y(r)$, is a relation on the attributes in Y consisting of the restrictions of the tuples of r to the attributes in Y:

$$\pi_Y(r) \stackrel{\text{def}}{=} \{t_{(Y)} \mid t \in r\} \ . \tag{2}$$

Example 2. Shown in Table 3 is a relation called `Priorities`, whose schema includes attributes `course` and `prerequisite`. `Priorities` is meant to specify the prerequisites of each course, that is, the courses a student is required to have passed before taking the exam of that course. For example, `Geometry` requires `Algebra` as prerequisite. `Software engineering` requires two prerequisites, namely `Programming languages` and `Databases`. Note that both attributes `course` and `prerequisites` are part of the key. In fact, given a course, there exist in general several different prerequisites for it. On the other hand, a course may be a prerequisite for several different courses. As a matter of fact, `Programming languages` is a prerequisite for three courses. This means that neither `course` nor `prerequisite` is a key by its own, but only the composition of both.

Table 4. Result of relational-algebra Expression (3)

course
Software engineering
Calculus
Algebra
Programming languages
Computers
Geometry
Databases

To find out the courses which are prerequisites for some courses we write the following expression:

$$\pi_{\text{prerequisite}}(\text{Priorities}) \tag{3}$$

which will give the result outlined in Table 4. Note that, generally speaking, some of the tuples may become identical when they are projected on a set of attributes: in this case duplicated tuples are deleted, so the resulting relation may actually contain less tuples than the operand. In our example, two of the three tuples relevant to **Programming languages** have been eliminated from the result, so that the resulting relation includes seven tuples instead of nine.   □

Selection. The selection operator is unary. Intuitively, the result of the selection is the subset of the tuples in the operand that satisfy a selection predicate expressed in terms of elementary comparisons of constants and attribute values plus logic connectives.

Propositional formula. Let r be a relation over the set of attributes X; a *propositional formula* $\wp$ over X is defined recursively as follows. *Atoms* over X have the form $A_1 \vartheta A_2$ or $A_1 \vartheta a$, where $A_1 \in X$, $A_2 \in X$, a is a constant, and ϑ is a comparison operator, $\vartheta \in \{=, <, >, \neq, \geq, \leq\}$. Every atom over X is a propositional formula over X; if $\wp_1, \wp_2$ are propositional formulas over X, then $\neg(\wp_1)$, $\wp_1 \wedge \wp_2$, and $\wp_1 \vee \wp_2$ are formulas over X. Parentheses can be used as usual. Nothing else is a formula. A propositional formula associates a Boolean value with each tuple in r.

Given a relation r over the schema R, the *selection* of r with respect to $\wp$, denoted by $\sigma_\wp(r)$, is a relation over the same schema R, containing the tuples of r that make $\wp$ true:

$$\sigma_\wp(r) \stackrel{\text{def}}{=} \{t \in r \mid \wp(t)\} \ . \tag{4}$$

Example 3. Considering Table 1, to find out the course names of the last two years we write:

$$\pi_{\text{course}}(\sigma_{\text{year} \geq 4}(\text{Courses})) \tag{5}$$

Table 5. Result of relational-algebra Expression (5)

course
Compilers
Operating Systems
Robotics
Software engineering

which is expected to produce the result outlined in Table 5. Note that the temporary result of the selection is the operand of the subsequent projection.   □

Product. It is a binary operator. Let r_1 and r_2 be two relations defined over the set of attributes X and Y, respectively, such that $X \cap Y = \emptyset$. The *product* of r_1 and r_2, denoted by $r_1 \times r_2$, is a relation on $X \cup Y$ consisting of all the tuples resulting from the concatenation of tuples in r_1 with tuples in r_2:

$$r_1 \times r_2 \stackrel{\text{def}}{=} \{t \text{ over } X \cup Y \mid (\exists t_1)(\exists t_2)(t_1 \in r_1, t_2 \in r_2, t_{(X)} = t_1, t_{(Y)} = t_2)\} \ . \tag{6}$$

Join. It is a binary operator that comes in two versions, usually referred to as *natural join* and *theta-join*. Let r_1 and r_2 be two relations defined over the set of attributes XY and YZ, such that $XY \cap YZ = Y$. The *natural join* of r_1 and r_2, denoted by $r_1 \bowtie r_2$, is a relation on $XYZ = XY \cup YZ$ consisting of all the tuples resulting from the concatenation of tuples in r_1 with tuples in r_2 that have identical values for the attributes Y:

$$r_1 \bowtie r_2 \stackrel{\text{def}}{=} \{t \text{ over } XYZ \mid (\exists t_1)(\exists t_2) (t_1 \in r_1, t_2 \in r_2, t_{(XY)} = t_1, t_{(YZ)} = t_2\} \ . \tag{7}$$

Example 4. To associate with ech course and corresponding teacher the relevant prerequisite we write:

$$\pi_{course,teacher,prerequisite}(Courses \bowtie Priorities) \tag{8}$$

which generates the relation outlined in Table 6.   □

Given two relations r_1 and r_2 over disjoint sets of attributes X and Y, a *theta-join* $r_1 \bowtie_\wp r_2$ is a relation over the set of attributes $X \cup Y$ containing tuples obtained by the concatenation of tuples of r_1 and r_2 that satisfy the propositional formula $\wp$, that is:

$$r_1 \bowtie_\wp r_2 \stackrel{\text{def}}{=} \{t \text{ over } X \cup Y \mid (\exists t_1)(\exists t_2) (t_1 \in r_1, t_2 \in r_2, t_{(X)}=t_1, t_{(Y)}=t_2, \wp(t))\} \ . \tag{9}$$

The theta-join can be expressed through the use of selection and product as follows:

$$r_1 \bowtie_\wp r_2 \equiv \sigma_\wp(r_1 \times r_2) \ . \tag{10}$$

Example 5. To retrieve, for each course, the department the corresponding teacher depends on we write:

$$\pi_{\texttt{course,department}} (\texttt{Courses} \bowtie_{\texttt{teacher=name}} \texttt{Teachers}) \tag{11}$$

which yields the relation outlined in Table 7. In this case we used a theta-join as the two linking attributes `teacher` and `name` have different identifiers.   □

Renaming. It is a unary operator that only changes the name of the attributes in the result, leaving the content of the relation unchanged. It is used to overcome difficulties with those operators for which attribute names are significant. Let r be a relation defined over a set of attributes X, and Y another set of attributes with the same cardinality of X, that is, $|X| = |Y|$. Besides, let $A_1, A_2, \ldots , A_k$ and $A'_1, A'_2, \ldots , A'_k$ be an order for attributes in X and Y respectively. The *renaming*

$$\rho_{A'_1,\ldots,A'_k \leftarrow A_1,\ldots,A_k}(r) \tag{12}$$

is a relation which includes a tuple t' for each tuple $t \in r$, defined as follows: t' is a tuple on Y and $t'_{(A'_i)} = t_{(A_i)}$, $i \in [1 \mathrel{..} k]$. In practice, only the renamed attributes

Table 6. Result of relational-algebra Expression (8)

course	teacher	prerequisite
Compilers	Jerri	Software engineering
Computers	Patricia	Calculus
Geometry	Martin	Algebra
Operating systems	Angelique	Programming languages
Programming languages	Jerri	Computers
Robotics	Richard	Geometry
Robotics	Richard	Programming languages
Software engineering	Angelique	Databases

Table 7. Result of relational-algebra Expression (11)

course	department
Algebra	Mathematics
Calculus	Mathematics
Compilers	Computer science
Computers	Electronics
Databases	Computer science
Geometry	Mathematics
Operating systems	Computer science
Programming languages	Computer science
Robotics	Electronics
Software engineering	Computer science

will be indicated within the two lists $A_1, \ldots, A_k$ and $A'_1, \ldots, A'_k$, namely those for which $A_i \neq A'_i$, rather than the complete list of attributes.

Since the renaming operator is not intended to change the domain of the attributes, we also require $\text{Dom}(A'_i) = \text{Dom}(A_i)$.

Example 6. We may formulate the same operation of Example 5 by replacing the theta-join in Expression (11) with a renaming and a natural join as follows:

$$\pi_{\text{course,department}}(\text{Courses} \bowtie (\rho_{\text{teacher} \leftarrow \text{name}}(\text{Teachers}))) \tag{13}$$

which yields the same relation displayed in Table 7.   □

It is worthwhile pointing out that a natural join can be expressed in terms of projection, theta-join, and renaming operations. Let r_1 and r_2 be two relations over schemas XY and YZ, respectively, where $\text{XY} \cap \text{YZ} = Y$. Then, the following equivalence holds:

$$r_1 \bowtie r_2 \equiv \pi_{\text{XYZ}}(r_1 \bowtie_{\wp_{Y,Y'}} (\rho_{A'_1, \ldots, A'_k \leftarrow A_1, \ldots, A_k}(r_2))) \tag{14}$$

where $Y = \{A_1, \ldots, A_k\}$, $Y' = \{A'_1, \ldots, A'_k\}$, $\text{XY} \cap Y'Z = \emptyset$, and $\wp_{Y,Y'}$ is a predicate defined as follows:

$$\wp_{Y,Y'}(t) \overset{\text{def}}{=} \begin{cases} true & \text{if } t_{(Y)} = t_{(Y')} \\ false & \text{otherwise .} \end{cases} \tag{15}$$

Set-theoretic operators. Since relations are special classes of sets, the meaning of the well-known set-theoretic operators can be retained for relations as well. Specifically, the *union*, *intersection*, and *difference* of two relations r and s are denoted by $r \cup s$, $r \cap s$, and $r - s$, respectively. However, since the result is expected to be defined over an expected schema, these operators require r and s to share the same schema, otherwise the tuples of the resulting relation would refer to different sets of attributes.

Example 7. To find the courses which either depend on some prerequisite courses or are themselves prerequisites for other courses we may formulate the following expression:

$$(\pi_{\text{course}}(\text{Priorities})) \cup (\rho_{\text{course} \leftarrow \text{prerequisite}}(\pi_{\text{prerequisite}}(\text{Priorities}))) \tag{16}$$

whose result is displayed in Table 8. Incidentally, all the courses in Courses are incorporated in Table 8. Duplicate courses have been removed.

Instead. if we are to retrieve the courses which are both prerequisite for some courses and depend on other courses we have just to change the union operator in Formula (16) with an intersection as follows:

$$(\pi_{\text{course}}(\text{Priorities})) \cap (\rho_{\text{course} \leftarrow \text{prerequisite}}(\pi_{\text{prerequisite}}(\text{Priorities}))) \tag{17}$$

Table 8. Result of relational-algebra Expression (16)

course
Algebra
Calculus
Compilers
Computers
Databases
Geometry
Operating Systems
Programming languages
Robotics
Software engineering

Table 9. Result of relational-algebra Expression (17)

course
Computers
Geometry
Programming languages
Software engineering

whose result is outlined in Table 9.

Finally, the following expression yields the courses which require some other courses, but are not prerequisite of any course:

$$(\pi_{\text{course}}(\texttt{Priorities})) - (\rho_{\text{course}\leftarrow\text{prerequisite}}(\pi_{\text{prerequisite}}(\texttt{Priorities}))) \tag{18}$$

whose result is outlined in Table 10.   □

Assignment. The assignment operator does not extend the expressive power of relational algebra. In other words, all the operations involving the assignment can be expressed as well using the operators defined above. However, the assignment operator is useful because it allows the specification of expressions in a more concise and modular fashion. An assignment has the following form:

$$w \leftarrow E(r_1, r_2, \dots, r_n) \tag{19}$$

Table 10. Result of relational-algebra Expression (18)

course
Compilers
Operating Systems
Robotics

where w is an identifier and E a relational expression. In general, a query expressed in relational algebra consists of a (possibly empty) sequence of assignments and a final relational expression Q:

$$w_1 \leftarrow E_1(r_1, r_2, \dots, r_n)$$
$$w_2 \leftarrow E_2(r_1, r_2, \dots, r_n, w_1)$$
$$\cdots \tag{20}$$
$$w_m \leftarrow E_m(r_1, r_2, \dots, r_n, w_1, w_2, \dots, w_{m-1})$$
$$Q(r_1, r_2, \dots, r_n, w_1, w_2, \dots, w_m) \ .$$

Each assignment instantiates a temporary relation w_i whose schema and instance are determined by the corresponding expression E_i. However, relations w_i are only visible within Q and disappear as soon as the execution of Q terminates.

Example 8. Consider the problem of determining the age of the youngest teacher. Note that relational algebra does not provide any aggregation function like min to determine a single value (e.g. maximum, minimum, or average) starting from a set of values. Consequently we have to specify the relevant query by means of relational operators only. This can be done as follows:

$$\text{Ages} \leftarrow \pi_{\text{age}}(\text{Teachers})$$
$$\text{NotMin} \leftarrow \pi_{\text{age}}(\sigma_{\text{age}>\text{age1}}(\text{Ages} \times (\rho_{\text{age1}\leftarrow\text{age}}(\text{Ages})))) \tag{21}$$
$$\text{Ages} - \text{NotMin} \ .$$

In the above sequence of expressions, the first two are assignments, while the third is the displayed result. Ages is a temporary relation incorporating the whole set of ages in Teachers (see Table 2). The other temporary relation NotMin represents the set of ages which cannot be the minimum value. This claim is supported by the evidence that each age is combined with each other age in Ages by means of the product (the renaming is necessary to avoid duplicated attribute names in the resulting schema). Then a selection is applied to the result of the product: only the ages which are greater than at least another age are selected. Consequently, NotMin will embody all the ages but the minimum, this being the age which is not greater than any other age. Hence, the final result is obtained by simply complementing NotMin with respect to Ages through the difference operation. The final result is therefore the singleton $\{36\}$, which incorporates the age of Patricia, the youngest teacher. □

3 Multisets in Relational Database Systems

The notion of a relation as a set of tuples is a simple and formal model of data. Being a set, a relation cannot include duplicates. Accordingly, all the relations involved in the examples of Sect. 2 do not incorporate any duplicate tuple. However, commercial relational database systems are almost invariably based on multisets instead of sets. In other words, tables are in general allowed to include duplicate tuples.

The main reason for considering relations as multisets in database systems is efficiency of query evaluation. Normally, base relations do not include duplicates. However, duplicate tuples can be generated within the evaluation of a relational-algebra expression which involves either projections or unions. For example, when we do a projection, if we want the result to be a set, we are required to compare each tuple in the result with all the other tuples in the result in order to be sure that tuples are not replicated. Generally speaking, duplicate removal is very expensive in time. Instead, if we are allowed to have a multiset as the result, then we merely project each tuple and add it to the result, regardless of other occurrences of the same tuple.

Likewise, when making the union of two relations r_1 and r_2, if we want the result to be a set, we must check that each tuple of one relation is not already included in the other. That is, if n_1 and n_2 are the number of tuples in r_1 and r_2, respectively, we are expected to make $n_1 \cdot n_2$ comparisons. Instead, if we accept a multiset as the result, then we just copy all the tuples of r_1 and r_2 into the answer, regardless of whether or not they appear in both relations, thereby making $n_1 + n_2$ operations without any comparison.

Sometimes, a hybrid approach is adopted: duplicate removals are delayed as much as possible in order to optimize efficiency whilst obtaining a set as the result. This means that, from the user point of view, relation are sets, while the evaluation system makes the best choices to produce the required result based on implementation-oriented criteria.

Example 9. With reference to Example 7, we may represent Expression (16) as a tree where leaves and internal nodes correspond to base relations and operators, respectively, as shown in Fig. 1.

A temporary relation is associated with each internal node, that implicitly represents the intermediate result of the expression inherent to the corresponding sub-tree.

Thus, the tree of Fig. 1 incorporates four internal nodes and two leaves, which incidentally happen to refer to the same base relation Priorities. Three nodes are virtually subjected to duplicate removals, these being the two projections

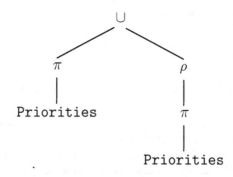

Fig. 1. Evaluation tree of Expression (16)

Table 11. Multiset result of Expression (16)

course
Compilers
Computers
Geometry
Operating Systems
Programming languages
Robotics
Robotics
Software engineering
Software engineering
Software engineering
Calculus
Algebra
Programming Languages
Computers
Geometry
Programming Languages
Programming Languages
Databases

and the union. If we allow the result to be a multiset, we will obtain the result displayed in Table 11.

Note that Table 11 incorporates 18 tuples while the set result of the same expression outlined in Table 8 includes 10 tuples only. However, the multiset result, although larger, can be computed more quickly.   □

The instance of a multiset m can be implemented in two different ways:

1. *Extensional representation*: different occurrences of the same tuple are physically replicated, such as in Table 11.
2. *Intensional representation*: the whole collection of occurrences of the same tuple t is physically implemented by a single tuple (t, ω), where $\omega = \text{Occ}(t, m)$.

Example 10. The intensional representation of the multiset displayed in Table 11 is shown in Table 12.   □

Note that the intensional representation requires ω to be updated whenever duplicates may be generated, thereby losing the advantages of the extensional representation in terms of computation efficiency. On the other hand, representing intensionally several tuples by means of a single tuple allows the database system to save space and even time in those operations which do not introduce duplicates, as a smaller set of tuples are processed[2].

---

[2] From the physical point of view, a multiset stored intensionally is in fact a set.

Table 12. Intensional representation of the multiset displayed in Table 11.

course	ω
Compilers	1
Computers	2
Geometry	2
Operating Systems	1
Programming languages	4
Robotics	2
Software engineering	3
Calculus	1
Algebra	1
Databases	1

3.1 Relational Algebra for Multisets

The operators of relational algebra can be straightforwardly extended for manipulating multisets. However, as it will be shown in Sect. 3.3, some of the algebraic properties which hold for sets hold no longer for multisets. In what follows, multisets are enclosed within square brackets. Furthermore, the number of occurrences of t within a multiset m is denoted by the function $\mathrm{Occ}(t, m)$. For example, if m is the multiset displayed in Table 11 and $t = (\texttt{Software engineering})$, we have $\mathrm{Occ}(t, m) = 3$.

In order to univocally define a multiset, it is not enough to specify its elements, it is also necessary to assign a cardinality to each of them. Thus, in what follows, the definition of the multiset resulting from each operation is usually twofold. Indeed, the specification of the cardinality of each element is a complete definition as, if the cardinality of an element is zero, then such an element will not belong to the multiset. Therefore, in some cases, only the definition of the cardinality is given.

Projection of multisets. Provided that multiple occurrences of the same tuple can be inserted into a multiset, the definition of projection of multisets is formally the same given for sets. Let m be a multiset defined over the schema M containing the set X of attributes, and let $Y \subseteq X$. The projection of m onto Y, $\pi_Y(m)$, is a multiset on the attributes in Y consisting of the restrictions of the tuples of m to the attributes in Y:

$$\pi_Y(m) \stackrel{\mathrm{def}}{=} [t_{(Y)} \mid t \in m] \ . \tag{22}$$

However, in contrast with Formula (2), duplicate restrictions of t on Y are not removed from the result. Specifically, the number of occurrences is given by the following formula:

$$\mathrm{Occ}(t, \pi_Y(m)) \stackrel{\mathrm{def}}{=} \sum_{t' \in m, t'_{(Y)} = t} \mathrm{Occ}(t', m) \ . \tag{23}$$

Selection of multisets. The selection operation on a multiset m simply applies the selection predicate to each tuple in m and retains in the result those tuples $t \in m$ for which the predicate evaluates to true:

$$\sigma_\wp(m) \stackrel{\text{def}}{=} [t \in m \mid \wp(t)] \ . \tag{24}$$

Note that the Boolean value $\wp(t)$ is the same for all the occurrences of t in m. Consequently, depending on the value of $\wp(t)$, either all the occurrences of t are selected or none or them, that is:

$$\mathrm{Occ}(t, \sigma_\wp(m)) \stackrel{\text{def}}{=} \begin{cases} \mathrm{Occ}(t, m) & \text{if } \wp(t) \\ 0 & \text{otherwise} \end{cases} \ . \tag{25}$$

This is due to the inability to distinguish different occurrences of t within m based on $\wp$. Furthermore, as for sets, the selection operator does not generate new duplicates. However, it preserves the selected duplicates in the result.

Product of multisets. The product operation can be extended for multisets in a natural way. Let m_1 and m_2 be two multisets defined over schemas X and Y, respectively, such that $X \cap Y = \emptyset$. The product $m_1 \times m_2$ is a multiset over $X \cup Y$ consisting of all the tuples resulting from the concatenation of tuples in m_1 with tuples in m_2:

$$m_1 \times m_2 \stackrel{\text{def}}{=} [t \text{ over } X \cup Y \mid (\exists t_1)(\exists t_2)(t_1 \in m_1, t_2 \in m_2, t_{(X)} = t_1, t_{(Y)} = t_2)] \ . \tag{26}$$

In other words, each tuple of m_1 is paired with each tuple of m_2, regardless of whether it is a duplicate or not. As a result, if tuple t_1 appears n_1 times in m_1 and tuple t_2 appears n_2 times in m_2, then, denoting with $t_1 \oplus t_2$ the concatenation of tuples t_1 and t_2, in the product $m_1 \times m_2$, the tuple $t_1 \oplus t_2$ will appear $n_1 \cdot n_2$ times, that is:

$$\mathrm{Occ}(t_1 \oplus t_2, m_1 \times m_2) \stackrel{\text{def}}{=} \mathrm{Occ}(t_1, m_1) \cdot \mathrm{Occ}(t_2, m_2) \ . \tag{27}$$

Join of multisets. Both natural join and theta-join can be extended for multisets. Let m_1 and m_2 be two multisets defined over disjoint schemas X and Z, respectively. Then, the theta-join of m_1 and m_2 is defined as follows:

$$m_1 \bowtie_\wp m_2 \stackrel{\text{def}}{=} [t \text{ over } X \cup Z \mid (\exists t_1)(\exists t_2) (t_1 \in r_1, t_2 \in r_2, t_{(X)} = t_1, t_{(Y)} = t_2, \wp(t))] \ . \tag{28}$$

The equivalence expressed in Equation (10) is required to hold for multisets as well, that is:

$$m_1 \bowtie_\wp m_2 \equiv \sigma_\wp(m_1 \times m_2) \ . \tag{29}$$

Therefore, based on the above equivalence, we can derive the number of occurrences of tuples in the theta-join result based on Equations (27) and (25) as follows:

$$\text{Occ}(t_1 \oplus t_2, m_1 \bowtie_{\wp} m_2) \overset{\text{def}}{=} \text{Occ}(t_1 \oplus t_2, \sigma_{\wp}(m_1 \times m_2))$$

$$= \begin{cases} \text{Occ}(t_1 \oplus t_2, m_1 \times m_2) & \text{if } \wp(t_1 \oplus t_2) \\ 0 & \text{otherwise} \end{cases}$$

$$= \begin{cases} \text{Occ}(t_1, m_1) \cdot \text{Occ}(t_2, m_2) & \text{if } \wp(t_1 \oplus t_2) \\ 0 & \text{otherwise} \end{cases} \quad (30)$$

The natural join for multisets can be defined as follows. Let m_1 and m_2 be two multisets defined over the set of attributes XY and YZ, respectively, where $\text{XY} \cap \text{YZ} = Y$. The natural join $m_1 \bowtie m_2$, is a multiset over $\text{XYZ} = \text{XY} \cup \text{YZ}$ consisting of all the tuples resulting from the concatenation of tuples in m_1 with tuples in m_2 that have identical values for the attributes Y:

$$m_1 \bowtie m_2 \overset{\text{def}}{=} \left[t \text{ over XYZ} \mid (\exists t_1)(\exists t_2) \ (t_1 \in m_1, t_2 \in m_2, t_{(\text{XY})} = t_1, t_{(\text{YZ})} = t_2 \right] . \quad (31)$$

We require the multiset version of Equivalence (14) to hold, that is:

$$m_1 \bowtie m_2 \equiv \pi_{\text{XYZ}}(m_1 \bowtie_{\wp_{Y,Y'}} (\rho_{A'_1,\dots,A'_k \leftarrow A_1,\dots,A_k}(m_2))) \quad (32)$$

where $\wp_{Y,Y'}$ is the predicate defined in Formula (15). Then, we make use of Formula (32) to compute the number of occurrences of tuples in the result. Let

$$m'_2 = \rho_{A'_1,\dots,A'_k \leftarrow A_1,\dots,A_k}(m_2) . \quad (33)$$

Then, based on Formulae (23) and (30), the number of occurrences of tuples in the natural join result can be derived as follows:

$$\text{Occ}(t, m_1 \bowtie m_2) \overset{\text{def}}{=} \text{Occ}(t, \pi_{\text{XYZ}}(m_1 \bowtie_{\wp_{Y,Y'}} m'_2)) =$$

$$\sum_{t' \in (m_1 \bowtie_{\wp_{Y,Y'}} m'_2), t'_{(\text{XYZ})} = t} \text{Occ}(t', m_1 \bowtie_{\wp_{Y,Y'}} m'_2) =$$

$$\sum_{t' \in (m_1 \bowtie_{\wp_{Y,Y'}} m'_2), t'_{(\text{XYZ})} = t, t'_{(\text{XY})} = t'_1, t'_{(Y'Z)} = t'_2} \begin{cases} \text{Occ}(t'_1, m_1) \cdot \text{Occ}(t'_2, m'_2) & \text{if } \wp_{Y,Y'}(t') \\ 0 & \text{otherwise} \end{cases} =$$

$$\sum_{t'_{(\text{XYZ})} = t, t'_{(\text{XY})} = t'_1, t'_{(Y'Z)} = t'_2, t'_1 \in m_1, t'_2 \in m'_2, t'_{(Y)} = t'_{(Y')}} \text{Occ}(t'_1, m_1) \cdot \text{Occ}(t'_2, m'_2) \quad (34)$$

where $Y'Z = Y' \cup Z$.

Renaming of multisets. From the instance point of view, the renaming operation is an identity function. In other words, the result differs from the operand in the schema whilst retaining the instance. As such, it can be applied to multisets based on the same semantics defined for relations. In particular, given a multiset m, the number of occurrences of each tuple in the result does not change, that is:

$$\text{Occ}(t, \rho_{A_1', A_2', \ldots, A_k' \leftarrow A_1, A_2, \ldots, A_k}(m)) \overset{\text{def}}{=} \text{Occ}(t, m) \ . \tag{35}$$

Set-theoretic operators for multisets. The set-theoretic operations defined for relations in Sect. 2.1, namely union, intersection, and difference, can be extended to multisets in a natural fashion. Let m_1 and m_2 be two multisets defined over the same schema. The union of m_1 and m_2 is defined as follows:

$$m_1 \cup m_2 \overset{\text{def}}{=} [t \mid t \in m_1 \lor t \in m_2] \ . \tag{36}$$

However, the peculiarity of the union definition lies in the fact that the resulting multiset is expected to incorporate all the occurrences of tuples, both in m_1 and m_2, that is:

$$\text{Occ}(t, m_1 \cup m_2) \overset{\text{def}}{=} \text{Occ}(t, m_1) + \text{Occ}(t, m_2) \ . \tag{37}$$

The intersection operation $m_1 \cap m_2$ will include a number of occurrences, of each shared tuple, equal to the minimum number of occurrences of the tuple in m_1 and m_2, respectively, that is:

$$\text{Occ}(t, m_1 \cap m_2) \overset{\text{def}}{=} \min(\text{Occ}(t, m_1), \text{Occ}(t, m_2)) \ . \tag{38}$$

In particular, if the tuple t is included in m_1 but not in m_2, it follows that, $\min(\text{Occ}(t, m_1), \text{Occ}(t, m_2)) = 0$, whatever $\text{Occ}(t, m_1)$. That is, as expected, t will not be part of the intersection. On the other hand, if t is both included in m_1 and m_2, the result will include the smallest multiset of t, that is, the minimum number of occurrences of t in m_1 and m_2.

The difference operation $m_1 - m_2$ will include a number of occurrences of each tuple t which depends on whether or not $\text{Occ}(t, m_1) \geq \text{Occ}(t, m_2)$. Specifically, if this relationship is met, then the number of occurrences in the result will be the difference $\text{Occ}(t, m_1) - \text{Occ}(t, m_2)$, otherwise the result will not include t. Formally,

$$\text{Occ}(t, m_1 - m_2) \overset{\text{def}}{=} \max((\text{Occ}(t, m_1) - \text{Occ}(t, m_2)), 0) \ . \tag{39}$$

Duplicate removal. Given a multiset m, we may introduce a special unary operator called *duplicate removal*, denoted by $\delta(m)$, which generates a multiset obtained by removing from m duplicate occurrences of tuples. Formally, let

$$M \overset{\text{def}}{=} \{T(t_1), T(t_2), \ldots, T(t_k)\} \tag{40}$$

be a partition of m where each part $T(t_i)$, $i \in [1 .. k]$, is the multiset of occurrences of tuple t_i. Then,

$$\delta(m) \stackrel{\text{def}}{=} [t_1, t_2, \dots, t_k] . \tag{41}$$

Clearly,

$$\text{Occ}(t, \delta(m)) \stackrel{\text{def}}{=} 1 . \tag{42}$$

3.2   Multiset Operations on Relations

Intuitively, each relation r can be viewed as a special multiset m in which each tuple happens to occur just once within the multiset, that is

$$(\forall t \in m) \, (\text{Occ}(t, m) = 1) . \tag{43}$$

We say that m is the *multiset mirror* of r, denoted by $\mathfrak{M}(r)$. There exists an isomorphism between the tuples of r and m. More precisely, r and m incorporate the same tuples. Consequently, we may apply to r both the operators of relational algebra for relations and those extended for multisets, as defined in Sect. 3.1. This is due to the fact that there exists an isomorphism between the operators for relations and those for multisets, that is, for each operator of relational algebra there exists a homonymous operator of the extended algebra for multisets and vice versa. In other words, the algebraic operators are polymorphic in nature, as they can be applied either to relations or multisets, even though, generally speaking, with different semantics.

Let r, r_1, and r_2 denote relations. Let ω denote an (either unary or binary) generic (polymorphic) algebraic operator. Then, it is worthwhile finding out which operators meet the following equivalences:

$$\begin{aligned} \mathfrak{M}(\omega(r)) &\equiv \omega(\mathfrak{M}(r)) \\ \mathfrak{M}(r_1 \, \omega \, r_2) &\equiv \mathfrak{M}(r_1) \, \omega \, \mathfrak{M}(r_2) \end{aligned} \tag{44}$$

If Equivalence (44) holds for an operator ω, it means that considering the operand as either a relation or a multiset is the same, that is, it leads to the same result.

You can verify that all of the unary operators but projection meet the first equivalence in (44), that is:

$$(\forall \omega \in \{\sigma, \rho\}) \, (\mathfrak{M}(\omega(r)) \equiv \omega(\mathfrak{M}(r))) . \tag{45}$$

The discrepancy of projection stems from the possible duplicate generation in multisets, which are instead removed in relations. Likewise, all of the binary operators but union meet the second equivalence in (44), that is:

$$(\forall \omega \in \{\times, \bowtie, \bowtie_\wp, \cap, -\}) \, (\mathfrak{M}(r_1 \, \omega \, r_2) \equiv \mathfrak{M}(r_1) \, \omega \, \mathfrak{M}(r_2)) . \tag{46}$$

To understand why union does not fulfill the equivalence, let us assume the case in which r_1 and r_2 contain the same tuple t. If so, the union of relations, $r_1 \cup r_2$, will contain just one instance of t, as the duplicate tuple is removed from the result. By contrast, the union of the corresponding multiset mirrors will embody two occurrences of t.

3.3   Algebraic Laws for Multisets

An algebraic law is an equivalence between two expressions of relational algebra. Each expression involves a number of variables denoting relations. The equivalence establishes that no matter what relations we replace for these variables, the two expressions give rise to the same result. However, an equivalence law which is valid for relations may happen not to hold when variables are interpreted as multisets.

A list of equivalence laws which are both valid for relations and multisets is given below, where x, y, and z denote collections (either relations or multisets) variables, A a subset of the attributes of the relevant projection operand, and $\wp_i$ a generic predicate on the tuples of the selection operand:

$$(x \cup y) \cup z \equiv x \cup (y \cup z) \tag{47}$$

$$(x \cap y) \cap z \equiv x \cap (y \cap z) \tag{48}$$

$$(x \bowtie y) \bowtie z \equiv x \bowtie (y \bowtie z) \tag{49}$$

$$x \cup y \equiv y \cup x \tag{50}$$

$$x \cap y \equiv y \cap x \tag{51}$$

$$x \bowtie y \equiv y \bowtie x \tag{52}$$

$$\pi_A(x \cup y) \equiv \pi_A(x) \cup \pi_A(y) \tag{53}$$

$$x \cup (y \cap z) \equiv (x \cup y) \cap (x \cup z) \tag{54}$$

$$\sigma_{\wp_1 \wedge \wp_2}(x) \equiv \sigma_{\wp_1}(x) \cap \sigma_{\wp_2}(x) \tag{55}$$

By contrast, the following algebraic laws hold for relations but not for multisets:

$$(x \cap y) - z \equiv x \cap (y - z) \tag{56}$$

$$x \cap (y \cup z) \equiv (x \cap y) \cup (x \cap z) \tag{57}$$

$$\sigma_{\wp_1 \vee \wp_2}(x) \equiv \sigma_{\wp_1}(x) \cup \sigma_{\wp_2}(x) \tag{58}$$

3.4   SQL and Multisets

Relational algebra provides a concise notation for defining queries. However, commercial database systems usually adopt a query language that is more user-friendly and even more powerful than relational algebra. The most commonly used concrete query language is SQL [13, 19, 37]. As a matter of fact, SQL has established itself as the standard relational-database language, which is not only a query language but also incorporates several other constructs, such as updating the database and defining security constraints.

Several different dialects (versions) of SQL exist. On the one hand, there are standards, among which are ANSI SQL and an updated standard called SQL2. There is also an emerging standard called SQL3 that extends SQL2 with several new capabilities, including recursion, triggers, and objects. On the other hand, there are many versions of SQL produced by the principal vendors of database management systems, which usually conform completely to ANSI SQL, but also, to a large extent, to SQL2. To a lesser extent, some of them, include some of the advanced capabilities of SQL3.

The data model of SQL is relational in nature, as well as the relevant operations. However, unlike relational algebra, the tables manipulated by SQL are not relations, but, rather, multisets. The reason for this peculiarity is twofold. First, as mentioned in Sect. 2.1, this is due to a practical reason: since SQL tables may be very large, duplicate elimination might become a bottleneck for the computation of the query result. Second, SQL extends the set of query operators by means of aggregate functions, whose operands are in general required to be multisets of values, as illustrated in the sequel.

SQL query paradigm. In its simplest form, the typical SQL query is expressed by the so-called *select-from-where* paradigm, generically defined as follows:

$$
\begin{aligned}
&\texttt{SELECT } A_1, \dots, A_k \\
&\texttt{FROM } M_1, \dots, M_n \\
&\texttt{WHERE } \wp
\end{aligned} \tag{59}
$$

where each M_i, $i \in [1..n]$, is a table (multiset of tuples) name, each A_j, $j \in [1..k]$, is an attribute name, and $\wp$ a predicate. In terms of algebraic operators for multisets, the above query can be interpreted as follows:

$$
\pi_{A_1,\dots,A_k}(\sigma_\wp(M_1 \times \cdots \times M_n)) \ . \tag{60}
$$

Therefore, the evaluation of the select-from-where statement is based on the following steps:

1. The tables listed in the FROM clause are combined through the product;
2. The corresponding result is filtered by the selection predicate $\wp$ specified in the WHERE clause;
3. The filtered tuples are projected on the list of attributes specified in the SELECT clause.

Example 11. With reference to the multiset mirrors of relations Courses and Teachers shown in Tables 1 and 2, respectively, we may display the name and the relevant department of those courses whose teachers are older than 40 as follows:

```
SELECT course, department
FROM Courses, Teachers
WHERE teacher = name AND age > 40
```

which yields the result outlined in Table 13. Incidentally, no duplicate tuples are generated. □

Table 13. Result of SQL query of Example 11

course	department
Calculus	Mathematics
Compilers	Computer science
Geometry	Mathematics
Operating systems	Computer science
Programming languages	Computer science
Robotics	Electronics
Software Engineering	Computer science

Example 12. With reference to the multiset mirrors of relations `Courses`, `Teachers`, and `Priorities`, displayed in Tables 1, 2, and 3, respectively, we may find out the courses which depend on some prerequisite course whose teacher is in the `Computer science` department as follows:

```
SELECT course
FROM Priorities, Courses, Teachers
WHERE prerequisite = course AND
      teacher = name AND department = 'Computer science'
```

which displays the result shown in Table 14. Note that, in contrast with the SQL query of Example 11, Table 14 is a multiset where the course `Software engineering` is replicated.   □

Duplicate removal in SQL. In order for a result of a select-from-where statement not to include duplicate tuples, we are required to follow the keyword `SELECT` by the keyword `DISTINCT`, which tells SQL to produce only one occurrence of any tuple. Consequently, the result is guaranteed to be duplicate-free. In terms of the algebraic operators for multisets defined in Sect. 3.1, the set-oriented select-from-where paradigm

$$\begin{aligned} &\texttt{SELECT DISTINCT } A_1, \dots, A_k \\ &\texttt{FROM } M_1, \dots, M_n \\ &\texttt{WHERE } \wp \end{aligned} \tag{61}$$

Table 14. Result of SQL query of Example 12

course
Compilers
Operating Systems
Robotics
Software engineering
Software engineering

can be interpreted by prefixing Expression (60) with the duplicate removal operator as follows:

$$\delta(\pi_{A_1,\dots,A_k}(\sigma_\wp(M_1 \times \cdots \times M_n)))\ . \tag{62}$$

In order to provide SQL queries with a set-oriented semantics, one might be tempted to place DISTINCT after every SELECT, on the assumption that it is harmless. On the contrary, it might be very expensive to perform duplicate removal on a table. Generally speaking, the table must be sorted so that the different occurrences of the same tuple appear next to each other. In fact, only by grouping the tuples in this way can we determine whether or not a given tuple should be eliminated. It is very likely that the time needed to sort the table is a good deal greater than the time it takes to execute the query itself. Consequently, the keyword DISTINCT should be used carefully.

Set-theoretic operations in SQL. SQL provides the usual set-theoretic operators, which can be applied to the results of queries, provided these queries produce tables with the same schema. Union, intersection, and difference operators are denoted in SQL by the keywords UNION, INTERSECT, and EXCEPT, respectively. However, unlike the SELECT statement, which preserves duplicates as a default and only eliminates them when instructed to by the DISTINCT keyword, the set-theoretic operators eliminate duplicates by default. We may change such a default by following UNION, INTERSECT, or EXCEPT, by the keyword ALL, thereby enabling the multiset semantics of such operators, as detailed in Sect. 3.1.

Example 13. We may rephrase in SQL Expression (16) of Example 7, which retrieves the courses which either depend on some prerequisite courses or are themselves prerequisites for other courses, as follows (the keyword AS allows us to change the name prerequisite into course, that is, to implement the renaming):

```
(SELECT course FROM Priorities)
    UNION
(SELECT prerequisite AS course FROM Priorities)
```

obtaining the same result displayed in Table 8. If, instead we write

```
(SELECT course FROM Priorities)
    UNION ALL
(SELECT prerequisite AS course FROM Priorities)
```

our result will preserve the duplicate tuples, as outlined in Table 15.

Compared with Table 8, the result in Table 13 includes duplicates for courses Robotics, Software engineering, Programming languages, Computers, and Geometry. However, apart from duplicates, Table 13 contains the same information of Table 8, even if with some redundancy. By contrast, it is worthwhile

Table 15. Result of multiset interpretation of Expression (16)

course
Compilers
Computers
Geometry
Operating Systems
Programming languages
Robotics
Robotics
Software engineering
Software engineering
Software engineering
Calculus
Algebra
Programming languages
Computers
Geometry
Programming languages
Programming languages
Databases

highlighting that the multiset counterpart of Expression (18) is inconsistent with the given query, that asks for the courses which require some other courses but are not prerequisite of any course:

```
(SELECT course FROM Priorities)
     EXCEPT ALL
(SELECT prerequisite AS course FROM Priorities)
```

Based on the multiset version of the difference operation defined in Sect. 3.1, the above SQL query gives rise to the result displayed in Table 16. Comparing Table 16 with Table 10, we note two discrepancies. First, in Table 16, `Robotics` is replicated, due to the projection within the first operand of the `EXCEPT` operator onto `course`. This is however only a redundancy. Second, Table 16 embodies the course `Software engineering`, which is instead not included in Table 10. Such a 'spurious' tuple stems from the semantics of difference operation for multisets,

Table 16. Result of multiset interpretation of Expression (18)

course
Compilers
Operating Systems
Robotics
Robotics
Software engineering

in which, according to Formula (39), the number of occurrences of each tuple t = "Software engineering" in the result is

$$\text{Occ}(t, m_1 - m_2) = \max((\text{Occ}(t, m_1) - \text{Occ}(t, m_2)), 0) = \max((2 - 1), 0) = 1 \ . \tag{63}$$

However, we cannot obtain the required result by simply removing the ALL qualifier from EXCEPT, thereby writing the following SQL query:

```
(SELECT course FROM Priorities)
    EXCEPT
(SELECT prerequisite AS course FROM Priorities)
```

In fact, as duplicate removal is activated after the difference operation, the above statement will simply remove the duplicate course Robotics, preserving however the spurious tuple of Software engineering. To implement the correct query, we should also eliminate duplicates from the first operand as follows:

```
(SELECT DISTINCT course FROM Priorities)
    EXCEPT
(SELECT prerequisite AS course FROM Priorities)
```

which yields the expected result displayed in Table 10.   □

Aggregate operations in SQL. Among other capabilities, SQL provides *aggregate functions*. An aggregate function takes a collection of values as input, which is in general a multiset, and returns a single value. Let a be the multiset of values of an attribute A of a table m, and $\mathcal{F}$ an aggregate function. Then,

$$\mathcal{F}(a) \tag{64}$$

is expected to return a scalar value based on a. SQL provides the following aggregate functions:

1. SUM, the sum of the values of a;
2. AVG, the average of values of a;
3. MIN, the least value in a;
4. MAX, the greatest value in a;
5. COUNT, the number of values in a.

Virtually, the computation of $\mathcal{F}(a)$ is performed by projecting a table m on an attribute A and by applying $\mathcal{F}$ to the resulting multiset a of A-values. Accordingly, from the syntactical point of view, in SQL the argument of the aggregate function is the identifier of an attribute A belonging to a table listed in the FROM clause.

Example 14. With reference to the multiset mirror of relation Teachers shown in Table 2, we may compute the average age of the teachers as follows:

```
SELECT AVG(age)
FROM Teachers
```

which gives rise to the value 45. Instead, if the average function would be applied to a set of values, we would obtain a result which is slightly greater than the actual average, owing to the elimination of one of the two occurrences of value 42, which is associated with both `Angelique` and `Martin`.   □

Not all of the aggregate functions are sensitive to duplicate elimination of elements within the operand. Specifically, both the least and the greatest values within a collection a are preserved after duplication removal. We can express this property by means of the δ operator introduced in Sect. 3.1 as follows:

$$(\forall \mathcal{F} \in \{\texttt{MIN}, \texttt{MAX}\}) \, (\mathcal{F}(a) \equiv \mathcal{F}(\delta(a))) \; . \tag{65}$$

Example 15. To determine the age of the youngest teacher in Table 2, we may rephrase the relational-algebra sequence of statements (21) of Example 8 by means of the following SQL query:

```
SELECT MIN(age)
FROM Teachers
```

which is expected to yield the value 36, relevant to `Patricia`. Note that the SQL version of the query is much simpler than its relational-algebra counterpart[3]. According to Equivalence (65), the same result would be obtained by prefixing age with the keyword DISTINCT.   □

By contrast, the other SQL aggregate functions return in general different results, depending on whether duplicates are removed or not, that is:

$$(\forall \mathcal{F} \in \{\texttt{SUM}, \texttt{AVG}, \texttt{COUNT}\}) \, (\mathcal{F}(a) \not\equiv \mathcal{F}(\delta(a))) \; . \tag{66}$$

To force duplicate removal, the attribute identifier must be preceded by the keyword DISTINCT.

Example 16. With reference to the multiset mirror of relation `Teachers` shown in Table 2, we may compute the number of departments as follows:

```
SELECT COUNT(DISTINCT department)
FROM Teachers
```

which produces the value 3. Note that the qualification DISTINCT is essential to tell SQL to discard duplicates before computing the number of departments.   □

---

[3] Among the SQL aggregate functions, only MIN and MAX can be implemented in relational algebra. Thus, the expressive power of SQL is greater than the expressive power of relational algebra.

Table 17. Result of SQL query of Example 17

department	numc
Computer science	5
Mathematics	3
Electronics	2

SQL allows aggregate functions to be applied to parts of a table, which are collected based on a grouping criterion, expressed by means of the GROUP BY clause. More precisely, the multiset of tuples of a table m is partitioned based on a set G of attributes, thereby obtaining a partition of m where each part is composed of all the tuples which share the same values for attributes in G. The aggregate function will be applied to each of these parts and, as a result, a set of values, one for each part, will be computed.

Example 17. With reference to the multiset mirrors of relations Courses and Teachers shown in Table 1 and Table 2, respectively, we may compute the number of courses relevant to each department as follows:

```
SELECT department, COUNT(course) AS numc
FROM Courses, Teachers
WHERE teacher = name
GROUP BY department
```

which yields the result displayed in Table 17. The special clause GROUP BY tells SQL to make a partition of the join of Courses and Teachers before applying the COUNT aggregate function to each part, that is, to each set of tuples relevant to each department. Accordingly, the GROUP BY clause is applied after the WHERE clause and before COUNT.   □

Even in the case of grouping, SQL aggregate functions are applied, by default, to multisets of values. The qualifier DISTINCT is used to force duplicate removal as usual.

Subqueries and test for absence of duplicates. SQL allows the WHERE clause to include SQL subqueries, whose result may represent the operand of a Boolean function included in the relevant predicate. For example, we may test for the membership of a value within a collection resulting from an SQL subquery by means of the IN Boolean function.

Example 18. Consider the multiset mirror of relation priorities in Table 3 and the problem of determining the courses which have some prerequisites but are not a prerequisite for any courses, which was already formulated in relational-algebra Expression (18). We may rephrase such a query in SQL either using the EXCEPT operator or by means of a selection in which the WHERE clause involves an SQL subquery as follows:

```
SELECT course
FROM Courses
WHERE course IN (SELECT course
                 FROM Priorities) AND
        course NOT IN (SELECT prerequisite
                       FROM priorities)
```

Each tuple in `Courses` is selected if and only if the relevant `course` appears within the set of courses in `Priorities` but not in the set of prerequisite courses. □

Example 19. To show how SQL allows us to test the absence of duplicates in a subquery, consider the multiset mirror of relation `Courses` displayed in Table 1. We may find out the courses who are taught by a teacher who teaches just that course as follows:

```
SELECT course
FROM Courses
WHERE UNIQUE (SELECT course
             FROM Courses AS C
             WHERE C.teacher = Courses.teacher
```

Note the use of the qualifier `AS` in the `FROM` clause, that renames `Courses` to C within the scope of the subquery. Besides, the pathname notation allows us to access tuples either in `Courses` or C. Thus, for each tuple of `Courses`, the predicate of the `WHERE` clause is evaluated by selecting on a copy of `Courses`, named C, the tuples matching the teacher name. Specifically, the result of each subquery returns the multiset of courses which are taught by the current teacher in the external query. Based on the `UNIQUE` Boolean function, the teacher is selected if and only if no duplicates are incorporated in the subquery, that is, if and only if the teacher teaches just one course, as required. Note that, if we substitute `UNIQUE` by `NOT UNIQUE`, we will retrieve the courses which are taught by a teacher who teaches several courses, that is, the complement of the previous query (the new predicate is in fact the logical negation of the previous one). □

Database modifications in SQL. Besides query-oriented language capabilities, SQL provides a means of changing the state of the database, specifically, by inserting tuples into a relation, by deleting certain tuples from a relation, or by updating the values of certain attributes in certain tuples.

Example 20. With reference to the multiset mirrors of relation `Priorities` shown in Table 3, we may insert the prerequisite `Algebra` for `Robotics` as follows:

```
INSERT INTO Priorities
VALUES ('Robotics', 'Algebra')
```

Conversely, we may remove all the prerequisites for `Robotics` as follows:

```
DELETE FROM Priorities
WHERE course = 'Robotics'
```

Finally, with reference to Table 1, we may move the courses of `Angelique` to one year upward as follows:

```
UPDATE Courses
SET year = year +1
WHERE teacher = 'Angelique'
```

$\square$

There are some remarkable points about how insertions and deletions interact with the duplicates included in SQL tables. On the one hand, an insertion adds the specified tuple irrespective of whether it is already included in the table. On the other, the deletion statement, which appears to describe a single tuple to be deleted, in fact removes in general several tuples, because there is no way in SQL to remove one occurrence only from a collection of identical tuples. This characteristics is bound to surprising effects: given a table m containing a tuple t, if we followed the insertion of another occurrence of t into m by the deletion of t, both occurrences of t would be removed from m. In other words, the insertion of t followed by the removal of t would leave the database in a different state from what it was before the two operations.

4   Nested Relational Model

The success of relational database systems was mainly due to the fact that they provide a satisfactory response to the typical needs of business applications, for which the idea of databases as large collections of persistent data to be handled in an effective, efficient, and reliable way was conceived.

The most successful features of the relational model are the ease of use of its query language, which is set-oriented in nature, compared with the procedural, navigational style of earlier proposals, together with the conceptual simplicity of the data model. In fact, the relational model is actually based on a single data structure, the relation. Each relation contains its own data, and connections between data of different relations are implicitly represented by means of equality of values. For this reason, as already remarked, the relational model is often qualified as value-oriented.

Business applications usually involve large amounts of data with a relatively simple structure. The relational model provides an effective and implementation-independent way of specifying this structure while allowing at the same time flexible and sophisticated querying capabilities through set-oriented operations that act on whole relations rather than a single tuple at a time. For all these reasons the relational model proved itself satisfactory with respect to the requirements of business applications, improving notably the productivity of software development in this area.

This success stimulated the adoption of the database technology in areas different from business applications, such as computer-aided design, computer-aided software engineering, knowledge representation, office systems, and multimedia systems. New applications, however, highlighted a number of shortcomings inherent to the relational database technology, among which are the following:

1. The involved data have a complex structure that cannot be expressed in a natural way in the relational model;
2. The relationships among data that derive from their semantics are very complex and cannot be efficiently stored in a value-oriented way;
3. Relational languages lack expressive power for most applications outside the business area.

The first step that was devised in the direction of widening the range of applicability of database systems was to extend the relational data model. This idea can quite naturally be understood starting from the consideration that two out of the above three limitations of relational systems arise from the simplicity of their data model. It looked reasonable to extend the data model, without losing its positive features, in order to explicitly represent data structures more complex than flat tuples of values. This would have also solved the second problem, that is, efficiently storing related data. As to the solution of the third problem, relational languages should have been extended in order to cope with more complex data structures, while retaining their set-oriented, declarative style: this extension, in the original idea, should have supplied the lacking expressive power.

Perhaps, the most noteworthy extension of the relational model was the *nested relational model*, according to which the assumption of atomic attributes (flat relations) is relaxed. Such an assumption, which excludes the possibility that an attribute value be a collection of other values, is called *First Normal Form* (1NF).

The standard relational model that derived implicitly from this assumption is therefore a *flat* relational model. Then, a relational database schema consists of relation schemas of the form:

$$R(A_1 : D_1, \dots, A_n : D_n) \tag{67}$$

where each D_i is an atomic domain. The easiest way of defining a data model that allows for the representation of complex data structures is the direct extnsion of the relational model obtained by relaxing the 1NF assumption. A *nested relation* is defined in terms of (possibly complex) attributes. A *complex* attribute is in turn a (possibly nested) relation. Nested relations can be manipulated by means of special purpose languages [30, 6, 22, 1, 18, 8, 3, 43, 14, 10], among which are various extensions of relational algebra. In the *nested* data model, also called *Non First Normal Form* ($\neg$ 1NF), attribute values can be nested relations themselves, with unbounded depth. A nested relation schema R could be expressed as follows:

$$R(A_1 : T_1, \dots, A_n : T_n) \tag{68}$$

where each T_j is either an atomic domain D or a nested relation schema of the form:

$$(A_{j1} : T_{j1}, \ldots, A_{jn_j} : T_{jn_j}) \ . \tag{69}$$

In the following, when irrelevant, atomic domains will be omitted.

As the notation suggests, the notion of a nested relation is the natural extension of the notion of a flat relation. A nested relation is a set of nested tuples in just the same way as a flat relation is a collection of flat tuples. A nested tuple associates a value from the corresponding domain with each attribute in the schema, as in the flat case. A flat tuple is a particular case of nested tuple, in which all the attributes are associated with an atomic domain. The value associated with an attribute is atomic if the attribute is simple, otherwise it is a nested relation. In the latter case the attribute is complex.

Even if the definition of nested relations is recursive, the schema is supposed to have a finite depth, because at each level every complex attribute must correspond to a new relation schema: it cannot be associated with a relation schema of the upper levels. Cyclical paths in the schemas are not allowed: the schema of a nested relation can therefore be conveniently represented by a tree, where the root is the name of the external nested relation, simple attributes are represented as leaves, and complex attributes correspond to internal nodes.

Nested relations and complex attributes have exactly the same properties as far as their structure is involved: we make use of the same definition for both of them. However, a nested relation is an instance consisting in a single set of nested tuples. By contrast, there are as many instances of complex attributes as tuples in its parent relation. Therefore, if the (external) nested relation instance consists of n tuples, a complex attribute corresponds to n sets of tuples. This asymmetry notably complicates the query language, even if it allows the expression of most queries more concisely.

Another consequence of this extension is that it gives a more complex semantics to elementary operations on tuples and attributes. For example, in the flat case, comparing two attribute values involves only a comparison of two atomic values, while in the nested model it requires a more complex comparison between sets when the attributes are complex.

Moreover, new operators are required for attribute comparison, such as in selection predicates. Besides the classical comparison operators, such as $=$, $>$, and $\geq$, it is necessary to introduce a number of relational operators for sets, such as $\supset$ (inclusion). Other set-oriented operations like $\cup$ (union) and $\cap$ (intersection) need also be included in the language, not only for relations, as in the flat case, but also for attributes.

Formally, an *extended relational database schema* S is a collection of rules. Each rule has the form

$$R = (R_1, \ldots, R_n) \tag{70}$$

where $R, R_1, \ldots, R_n$, which are called *names*, are distinct and there is no ordering on $R_1, \ldots, R_n$. The names on the right-hand side of the rule R form a set, which is denoted by E_R. Each rule has a different name on the left-hand side.

A name is a *higher-order* name if it occurs on the left-hand side of some rule; otherwise, it is *zero order*, or *attribute name*. Higher order names correspond to nested relations or attributes with a nested structure, while zero order names are ordinary, atomic attributes. Rules in the database schema associate each name with its structure. Since the structure is expected to have a finite depth, the structure of rules cannot be cyclic: this requirement will be specified below.

A name is *external* if it occurs only on the left-hand side of some rule; otherwise, it is *internal*. External names correspond to complex relations of the database, while internal names are attributes, either simple or complex.

Given an external name R in an extended relational database schema S, consider the smallest subset S' of S including:

1. The rule with R on the left-hand side;
2. For each higher-order name R_k on the right-hand side of some rule in S', the rule with R_k on the left-hand side.

S' is called the (nested) *relation schema* corresponding to R. The set S' corresponds to the rules in S that are accessible from R. We will normally identify a relation schema by its external name R rather than by listing explicitly the set S' of rules.

Given a relation schema R, we can define a unique *schema tree* of R, written G_R. Thus the schema tree of a relation schema S' with external name R is a tree rooted in R. The internal nodes of the tree will be the left-hand sides of other rules in S', and the leaves of the tree will be zero-order objects in the rules of S', representing basic attributes. Each non-leaf node of a relation schema tree represents a collection of tuples, each composed of the children of the node. The nodes of G_R are exactly the names in the rule R. G_R contains a directed edge from R to R' if and only if $R' \in E_R$.

An extended relational database schema S consists of one or several relation schemas, possibly sharing some attributes. A unique *schema graph* of S, written G_S, can be defined, which is a graph resulting from the merging of the schema trees of all the relation schemas included in S.

An extended relational database schema S is *valid* if and only if G_S is a directed acyclic graph: this implies that the hierarchical structure of a nested relation has an unlimited but finite depth.

Two valid nested relation database schemas are *equivalent* if their schema graphs are isomorphic, in other words, if the graphs are equal up to renaming of internal (non-leaf) nodes.

Example 21. Considering the relations Courses, Teachers, and Priorities displayed in Tables 1, 2, and 3, respectively, we may define an extended database schema S as follows:

$$
\begin{aligned}
\text{Departments} &= (\text{department}, \text{teachers}) \\
\text{teachers} &= (\text{name}, \text{age}, \text{courses}) \\
\text{courses} &= (\text{course}, \text{year}) \\
\text{Priors} &= (\text{course}, \text{prerequisites}) \\
\text{prerequisites} &= (\text{prerequisite})
\end{aligned}
\tag{71}
$$

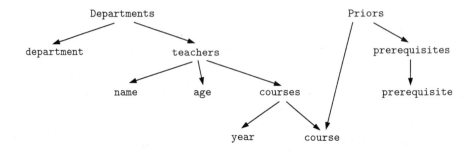

Fig. 2. Schema graph G_S of the extended database schema defined by Rules (71)

According to our definitions, this schema embodies two external names only, namely Departments and Priors. Higher-order names include Departments, teachers, courses, Priors, and prerequisites. Zero-order names are name, age, course, department, year, and prerequisite. The relevant schema graph G_S, which happens to be valid, is shown in Fig. 2.   □

Having defined an extended relational database schema, we now turn to the problem of defining an instance of a complex relation. We want to define instances so that they are independent of the order of columns. We have simply to associate with each atomic attribute a single value taken from an appropriate domain, and to each higher-order name a set of tuples of the appropriate type. Since the ordering of names is immaterial, each value must be labeled by its name. Like for the flat relational model, we assume the existence of a function Dom that associates with each zero-order name its domain.

An instance of a name R, denoted by r, is an ordered pair $\langle R, V_R \rangle$, where V_R is a *value* for name R. If R is a zero order name, a value is an element of $\text{Dom}(R)$. If R is a higher-order name, a value is a set $\{t\}$ of tuples t, where t contains a component $\langle R_i, V_{R_i} \rangle$ for each $R_i \in E_R$.

Note that two different complex relations can share part of the schema, that is, they can have some inner attribute with the same name and schema; however, they cannot share part of the instance: every relation is completely independent of any other relation. That is, nested tuples cannot share common sub-objects, and every update on a tuple or relation is confined to the tuple or relation.

The schema and instance associated with the same external name R form a *structure*. A structure is therefore a pair (R, r), where R is an external name and r an instance of R. A *database structure* (S, s) is a database schema together with an instance for its external names.

Example 22. With reference to Example 21, instances of the nested relations Departments and Priors are shown in Tables 18 and 19, respectively.

The instance of the nested relation Departments is composed of three tuples, each of which refers to a specific department. Associated with each department is a set of teachers, each of which is characterized by a name, an age, and a set of courses. A course is described by a name and the year in which it is taught.

Table 18. Instance of the nested relation `Departments`

department	name	age	course	year
			teachers()	
			courses()	
Computer science	Angelique	42	Operating systems	4
			Software engineering	4
	Carol	38	Databases	3
	Jerri	54	Compilers	5
			Programming languages	3
Electronics	Patricia	36	Computers	2
	Richard	45	Robotics	5
Mathematics	Gregory	63	Calculus	1
	Martin	42	Geometry	2
	Sheila	40	Algebra	1

As such, the instance of `Departments` incorporates in a hierarchical fashion all the information contained in Tables 1 and 2. Similar considerations apply to Table 19, which represents the instance of the nested relation `Priors`. In fact, each course is associated with all its preceding courses by means of the complex attribute `prerequisites`. □

4.1 Nested Relational Algebra

A meaningful amount of work on the extended relational model was devoted to the definition of query languages. Almost all paradigms of languages for the relational model have been extended to the nested case. The advantage of the extended relational model is here particularly evident: in principle it is not necessary to design new query languages, all is needed is the freedom to apply old constructs of relational query languages to more complex data structures. More formally, we need a language design that is fully *orthogonal*: since relations can now occur not only at the outermost level as external operands but even as complex attributes, the same operators should be applicable at the attribute level as well.

Due to the increased complexity of the data model, the algebraic approach has become more popular for the nested model than it was for the classical flat one, imposing itself as the predominant stream of research on languages for nested relations. An operational approach (as opposed to the declarative approach of the calculus-based and rule-based languages) is more appropriate, for

Table 19. Instance of the nested relation `Priors`

course	prerequisites() prerequisite
Compilers	Software engineering
Computers	Calculus
Geometry	Algebra
Operating systems	Programming languages
Programming languages	Computers
Robotics	Geometry Programming languages
Software engineering	Programming languages Databases

example, for specifying the schema-restructuring operations that are particularly relevant in the extended relational model.

This is accomplished through the introduction of two new algebraic operators, namely *nest* and *unnest*, defined later in this section, that allow creation and deletion of complex attributes starting from atomic domains.

Actually, the first proposals inherent to nested algebra involved these operators only, and did not discuss the extension of algebraic operators to the attribute level [22, 30, 42]. The idea was that, whenever relation-valued attributes are to be manipulated, one could first unnest, apply the standard relational operators and finally re-nest to obtain the desired result. However it was soon noticed that this cannot work in general, since unnest may not be reversible by nesting operations; moreover, this is neither an efficient nor a natural way of computing the results.

Subsequent research on algebraic languages for nested relations was then focused on investigating the expressive power of the algebra extended with the nest/unnest operators and on designing languages that were suited for efficient implementation and that allow for the manipulation of complex attributes without unnesting them first.

Algebras for manipulating complex attributes have been proposed by many authors [2, 17, 21, 29, 39, 40, 41]. In these algebras, ordinary algebraic operators are extended to accept nested relations as their operands. Certain algebras define set operators (union, difference, etc.) that apply recursively to all complex attributes of their operands.

The approach we follow in this section is to proceed from the more immediate and straightforward extensions to the more complex and specific ones. Specifically, we introduce the following class of extensions to relational algebra:

1. Set-theoretic operations and product extended to nested operands;
2. Nest and unnest;
3. Operators involving a predicate;
4. Extended projection;
5. Extended selection;
6. Expressions involving nested applications of operators.

Extension to set-theoretic operations and product. The basic operators on sets, namely union, difference, and intersection, are defined exactly as for flat relations: the only difference is that domains of attributes may now be either atomic or set-valued. The set operations are always performed at the most external level, that is, on tuples of external names, in other words, we cannot perform set operations on tuples of nested attributes.

Set-theoretic operations usually require operands to have the same domain. However, we can relax this requirement on relation schemas: the two schemas may also be equivalent (that is, isomorphic up to a renaming), if we give some rule to determine the names for the attributes of the result. To this end, we can introduce the following rule: when two schemas are equivalent but do not have the same attribute names, the result inherits the names of the first schema. Alternatively, an explicit renaming operator can be introduced as in the case of relational algebra.

Also the extension of the product is straightforward, since it only involves the extension of the schema at the external level.

Nest and unnest. These two operators produce a result obtained as a modification of both schema and instance of the operand. Informally, *nest*, denoted by ν, builds a higher-order attribute within an external relation starting from one or more atomic attributes, thus creating a further level of nesting. On the other hand, *unnest*, denoted by μ, deletes a higher-order attribute. When nesting, a set is created containing all the tuples of nested attributes having the identical values on the non-nested attributes. For unnesting, each tuple of the unnested attribute is concatenated with the external tuple containing the unnested attribute, thus resulting in a sort of tuple-oriented product. As remarked in [30], nest and unnest are actually the inverse of each other. However, while an unnest can always restore the situation previous to a nesting, the inverse is not in general true.

Unnest. The definition of μ can be formalized as follows. Given a database schema S, let r be a relation with schema R in S. Assume B is some higher-order name in E_R with an associated rule $B = (B_1, \ldots, B_m)$. Let $\{C_1, \ldots, C_k\} = E_R - B$. Then

$$\mu_B(r) \tag{72}$$

is a relation r' with schema R' where:

Table 20. Result of the unnesting in Expression (73)

department	name	age	teachers() course	year
Computer science	Angelique	42	Operating systems	4
	Angelique	42	Software engineering	4
	Carol	38	Databases	3
	Jerri	54	Compilers	5
	Jerri	54	Programming languages	3
Electronics	Patricia	36	Computers	2
	Richard	45	Robotics	5
Mathematics	Gregory	63	Calculus	1
	Martin	42	Geometry	2
	Sheila	40	Algebra	1

1. $R' = (C_1, \ldots, C_k, B_1, \ldots, B_m)$ and the rule $B = (B_1, \ldots, B_m)$ is removed from the set of rules in S if it does not appear in any other relation schema;
2. $r' = \{t \mid u \in r, t_{(C_1, \ldots, C_k)} = u_{(C_1, \ldots, C_k)}, t_{(B_1, \ldots, B_m)} \in u_{(B)}\}$.

Example 23. With reference to the nested relation Departments displayed in Table 18, the unnesting of collection courses:

$$\mu_{\text{courses}}(\text{Departments}) \tag{73}$$

is expected to yield the nested relation whose schema and instance are shown in Table 20. Owing to the unnest of courses, the result incorporates one tuple of teachers for each department.   □

Nest. The definition of ν can be formalized as follows. Given a database schema S, let r be a relation with schema R in S. Let $\{B_1, \ldots, B_m\} \subset E_R$ and $\{C_1, \ldots, C_k\}$ $= E_R - \{B_1, \ldots, B_m\}$. Assume that B does not occur in the left-hand side of any rule in S. Then

$$\nu_{B=(B_1, \ldots, B_m)}(r) \tag{74}$$

is a relation r' with schema R' where:

1. $R' = (C_1, \ldots, C_k, B)$, where rule $B = (B_1, \ldots, B_m)$ is appended to the set of rules in S;
2. $r' = \{t \mid u \in r, t_{(C_1, \ldots, C_k)} = u_{(C_1, \ldots, C_k)}, t_{(B)} = \{v_{(B_1, \ldots, B_m)} \mid v \in r, v_{(C_1, \ldots, C_k)} = t_{(C_1, \ldots, C_k)}\}\}$.

Example 24. To show the use of the nest operator, we show how we can generate the nested relation `Priors` displayed in Table 19 starting from the relation `Priorities` outlined in Table 3 as follows:

$$\text{Priors} \leftarrow \nu_{\text{prerequisites}=(\text{prerequisite})}\text{Priorities} . \tag{75}$$

In the result, prerequisite courses relevant to the same course are grouped into the complex attribute `prerequisites` specified in the ν operator.   □

Operators requiring a predicate. This class of operators includes selection and join. Predicates are more difficult to define in the extended relational model, due to the possibility of different nesting depths for the attributes. The problem can be illustrated as follows. Assume we have the following database schema: $R = (A, M)$, $M = (B, C)$, $V = (D, N)$, $N = (E, F)$, and consider the following operation:

$$R \bowtie_{C=E} V . \tag{76}$$

It is not clear at which level the tuples of the product of R and V should be selected. Specifically, should all the combinations of tuples in the inner collections be verified? Or the selection is taking place at the outermost level, selecting those tuples whose inner collections agree on the value of attributes C and E for all their tuples?

This semantical ambiguity (arising only in specific comparisons among attributes) depends on the (so called) *quantification level* of the attributes involved in the predicate. The general solution to this problem will be presented later. At the moment, we confine ourselves to extending the form of predicates in order to account for set-valued nested attributes. Thus, in this section, we consider selections and joins whose predicate contains attributes belonging only to the outermost level, that is, to the schema rule of the external name involved in the operation.

Extended propositional formula. Let R be a relation schema. A *propositional formula* $\wp$ over E_R is defined recursively as follows. *Atoms* over E_R have the form $A_1 \theta A_2$ or $A_1 \theta a$, where both A_1 and A_2 are in E_R, a is a constant, which can be set-valued, and θ is a comparison operator, that is, $\theta \in \{=, <, >, \neq, \geq, \leq, \supset, \supseteq, \subset, \subseteq, \in\}$. Every atom over E_R is a propositional formula over E_R; if $\wp_1, \wp_2$ are propositional formulas over E_R, then $\neg(\wp_1)$, $\wp_1 \wedge \wp_2$, and $\wp_1 \vee \wp_2$ are formulas over E_R. Parentheses can be used as usual. Nothing else is a formula. A propositional formula associates a Boolean value with each tuple in the instance r of R.

Selection. Given a relation r over the schema R, the *selection* of r with respect to $\wp$, denoted by $\sigma_\wp(r)$, is a relation over the same schema R, containing the tuples of r that make $\wp$ true:

$$\sigma_\wp(r) \stackrel{\text{def}}{=} \{t \in r \mid \wp(t)\} . \tag{77}$$

The only changes introduced with respect to relational algebra are set comparison.

Example 25. With reference to the nested relation `Priors` displayed in Table 19, to find out the courses which require `Programming languages` as a prerequisite we write:

$$\sigma_{\text{`Programming languages'}\, \in\, \text{prerequisites}}(\texttt{Priors}) \,. \tag{78}$$

where '`Programming languages`' is a constant. This query will select the fourth, sixth , and seventh tuple of `Priors`.   □

Join. The theta-join between nested relations can be defined based on Equivalence (10), that is, as a selection of a product:

$$r_1 \bowtie_{\wp} r_2 \equiv \sigma_{\wp}(r_1 \times r_2) \,. \tag{79}$$

where r_1 and r_2 are nested relations and $\wp$ an extended propositional formula, as defined above. The extension of the natural join to nested relations comes without any surprise, once considered that the implicit predicate of equalities between homonymous attributes may in general involve complex attributes.

Extended projection. The operators introduced so far, with the exception of nest and unnest, are a direct extension of relational algebra, but their expressive power is inadequate, since they lack the capability of accessing inner collections. In order to perform even simple queries, a lot of nesting and unnesting is therefore needed.

Example 26. Considering `Departments` in Table 18, to select the departments which refer to (at least) a course of the last two years, we cannot access directly attribute `year` in the selection predicate since it belongs to the inner relation `courses`. Thus we need to unnest in cascade `courses` and `teachers` so as to apply the selection on the resulting flat relation:

$$\pi_{\text{department}}(\sigma_{\text{year} \geq 4}(\mu_{\text{teachers}}(\mu_{\text{courses}}(\texttt{Departments})))) \tag{80}$$

□

The solution to this problem is to further extend relational operators by allowing the manipulation of inner collections. We first introduce projection so as to make it possible to project also on inner collections. Then we extend set operations so that they can perform union, difference, and intersection of inner attribute

Given an external name R, consider the set S' of rules corresponding to its relation schema. The projection $\pi_{A_1,\dots,A_n}(R)$ defines the projection list $A_1,\dots,A_n$: each name A_i must occur in the right-hand side of just one rule in S'. This is called the uniqueness constraint. The result of the projection has a relation schema S'' obtained by replacing the rules of S' with their projection on the projection list as follows:

1. Include in S'' the rules of S' that contain in their right-hand side one or several names of the projection list, limiting the right-hand side to names of the projection list;

2. Include in S'' the rules of S' that have on the left-hand side a name appearing in the rules of S'' that does not already occur on the left-hand side of a rule in S'';
3. Include in S'' the rules of S' that contain in their right-hand side a name appearing in S'', limiting the right-hand side to names appearing in S'';
4. Apply Steps 2 and 3 until no more rules are added to S''.

The instance r' of the result is obtained from the instance r of R by projecting each tuple on the new relation schema S''

We have a problem with the formalism: the 'projected' rules are valid for the result of the projection, but they are also used in the operand schema, so we should keep also the original, non-projected version. The simplest solution is to assume that, as in the flat relational model, each algebraic operation builds a temporary (nested) relation, that can be used as the operand of another operation or stored as a persistent relation. In both cases, the schema of the relation is defined by new rules, that are added and not substituted to the original version. Names should be changed accordingly, so as to respect the uniqueness constraint.

Example 27. With reference to **Departments** outlined in Table 18, the following projection:

$$\pi_{\text{department,name,courses}} (\textbf{Departments}) \tag{81}$$

will result in the nested relation shown in Table 21. □

Extended selection. In our previous version of the selection operator of nested relational algebra, selection was only possible at the outermost level, the level of tuples of the external relation. Informally, in order to be able to use inner attributes in the predicate, we need a way to decide:

1. On which collection the selection is performed;
2. Which values should be compared in a predicate atom.

We first formalize the concept of a *quantification level*, informally introduced above: constants and external names have quantification level 0 (that is, there is only one instance for them in the database); the quantification level of each name occurring in the right-hand side of a rule, for which the left-hand side name has quantification level i, is $i + 1$.

We now extend the notion of an atom in a selection predicate. Let R be a relation schema; atoms over E_R have the form $A_1 \theta A_2$ or $A_1 \theta a$, where a is a constant, θ is a comparison operator, $\theta \in \{=, <, >, \neq, \geq, \leq, \supset, \supseteq, \subset, \subseteq, \in\}$, and A_1, A_2 are names occurring in the relation schema such that:

(α) The quantification level of $A_1 \theta a$ is that of the parent of A_1 and the atom refers to its tuples;
(β) If A_1 and A_2 are siblings (they have a common parent) in G_R (the schema tree corresponding to R), then the quantification level of $A_1 \theta A_2$ is that of the common parent, and the atom refers to its tuples;

Table 21. Result of the extended projection (81)

department	teachers()		
	name	courses()	
		course	year
Computer science	Angelique	Operating systems	4
		Software engineering	4
	Carol	Databases	3
	Jerri	Compilers	5
		Programming languages	3
Electronics	Patricia	Computers	2
	Richard	Robotics	5
Mathematics	Gregory	Calculus	1
	Martin	Geometry	2
	Sheila	Algebra	1

(γ) If a sibling of A_1 is an ancestor of A_2 (or vice versa) in G_R, then the quantification level of $A_1 \theta A_2$ is that of the parent of A_2 (respectively A_1) and the atom refers to its tuples.

According to the above axioms, external constants and external names are considered as siblings of any external name (the quantification level is 0).

Propositional formulas of predicates are built by means of atoms. For each atom involved in a propositional formula, we consider the tuples to which the atom refers. Operands of binary logical connectives must be atoms or other formulas whose corresponding tuples still obey to the above axioms. The logical connective will refer to tuples of the lower quantification level, unless they belong to the same parent collection.

Therefore, we have given axioms to recursively determine the quantification levels and tuples to which the predicate refers; the selection is performed only on these tuples.

Example 28. Based on the above axioms, the following extended selection:

$$\sigma_{\text{department}=\text{'Computer science'} \land \text{age}>40}(\textbf{Departments}) \tag{82}$$

will (somewhat surprisingly) result in the nested relation displayed in Table 22. With reference to Axioms (α), (β), and (γ), the quantification level of the selection predicate $\wp$ within Expression (82) is computed as described in Table 23.

Table 22. Result of the extended selection (82)

department	teachers()			
	name	age	courses()	
			course	year
Computer science	Angelique	42	Operating systems	4
			Software engineering	4
	Jerri	54	Compilers	5
			Programming languages	3
Electronics				
Mathematics				

Table 23. Steps to yield the quantification level of the selection predicate $\wp$ in (82)

Predicate	Description	Axiom	Quantification level
$\wp_1$	department = 'Computer science'	(α)	Departments
$\wp_2$	age > 40	(α)	teachers
$\wp$	$\wp_1 \wedge \wp_2$	(γ)	teachers

Accordingly, tuples of `teachers` are selected based on the value of predicate $\wp$. Within Table 22, the nested attribute `teachers` is empty for both departments `Electronics` and `Mathematics`, because every tuple of them does not satisfy predicate $\wp_1$ and, consequently, $\wp$. Besides, removing a tuple from `teachers` causes the removal of all its attributes, specifically, `courses`.   □

Note that, generally speaking, when the quantification level of a selection is internal, that is, the selection is applied to an internal relation R (in Example 28, $R = $ `teachers`), duplicates may be generated among the tuples R belongs to (in Example 28, the tuples of `Departments`). In fact, in this case, the selection operates as a modification of attribute values, which is bound to generate duplicates.

Expressions involving nested applications of operators. Since operands are nested relations, it is worth nesting the operators. For example, the selection predicate could consist of a comparison between the results of the selections on two inner attributes. This way, relational operators can be applied to inner attributes as well, thereby extending the expressiveness of the algebra.

Nested expressions are used also in languages for the flat relational model: for example, SQL allows nested queries, that are generally used as an alternative to joins between relations. When the structure of relations is more complicated, the need for nested expressions becomes a natural requirement.

Nested expressions in the context of extended algebras were first studied in [29] (where they are called *recursive* expressions) and [41].

Essentially, nested algebraic operations can be defined on the basis of the following principles:

1. A relational expression can occur wherever a relation name is needed (this was implicitly assumed in some of our previous examples).
2. A relational expression can occur wherever an attribute name is expected.

Example 29. To find out the departments that include (at least) a teacher younger than 40 we may write:

$$\pi_{\text{department}}\left(\sigma_{(\sigma_{\text{age}<40}(\text{teachers}))\neq\emptyset}(\text{Departments})\right) \tag{83}$$

In Expression (83), a selection is applied to Departments, whose predicate involves a complex comparison between a nested selection on teachers and the empty set. Intuitively, each tuple of Departments is selected if and only if the selection of the corresponding attribute teachers based on the simple comparison age < 40 yields at least a tuple, in other words, when the department includes a teacher younger than 40. This holds for Computer science and Electronics only, which are in fact the departments displayed after the final projection.   □

Example 30. To find out the departments that include (at least) a teacher older than 50 and teaching a course of the last two years we may write:

$$\pi_{\text{department}}\left(\sigma_{\left(\sigma_{\text{age}>50\wedge(\sigma_{\text{year}\geq4}(\text{courses})\neq\emptyset)}(\text{teachers})\right)\neq\emptyset}(\text{Departments})\right) \tag{84}$$

In Expression (84), a selection is applied to Departments, whose predicate involves a complex comparison between a nested selection on teachers and the empty set. Besides, the nested selection on teachers involves a complex comparison between a nested selection on courses and the empty set. The latter selection is based on the simple comparison year ≥ 4. This way, the selection operation is propagated through the nested schema of the operand Departments. The final result is expected to include the singleton {(Computer science)}.   □

5   Multisets in Nested Relational Database Systems

In Sect. 3, we showed how the relational model may be extended to deal with multisets. In particular, in Sect. 3.1, the operators of relational algebra have been extended to manipulate multisets. Then, in Sect. 4 we showed how the relational model can be extended to incorporate nested relations and how these nested relations can be manipulated by means of a nested relational algebra.

It is natural at this point to combine the two extensions, namely multisets and nested relations, into a new data model in which multisets can be nested within *complex relations*. Intuitively, a complex relation is like a nested relation incorporating both sets and multisets[4].

We may easily extend the formal definition of a nested relational database schema given in Sect. 4 to a complex relational database schema by making rules polymorphic, that is, a rule R may be either

$$R = (R_1, R_2, \ldots, R_n) \tag{85}$$

or

$$R = [R_1, R_2, \ldots, R_n] \tag{86}$$

where, intuitively, Rule (85) defines a set, while Rule (86) a multiset.

Example 31. Shown in Table 24 is a complex relation called `Products`, whose schema is defined by the following rules:

$$\begin{aligned} \texttt{Products} \ &= \ (\,\texttt{product}, \texttt{components}, \texttt{time}\,) \\ \texttt{components} \ &= \ [\,\texttt{component}\,] \end{aligned} \tag{87}$$

Each `product` is associated with a multiset `components`[5] and a `time` needed to assemble the product by means of the collection of components. Thus, according to the simplified information stored in `Products`, a `car` is composed of a `body`, an `engine`, four occurrences of `wheel` , two occurrences of `light`, and a `tank`. Then, each component is itself a product, possibly composed of other components, and so on, until reaching atomic components, such as `light` or `tank`.   □

5.1   Nested Relational Algebra for Complex Relations

Intuitively, we may define a nested relational algebra for complex relations by combining the concepts inherent to the relational algebra for multisets introduced in Sect. 3.1, and the relational algebra for nested relations presented in Sect. 4.1. In fact, there is almost nothing surprising in integrating the two approaches into a uniform operational framework. Among other papers, the book [10] presents an extensive description of a nested relational algebra for complex relations. However, in this section, we only focus upon a limited variety of computational aspects relevant to the manipulation of complex relations.

Type transformation. Generally speaking, a complex relation is polymorphic in nature, as it may be either a set or a multiset and include both set and multiset attributes. For example, the complex relation `Products` shown in Table 24 is a

---

[4] Complex relations may include other kinds of collections, like *lists* and *arrays*. However, according to the scope of this paper, we confine our discussion to multisets only.

[5] Within the schema, the name of a multiset, such as `components`, is followed by square brackets to distinguish it from a set.

Table 24. Instance of the complex relation Products

product	components[] component	time
car	body engine wheel wheel wheel wheel light light tank	32
body		0
engine	block block carburetor	18
block	cylinder piston valve	26
wheel	tyre tube	4
light		0
tank		0
carburettor		0
cylinder		0
piston		0
valve		0
tyre		0
tube		0

set, but the internal collections relevant to attribute components are multisets. That is, the schema of Products involves both set and multiset constructors. Since these two collection types may coexist within the same complex relation, it makes sense to define an operator which allows us to change the set into a multiset and vice versa[6].

---

[6] Such a change, however, must always be interpreted within a functional framework, where operands never change their state but, rather, a new complex relation is generated with the changed schema.

Let r be a complex relation, and $\ell = A_1, \ldots, A_n$ a (possibly empty) set of attribute names of r. Then, the following expression:

$$\tau_\ell(r) \tag{88}$$

denotes the type transformation of r, which results in a complex relation r' such that:

1. If $\ell = \emptyset$, then the type of r' is changed with respect to the type of r;
2. If $\ell \neq \emptyset$, then the types of attributes of r' in ℓ are changed with respect to the corresponding types in r;
3. The instance of r' is obtained from the instance of r in accordance with the type transformations performed in Points 1 and 2.

Example 32. With reference to Table 24, to generate a new complex relation obtained by transforming `Products` into a multiset we may write:

$$\tau(\texttt{Products}) \ . \tag{89}$$

Instead, to change the nested multiset `components` into a set we write:

$$\tau_{\texttt{components}}(\texttt{Products}) \ . \tag{90}$$

Consequently, the duplicates originally contained in multiset `components` (see Table 24) will be removed from the result. □

Type transformations are not only useful but even necessary to specify queries appropriately. In set-theoretic operations, since we require the equality of the operand schemas, a renaming might be in general not sufficient if one operand is a set and the other a multiset. Moreover, we require as well that product and join operate on complex relations of the same type. Thus, type transformation might be needed in this case too[7].

Nest and unnest. Nest and unnest operations were introduced in Sect. 4.1 for nested relations. Instead of extending such operations formally, we prefer providing some intuitive guidelines.

Unnest. The unnest operation creates a complex relation by reducing by one level the depth of the operand schema tree. Given an instance r of a complex relation schema R defined as follows:

$$R\langle A_1, \ldots, A_m, \beta\langle B_1, \ldots, B_n\rangle\rangle \tag{91}$$

---

[7] Another viable approach might be to have implicit casting rules (as in general-purpose programming languages in which, for example, an integer is transformed into a real), where the type of a complex relation is automatically transformed into another type based on the specific computational context.

Table 25. Result of Expression (94)

product	component	time
car	body	32
car	engine	32
car	wheel	32
car	light	32
car	tank	32
engine	block	18
engine	carburetor	18
block	cylinder	26
block	piston	26
block	valve	26
wheel	tyre	4
wheel	tube	4

where angles denote either set or multiset collections, the following operation:

$$\mu_\beta(r) \tag{92}$$

generates a complex relation r' with the following schema R':

$$R'\langle A_1, \ldots, A_m, B_1, \ldots, B_n \rangle \tag{93}$$

Since more than one tuple on $B_1, \ldots, B_n$ exists in R for each tuple $A_1, \ldots, A_m$, r' will be built joining the tuples together. The type of R' is determined from the type of R and not from the type of β. Accordingly, the tuples resulting from the concatenation of the tuples of R and β will be considered as belonging to a complex relation of the type of R. Two cases are possible:

1. R is a set: the multiple occurrences of the tuples on β will be reduced to one;
2. R is a multiset: the multiple occurrences of the tuples of β will generate multiple occurrences in r' too.

Example 33. With reference to the complex relation Products shown in Table 24, the following operation:

$$\mu_{\text{components}}(\text{Products}) \tag{94}$$

will result in the (flat) set displayed in Table 25. Note that the unnesting of empty collections give rise to the removal of the corresponding tuple. □

Nest. Given an instance r of a complex relation schema R defined as follows:

$$R\langle A_1, \ldots, A_m, B_1, \ldots, B_n \rangle \tag{95}$$

where each A_i and B_i can be a complex attribute, the following operation:

$$\nu_{\beta=\langle B_1, \ldots, B_n \rangle}(r) \tag{96}$$

generates a complex relation r' with the following schema:

$$R'\langle A_1, \ldots, A_m, \beta\langle B_1, \ldots, B_n\rangle\rangle \ . \tag{97}$$

The instance of r' is built as follows: tuples having the same projection on attributes $A_1, \ldots, A_m$ are grouped together and for each group we consider the corresponding parts B_i which form a set or a multiset, according to the type of β. Two cases are possible:

1. R is a set: only one tuple is maintained in r', which is obtained through the concatenation of the A_i, $i \in [1 \mathbin{..} m]$, and β;
2. R is a multiset: a tuple of r' is created for each tuple of r; within the tuples of the same group the part β is repeated.

Example 34. Shown in Table 26 are a multiset **Flat** (left) and the result of the following nest operation (right):

$$\nu_{\texttt{components}=(\texttt{component})}(\texttt{Flat}) \ . \tag{98}$$

Expression (98) generates a complex relation of type multiset with a nested set called **components**. Therefore, according to the above rules, for each tuple in **Flat**, a tuple is created in the result, where, within the tuples of the same group, the instance of the newly created attribute is repeated. Note, however, that, since **components** is defined as a set, duplicates are removed from the nested collections. □

Table 26. Multiset **Flat** (left) and result of Expression (98) (right)

product	component
car	wheel
car	wheel
car	engine
engine	block
engine	block
engine	carburetor

product	components() component
car	wheel engine
car	wheel engine
car	wheel engine
engine	block carburetor
engine	block carburetor
engine	block carburetor

Table 27. Result of Expression (101)

component
body
engine
wheel
light
tank
block
carburetor
cylinder
piston
valve
tyre
tube

Complex aggregate operations. We have already introduced the notion of an aggregate operation in Sect. 3.4 within the context of SQL. Considering complex relations, it is possible to introduce new kinds of aggregate operations that are applied to collections of complex (rather than simple) values. Specifically, these are *complex union* and *complex intersection*, denoted by $\cup^c$ and $\cap^c$, respectively.

Let r be a complex relation involving an attribute A, either of type set or multiset[8]. The complex union of A within r is defined as follows:

$$\cup_A^c (r) \stackrel{\text{def}}{=} \cup_{t \in r} (t_{(A)}) \ . \tag{99}$$

Likewise, the complex intersection of A within r is so defined:

$$\cap_A^c (r) \stackrel{\text{def}}{=} \cap_{t \in r} (t_{(A)}) \ . \tag{100}$$

Note that Definitions (99) and (100) can be generalized to deal with an attribute A which is at any level within the schema of r, not necessarily at the level of the tuples of r.

Example 35. With reference to the complex relation **Products** displayed in Table 24, we might ask for the set of products that are components of some other product as follows:

$$\tau(\cup_{\text{components}}^c(\textbf{Products})) \tag{101}$$

which gives rise to the set displayed in Table 27. Since both union and intersection of multisets result in a multiset, the final type transformation is needed in order to remove duplicated components.

Note that, if we replaced in (101) $\cup^c$ with $\cap^c$, we would obtain the empty set. □

---

[8] Attribute A can be itself complex, that is, it may be defined in terms of other complex attributes.

function FixedPoint(r): relation resulting from $\varphi[\Delta \leftarrow \mathcal{F}_t(r); \mathcal{F}_p(\Delta, r); \mathcal{F}_c(\Delta, r)](r)$
  input
    r: the operand relation;
  begin
    $r' := r$;
    stop := **false**;
    repeat
      $\Delta := \mathcal{F}_t(r')$;
      if $\mathcal{F}_p(\Delta, r')$ **then**
        stop := **true**
      else
        $r' := \mathcal{F}_c(\Delta, r')$
    until stop;
    return r'
  end.

Fig. 3. Semantics of the fixedpoint operation defined in Formula (102)

Fixedpoint operation. An interesting extension to the algebra for complex relations is the *fixedpoint* operator, which allows recursive queries to be expressed in an algebraic style without the use of control structures or recursion. Such an operation can be seen as a generalization of the *closure* operation introduced in various advanced query languages.

The fixedpoint operation applies repeatedly a function $\mathcal{F}_t$, called the *transformer*, to a complex relation r, combining the result with r using another function $\mathcal{F}_c$, called the *combiner*, which generates the new relation to be used in place of r in the next iteration. The loop continues until the *predicate* $\mathcal{F}_p$ becomes true.

From the algebraic point of view, the fixedpoint operator, denoted by φ, is a unary operator computing a result which has the same schema of the operand (but, in general, different instance). The generic form of the fixedpoint operation applied to a complex relation r is the following:

$$\varphi \left[\Delta \leftarrow \mathcal{F}_t(r); \; \mathcal{F}_p(\Delta, r); \; \mathcal{F}_c(\Delta, r) \right] (r) \tag{102}$$

where the involved functions $\mathcal{F}_t$, $\mathcal{F}_c$, and $\mathcal{F}_p$ are specified within square brackets, between the operator symbol and the operand. The semantics of the fixedpoint operator can be specified by the pseudo-code of a *FixedPoint* function, which takes as input the operand r and computes the result r', as illustrated in Fig. 3. The meaning is the following: r is the name of the φ operand and r' is a copy of r. First, r' is taken as the argument of the transformation function $\mathcal{F}_t$ (possibly with other arguments) that calculates the partial value Δ. Before going to the combination function, the predicate $\mathcal{F}_p(\Delta, r')$ is calculated. This predicate may possibly contain further arguments beside Δ and r'. If the predicate evaluates to true, the loop is broken and the current instance of r' is the result.

The quantification level of the predicate $\mathcal{F}_p$ must be related with the quantification level of the operand according to the axioms given in Sect. 4.1. If the predicate $\mathcal{F}_p(\Delta, r')$ evaluates to false, the result of the combination function

Table 28. Query result of Example 36

product
cylinder
cylinder
cylinder
cylinder
piston
piston
piston
piston
valve
valve
valve
valve
carburetor
carburetor
tyre
tube

$\mathcal{F}_c(\Delta, r')$, possibly with further arguments, is assigned to r', and the body of the loop is repeated [9]. When the loop terminates, the instance of r' is the returned result.

Example 36. Consider the complex relation **Products** displayed in Table 24. We might be interested in finding out the multiset of atomic components needed to assemble a multiset of given products. For example, to assemble two engines we need four cylinders, four pistons, four valves, and two carburetors.

The multiset containing the products for which we ask atomic components is called **Input**, which is characterized by a single attribute called **product**. For example, **Input** might include two engines and one wheel. Here is our query:

$$\texttt{Input} \leftarrow [(\texttt{engine}), (\texttt{engine}), (\texttt{wheel})]$$

$$\varphi[\ \texttt{Comps} \leftarrow \rho_{\texttt{product}}(\cup^c_{\texttt{components}}(\texttt{Input} \bowtie (\tau(\texttt{Products}))));$$
$$\texttt{Comps} = \emptyset;$$
$$\texttt{Comps} \cup (\pi_{\texttt{product}}(\sigma_{\texttt{components}=\emptyset}(\texttt{Input} \bowtie (\tau(\texttt{Products})))))$$
$$](\texttt{Input})$$

The first statement simply instantiates the multiset **Input**. The actual query is expressed by the next expression, where the fixedpoint operator is applied to **Input**. According to the terminology introduced in Formula (102), the transformer first makes a natural join between **Input** and the multiset mirror of **Products**, then makes the complex union of the relevant components, and finally performs a renaming of the result. This way, **Comps** is expected to contain

---

[9] Note that the function will not terminate if the condition $\mathcal{F}_p(\Delta, r')$ never evaluates to true.

the multiset of components (at the first level) of the products in Input. The predicate tests for the emptiness of Comps. Then, the combiner makes the union of Comps and the atomic products contained in the current instance of Input. The result of φ is displayed in Table 28.   □

6 Related Work

Various database systems and query languages support multisets in the relevant data model [15, 47, 28, 45, 36, 44, 23, 12, 10]. For instance, a query written in DAPLEX [45] supports a flexible control over creation and elimination of duplicates in the query result. Therefore, starting from the 1980s, multisets were proposed as an additional type constructor in order to formally deal with duplicates in databases.

Some research has been devoted to extending the set-theoretic operations to multisets, as well as to developing techniques for algebraic query optimization for data models supporting multisets. The work presented in [20] extends the relational model to include *multiset relations*, i.e. relations with duplicated tuples. A framework for query optimization is also defined. Both the operands and the results of queries are in general multiset relations.

A different approach to formalize multisets in relational databases can be found in [31], wherein, at the conceptual level, a relation with duplicates, called *multirelation*, is conceived as a subset of the columns of a larger relation with no duplicates, by keeping hidden those columns that yield the uniqueness of all the tuples in the relation. Consequently, relations or views with duplicated tuples are not allowed at the conceptual level, and there is no need for the introduction of any explicit multiset semantics.

In [27] it is investigated how the use of multisets in the nested relational algebra extends its expressive power and increases its computational complexity.

In [49] a framework for algebraic query optimization is developed in the context of an object-oriented data model that supports multisets. In the quoted work a number of specific algebraic rules for query transformation are presented. The emphasis is posed on rules for manipulating the instances of types in a super-type/sub-type lattice, with complex type constructors. Union, intersection, and difference are defined for multisets, and some transformation rules for these operators are given.

The work described in [38] deals with recursion and aggregates according to a proposed declarative multiset semantics for deductive databases. Such databases have been enriched with aggregate operators and predicates that may return multisets of tuples. A formal basis for efficient evaluation of queries when multisets are generated as intermediate results in a query evaluation process is provided.

6.1 Multisets and Views

Given a relational database, views are data derived from it that can be materialized (that is, stored permanently in the database) and subsequently queried against. The use of views in order to answer queries is common to different database areas and it is very relevant in data warehouse applications [32]. In fact, in such applications, queries against data typically involve the computation of aggregate operations, which, as known, have to handle duplicated data to be correctly evaluated.

In [46] the problem of transforming an SQL query that includes grouping and aggregation into an equivalent one, that can be answered more efficiently, is considered. Such a rewriting has to fulfill two requirements:

1. The rewritten query has to be posed against available materialized views (conditions under which this is possible are given);
2. The multiset semantics of queries has to be preserved, that is, the rewritten query computes the same multiset of tuples for any given database.

The problem of view maintenance occurs when some of the base relations change and, consequently, materialized views must be recomputed to ensure that answers to queries posed against the views be correct. The recomputation of a view, after a change has occured in base relations, is usually computationally expensive. An alternative approach to complete recomputation consists in determining the changes that must be applied to the view, given the changes to base relations and the expression that define the view. In [25] an algorithm for view maintenance that follows this alternative approach is proposed. Views are regarded as multisets of tuples defined by means of expressions of a multiset algebra. The algorithm for view maintenance gives solutions that satisfy some minimality requirements, and, moreover, that are proven to be correct with respect to the aggregate operations computed over the changed view.

6.2   Multisets in Web Information Discovery

Web data mining [35] is defined as the discovery and analysis of useful knowledge from Web data. An approach to Web data mining is proposed in [7], according to which queries are posed against the Web by specifying an entry point Web address and some conditions about paths and data attributes. Once retrieved, data matching the query are stored in a special relation. Each tuple of such a relation consists of interconnected directed graphs, which represent Web documents and hyperlinks. In order to isolate the columns of interest from a table, a *Web projection* operator can be used, which does not eliminate identical tuples automatically. The result of a Web projection is a *Web bag*, that is a relation that may contain multiple occurences of the same tuple. A few differences between Web bags and multisets in relational databases can be identified:

1. Different nature of data: a Web bag contains unstructured data derived from the Web, while each piece of data in a multiset within a relational database has a known domain;
2. Different purposes: Web bags are useful for discovering knowledge, while multisets in relational databases mainly focus on efficiency and on semantics of aggregate operators;

3. Different origin: Web bags can only be obtained as the result of a projection, while multisets in relational databases can be obtained also as the result of other operations.

The exploitation of Web bags for discovering useful knowledge from Web data is discussed in the same contribution.

6.3   Query Languages for Multisets

In this section we will describe in some detail two meaningful works in the area of database query languages for multisets. The first [5] is a conference paper that historically opened the way to a research stream. The second work [33, 26, 34], carried out over several years, moved many further steps in this stream. In both works the name *bag* is preferred rather than *multiset* and we will adhere to this preference. In each description we will retain the notation used by the authors of the considered works. Different authors adopt different notations, that is, they may use different symbols for denoting the same operator or, vice versa, they may use the same symbol for denoting distinct operators. Thus, when faced with algebraic properties, the reader has to recall the context-specific meaning of the involved operators. In order to help the reader, while providing any operator definition given in a work, we have explicitly stated the correspondence of the current operator with operators defined by other authors and/or introduced in previous sections.

Algebraic properties of bags. Albert [5] defines some bag operations and investigates to what extent the typical algebraic properties of set operations are obeyed when bags are assumed as operands. The proposed semantics of operators agree with [20]. Nested bags are not dealt with. The ultimate goal is to provide a formal basis for database query optimization.

Constants. The notation [*list of elements*] is used for representing a bag, where each element belongs to a countable set of primitive objects. The symbol $\emptyset$ denotes an empty bag.

Membership and multiplicity. The expression $x \in B$ denotes that bag B contains x, and $x \in\in B$ represents the number of copies of x in B. If $x \in B$, then $x \in\in B > 0$, and vice versa. For example, given bag $B = [a, a, b, b, b]$, both $a \in B$ and $b \in B$ hold; in particular, $a \in\in B$ equals 2, and $b \in\in B$ equals 3.

Containment. Given two bags A and B, by definition

1. $A \subseteq B$ means that $(\forall x \in A) \, ((x \in\in A) \leq (x \in\in B))$;
2. $A \subset B$ means that $A \subseteq B$ and $A \neq B$.

In either case, A is a *subbag* of B. A theorem is proven, according to which $\subset$, as defined above, is a partial order relation (i.e. it is an irreflexive transitive relation) on any set of bags.

Powerset. Given a bag A, by definition, the powerset of A is

$$\mathcal{P}(A) \stackrel{\text{def}}{=} \{B \mid B \subseteq A\} \ . \tag{103}$$

Union and intersection. Let A and B be bags. By definition

1. $A \cup B$ is the smallest bag C such that $A \subseteq C$ and $B \subseteq C$;
2. $A \cap B$ is the largest bag C such that $C \subseteq A$ and $C \subseteq B$.

The notion of *largest* and *smallest* used in the above definitions refer to the partial order relation $\subseteq$.

The definitions of $\cup$ and $\cap$ yield the usual set union and intersection when restricted to sets. Besides, these operations, as defined on bags, share the same algebraic properties as the standard set-theoretic ones. This is proven by means of a theorem which states that, given a bag U, $(\mathcal{P}(U), \cup, \cap)$ is a distributive lattice, called *subbag lattice*, induced by the partial order $\subseteq$. This means that, for the binary operations $\cup$ and $\cap$, as applied to any possible subbag of a given (universal) bag U, all the axioms of Boolean algebra wherein complement is not referenced hold, with U corresponding to one and $\emptyset$ to zero.

Note that the union operation, as defined above for bags, is neither that defined in Sect. 3.1 nor that adopted by SQL for multisets. As to the intersection operator, it corresponds to both the intersection as defined in Sect. 3.1 and the SQL operator `INTERSECT ALL` (see Sect. 3.4).

By means of a theorem, it is shown that the above operations are well-defined, that is, for any given pair of bags, there is a unique bag satisfying the definition of union/intersection of such bags. In particular, given three bags A, B, and C, for union it is proven that

1. $x \in\in A \cup B = \max(x \in\in A, x \in\in B)$;
2. $A \subseteq C$ and $B \subseteq C \Rightarrow A \cup B \subseteq C$.

Likewise, for intersection it is proven that

1. $x \in\in A \cap B = \min(x \in\in A, x \in\in B)$;
2. $C \subseteq A$ and $C \subseteq B \Rightarrow C \subseteq A \cap B$.

Concatenation. Given two bags A and B, and the countable set of primitive objects *Obj* that is the domain of any element of A and B, by definition the concatenation of A and B, $A \sqcup B$, is the bag C such that

$$x \in\in C = (x \in\in A) + (x \in\in B) \ . \tag{104}$$

This operator becomes the usual set union when bags are turned into their set counterparts. SQL use the keyword `UNION ALL` for this operator.

Difference. Let A and B be bags, and *Obj* the countable set of primitive objects that is the domain of any element of A and B. By definition, the difference $A \setminus B$ is the bag $C \subseteq A$ such that

$$(\forall x \in Obj) \ (x \in\in C = \max((x \in\in A) - (x \in\in B), 0)) \ . \tag{105}$$

This operator matches the usual notion of set difference when restricted to sets. It corresponds to the difference operator for multisets defined in Sect. 3.1 as well as to the EXCEPT ALL operator of SQL (see Sect. 3.4). However, taking any bag U that is not a set as a universe, this operator, if used as a unary operator, does not induce a complement operator relative to this universe. Operator $\setminus$ would be a unary complement operator if it would satisfy the following two axioms:

$$(\forall A \in \mathcal{P}(U)) \, (A \cup (\setminus A) = U)$$
$$(\forall A \in \mathcal{P}(U)) \, (A \cap (\setminus A) = \emptyset) \; . \tag{106}$$

However, these two axioms does not hold for operator $\setminus$ as defined above. Still more, Albert demonstrates a theorem according to which no such an operator exists and, therefore, looking for a different semantics for bag difference cannot solve the problem.

Given a complement operator defined on a universe and two operators that form a distributive lattice on the same universe, the three operators altogether are a Boolean algebra. Thus $(\mathcal{P}(U), \cup, \cap, \setminus)$ is not a Boolean algebra for bags, whereas it is for sets. Therefore, while all the properties of a Boolean algebra which involve the difference operator hold for sets, there are some of them which fail for bags, as illustrated in the following example, taken from [5].

Example 37. Let's consider the following algebraic transformations inherent to the set-theoretic operations of union, intersection, and difference:

$$A \setminus (B \cup C) = A \cap (\overline{B \cup C}) = A \cap (\bar{B} \cap \bar{C}) = (A \cap \bar{B}) \cap \bar{C} = (A \setminus B) \setminus C \; . \tag{107}$$

The equivalence of set difference and the intersection with the complement has been exploited in the first and last transformations, in the second transformation De Morgan's law has been applied, and, in the next one, the associativity of intersection.

The identity obtained between the initial and final expressions holds if A, B, and C are sets, however, in general, it does not hold for bags. In fact, let's suppose, for instance, that $A = [x, x, x, x]$, $B = [x, x]$, and $C = [x]$. Then $A \setminus (B \cup C) = [x, x]$, while $(A \setminus B) \setminus C = [x]$. $\square$

Duplicate elimination. A bag B is transformed into a set by function δ. By definition, $\delta(B) = \{x \mid x \in B\}$. δ looks like the opposite function of the mirror function $\mathfrak{M}$ defined in Sect. 3.2.

(Boolean) selection. Let A be a bag, Obj the domain of its elements, and ψ a quantifier-free predicate, either atomic or built up from a set of atomic predicates, whose domain includes $\delta(A)$. The selection of A based on ψ selects all and only the elements of A which satisfy the predicate ψ, that is, by definition, $\sigma_\psi(A)$ is the bag C such that

$$(\forall x \in Obj) \left(x \in C = \begin{cases} x \in A & \text{if } \psi(x) \\ 0 & \text{otherwise} \end{cases} \right) \; . \tag{108}$$

This operator is the same as the selection operator for multisets defined in Sect. 3.1, which can be expressed in SQL by means of the select-from-where paradigm (see Sect. 3.4).

A theorem [5] states that, given any predicates ψ and φ whose domains include $\delta(A)$, the following equalities hold:

$$\sigma_{\psi \vee \varphi}(A) = \sigma_\psi(A) \cup \sigma_\varphi(A)$$
$$\sigma_{\psi \wedge \varphi}(A) = \sigma_\psi(A) \cap \sigma_\varphi(A)$$
$$\sigma_{\neg \varphi}(A) = A \setminus \sigma_\varphi(A) \tag{109}$$
$$\delta(\sigma_\varphi(A)) = \sigma_\varphi(\delta(A))$$
$$x \in\in \sigma_\varphi(A) = (x \in\in A) \cdot (x \in\in \sigma_\varphi(\delta(A))) \ .$$

The first three properties above imply that the semantics of $\vee$, $\wedge$, and $\neg$ with respect to Boolean selection correspond to $\cup$, $\cap$, and $\setminus$, respectively.

The next two properties formally state the *all-or-nothing* semantics of the selection operator, that is, either all copies of some $x \in A$ are selected, or none. The fourth property exhibits a slight abuse of notation on the right-hand side since the Boolean selection operator defined over bags is applied to a set. Indeed, Albert makes no clear distinction between a set and a bag containing just an occurrence of everyone of its element.

Albert proves a theorem stating that, with respect to selection operations, bags behave like sets. Formally, let E_1 and E_2 be two compositions of selection operators, that is, $E_1 = \sigma_{\psi_1} \circ \sigma_{\psi_2} \cdots \circ \sigma_{\psi_n}$ and $E_2 = \sigma_{\varphi_1} \circ \sigma_{\varphi_2} \cdots \circ \sigma_{\varphi_k}$. If, for every set A, $E_1(A) = E_2(A)$, then also $E_1(B) = E_2(B)$ for every bag B.

Complement. Let U be a bag and *Pred* the smallest set of predicates that includes a given set of atomic predicates, and is closed under the propositional connectives. Let $S(U)$ be the set of subbags of U obtained by applying any selection operator to U, that is,

$$S(U) = \{A \subseteq U : (\exists \varphi \in Pred)(A = \sigma_\varphi(U))\} \ . \tag{110}$$

The unary complement operator $-$ is defined on $S(U)$. By definition, for any $A \in S(U)$, $-(A) = U \setminus A$. Albert proves that $(S(U), \cup, \cap, -)$ is a Boolean algebra.

Operators and query languages. By means of the following properties, Albert proves that $\sqcup$ and $\setminus$ are sufficient to obtain $\cup$ and $\cap$.

$$A \cup B = (A \setminus B) \sqcup B$$
$$A \cup B = (A \sqcup B) \setminus (A \cap B)$$
$$A \cap B = A \setminus (A \setminus B) \tag{111}$$
$$A \cap B = (A \sqcup B) \setminus (A \cup B) \ .$$

The converse, instead, does not hold, that is, neither $\sqcup$ nor $\setminus$ can be constructed out of the three remaining operations, as stated by means of two theorems.

The implication of these results is that $\sqcup$ and $\setminus$ are strictly more general operations than $\cup$ and $\cap$. This justifies the presence in SQL of UNION ALL and

EXCEPT ALL, corresponding respectively to operators $\sqcup$ and $\backslash$, along with the absence of any operator corresponding to bag union $\uplus$ and intersection $\cap$ (as defined by Albert).

Since a Boolean algebra $(S(U), \cup, \cap, -)$ has been defined on the set of the results of any Boolean selection as applied to some universal bag, the usual algebraic properties for sets hold within this domain. In particular, the algebraic transformations that are applied to sets for query optimization keep being valid when the operands are bags which are formed by applying a Boolean selection to some universal bag.

Properties of query languages for bags. The research described in [33, 26, 34] focuses on how database query languages are affected by the use of duplicates.

Many query languages have traditionally been developed based on relational algebra and relational calculus. Relational algebra is the standard language for sets, and (in its flat form) an algebraization of first-order logic. However, structures such as bags, along with nested structures in general, are not naturally supported by first-order logic-based languages. Thus, in the work at hand an approach to language design was exploited [11] that, instead of using first-order logic as a universal platform, suggests to consider the main type constructors (such as sets, bags, and records) independently and to turn the universal properties of these collections into programming syntax. This approach was adopted for the development of a query language with the aim of investigating the theoretical foundations for querying databases based on bags. The expressive power and complexity of this language, which can be regarded as a rational reconstruction of SQL, and its relationships with both relational algebra and a nested relational algebra [3] have been investigated.

The operations defined in this work are intended for *flat* bags (bags of tuples with attributes of basic types) as well as for *nested* bags (where tuple attribute values can also contain nested bags).

Types and domains. A distinction between types and domains is made. $[T_1, \ldots, T_n]$ is a tuple type, whose domain is the set of tuples over $T_1, \ldots, T_n$, that is, $\text{Dom}([T_1, \ldots, T_n]) = \text{Dom}(T_1) \times \ldots \times \text{Dom}(T_n)$. $\{|T|\}$ is a bag type, whose domain is the set of finite bags of objects of type T.

Operations on records and tuples. The *tupling* function τ transforms a record into a tuple. Formally, $\tau(o_1, \ldots, o_k) = [o_1, \ldots, o_k]$, where $o_1, \ldots, o_k$ are k attributes of type $T_1, \ldots, T_k$, respectively, and $[o_1, \ldots, o_k]$ is a k-ary tuple, of type $[T_1, \ldots, T_k]$, containing o_i ($i = 1 \ldots k$) in its i-th attribute.

The *attribute projection* function α_i returns the i-th attribute of a given tuple. Formally, $\alpha_i([o_1, \ldots, o_k]) = o_i$.

Membership test and multiplicity. An element n-belongs to a bag if it belongs to that bag and has exactly n occurrences. Function *member* of type $T \times \{|T|\} \to Boolean$ returns *true* on a pair (o, B) iff o p-belongs to B, $p > 0$.

In other words, in order to relate [26], [5], and Sect. 3.1, given a bag B and an element b,

1. b n-belongs to $B \Leftrightarrow b \in\in B = n \Leftrightarrow Occ(b, B) = n$;
2. $member(b, B) = true \Leftrightarrow b \in B$.

Constants. $\{|\cdot, \ldots, \cdot|\}$ is a bag; constant $\{|\,|\}$ denotes the empty bag, even if also symbol $\emptyset$ is used very often with the same meaning.

Bagging (or bag singleton). Given an element o, by definition $\beta(o) = \{|o|\}$ is a bag containing o as a single element, i.e. o 1-belongs to $\beta(o)$.

Additive union. Bags are built by using the additive union operation $\uplus$ starting from the empty bag and singleton bags. That is, each bag is either $\{|\,|\}$, or a singleton $\{|x|\}$, or the additive union of two bags $B \uplus B'$. Given two bags B and B' of type $\{|T|\}$, by definition $B \uplus B'$ is a bag of type $\{|T|\}$ such that o n-belongs to $B \uplus B'$ iff o p-belongs to B and q-belongs to B' and $n = p + q$.

This operator is the same as the concatenation operator $\sqcup$ defined by [5], here extended also to nested bags.

Extension. If f is a function of type $T \to \{|T'|\}$, the extension of f extends f to a function of type $\{|T|\} \to \{|T'|\}$ as follows:

$$EXT_f(\{|x_1, \ldots, x_n|\}) \stackrel{\text{def}}{=} f(x_1) \uplus \ldots \uplus f(x_n) . \tag{112}$$

MAP_g is a concise form for $EXT_{\beta \circ g}$, where $\beta \circ g$ is the composition of function β (bagging) and function g, that is, $(\beta \circ g)(d) = \beta(g(d))$.

Cartesian product. If B and B' are bags containing tuples of arity k and k', respectively, by definition $B \times B'$ is a bag containing tuples of arity $k + k'$ such that $o = [a_1, \ldots, a_k, a_{k+1}, \ldots, a_{k+k'}]$ n-belongs to $B \times B'$ iff $o_1 = [a_1, \ldots, a_k]$ p-belongs to B, $o_2 = [a_{k+1}, \ldots, a_{k+k'}]$ q-belongs to B' and $n = p \cdot q$.

The Cartesian product operator is the same as the product for multisets defined in Sect. 3.1, which can be expressed in SQL by means of the select-from-where paradigm. Such an operator was naturally extended to nested multisets in Sect. 4.1.

Subtraction. Given two bags B and B' of type $\{|T|\}$, by definition $B - B'$ is a bag of type $\{|T|\}$ such that o n-belongs to $B - B'$ iff o p-belongs to B and q-belongs to B' and $n = \max(0, p - q)$.

This operator, called also *difference* [34], is the same as the difference operator $\setminus$ defined by [5] and outlined in Sect. 3.1. Such an operator was naturally extended to nested multisets in Sect. 4.1.

Maximal union. Given two bags B and B' of type $\{|T|\}$, by definition $B \cup B'$ is a bag of type $\{|T|\}$ such that o n-belongs to $B \cup B'$ iff o p-belongs to B and q-belongs to B' and $n = \max(p, q)$.

This operator is the same as the union operator $\cup$ defined by [5], here extended also to nested bags.

Intersection. Given two bags B and B' of type $\{|T|\}$, by definition $B \cap B'$ is a bag of type $\{|T|\}$ such that o n-belongs to $B \cap B'$ iff o p-belongs to B and q-belongs to B' and $n = \min(p, q)$.

This operator, also called *minimum intersection* [34], is the same as the intersection operator $\cap$ defined by [5] and outlined in Sect. 3.1. Such an operator was naturally extended to nested multisets in Sect. 4.1.

Duplicate elimination. Given a bag B of type $\{|T|\}$, $\epsilon(B)$ is a bag of type $\{|T|\}$ containing exactly one occurrence of each object of B. Formally, an object o 1-belongs to $\epsilon(B)$ iff o p-belongs to B for some $p > 0$, and 0-belongs to $\epsilon(B)$ otherwise.

Function ϵ, also called *unique* [34], becomes the identity function when it is turned from an operation on bags into its set analog.

Equality test. Function *eq* of type $T \times T \to Boolean$ returns *true* on a pair (o, o') iff o and o' are equal objects.

Subbag test. Function *subbag* of type $\{|T|\} \times \{|T|\} \to Boolean$ returns *true* on a pair (B, B') iff whenever o p-belongs to B, then o p'-belongs to B' for some $p' \geq p$.

In other words, relating [26] and [5], $subbag(B, B') = true \Leftrightarrow B \subseteq B'$. However, here the subbag test is inherent also to nested bags.

Complexity and expressive power. Complexity and expressive power are considered with respect to the language defined by operators τ, α_i, β, $\uplus$, EXT, $\times$, together with the empty bag constant. From the complexity point of view, each of the subtraction, union, intersection, duplicate elimination, subbag, membership, and equality test operators has polynomial time complexity with respect to the size of the input [34]. From the expressive power point of view, the same operators are related as follows [26, 34]:

1. $-$ can express all primitives other than ϵ;
2. ϵ is independent of the rest of the primitives;
3. $\cap$ is equivalent to *subbag* and can express both $\cup$ and *eq*;
4. *member* and *eq* are interdefinable, both are independent of $\cup$, and each of them, together with $\cup$, can express $\cap$.

Thus, the strongest combination of primitives among $-$, $\cup$, $\cap$, ϵ, *subbag*, *member* and *eq*, is $-$ and ϵ.

Bag query languages. Operators τ, α_i, β, $\uplus$, EXT, $\times$, $-$ and ϵ, together with the empty bag constant, are the primitives of *BALG (standard bag algebra)* [26], also referred to as *BQL (bag query language)* in [34].

BALG can express many operations commonly found in database languages. For instance, $MAP_{\lambda x.[\alpha_2(x), \alpha_3(x)]}$ denotes the projection of a tuple type (the type of the elements of the bag which is the operand) on its second and third arguments. Following [26], for the sake of brevity, the map projecting the attributes $i_1, \ldots, i_n$ will below be denoted by $\pi_{i_1, \ldots, i_n}$.

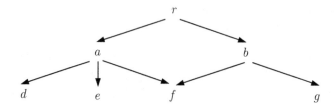

Fig. 4. The directed graph of Example 39

Likewise, $MAP_{\lambda x.\alpha_i(x)=a}$ denotes the selection within a bag of tuples of all and only the tuples wherein the i-th argument equals a. A shorthand for this query is $\sigma_{i=a}$.

Assumed to represent an integer i by means of a bag containing i occurrences of the same element (and nothing else), BALG allows the definition of several fundamental database primitives. For example, bags can be used to simulate SQL aggregate functions, such as SUM and COUNT. In fact, if B is a bag of tuples, $COUNT(B) = \pi_1(\{|a|\} \times B)$.

Example 38. A relation can be represented in BALG by means of a bag with no duplicates, such as $R = \{|7, 5, 3, 9, 4, 2|\}$. The *parity* of the cardinality of a relation becomes definable in BALG in the presence of an *order* on the domain. The following Boolean expression is *true* iff the parity of the cardinality of relation R is even:

$$\sigma_{\lambda x.\left(MAP_{[a]}(\sigma_{\lambda y.(y \le x)}R)=(MAP_{[a]}(\sigma_{\lambda y.(x<y)}R))\right)}(R) \neq \{||\} \ . \tag{113}$$

In fact, the inequality holds only if there exists an x such that the number of elements smaller than or equal to x equals the number of elements strictly bigger than x. For instance, the above query can check the parity of the sample relation R by exploiting the order of integers. Based on Table 29, it emerges that the result of the left-hand side is $\{|4|\}$, thus the query gives *true*.

This example is quite simple as it involves a relation among integers. However, the same considerations hold also for relations among tuples, provided that there is at least one attribute whose domain is ordered. □

Example 39. Consider a directed graph whose edges are recorded in a binary relation G, such as the one displayed in Table 30 inherent to the graph of Fig. 4.

Formally, in BALG a graph is an object of type $\{|t \times t|\}$, where t is a base type with a countably infinite domain of uninterpreted constants. For instance, G is represented by the bag of tuples $\{|[r, a], [r, b], [a, d], [a, e], [a, f], [b, f], [b, g]|\}$.

Note that the domain of a graph is unordered, since the domain of t is unordered, that is, only equality test is available for constants of type t.

The following is a Boolean query on the graph, that is, a query defined on $\{|t \times t|\} \rightarrow Boolean$,

$$(\pi_2(\sigma_{2=a}(G))) - (\pi_1(\sigma_{1=a}(G))) \neq \{||\} \ . \tag{114}$$

Table 29. Parity of the cardinality of an ordered relation represented as a bag.

x	$\sigma_{\lambda y.(y \leq x)} R$	$MAP_{[a]}(\sigma_{\lambda y.(y \leq x)} R)$	$\sigma_{\lambda y.(x < y)} R$	$MAP_{[a]}(\sigma_{\lambda y.(x < y)} R)$
7	$\{\|7,5,3,4,2\|\}$	$\{\|[a],[a],[a],[a],[a]\|\}$	$\{\|9\|\}$	$\{\|[a]\|\}$
5	$\{\|5,3,4,2\|\}$	$\{\|[a],[a],[a],[a]\|\}$	$\{\|7,9\|\}$	$\{\|[a],[a]\|\}$
3	$\{\|3,2\|\}$	$\{\|[a],[a]\|\}$	$\{\|7,5,9,4\|\}$	$\{\|[a],[a],[a],[a]\|\}$
9	$\{\|7,5,3,9,4,2\|\}$	$\{\|[a],[a],[a],[a],[a],[a]\|\}$	$\{\|\|\}$	$\{\|\|\}$
4	$\{\|3,4,2\|\}$	$\{\|[a],[a],[a]\|\}$	$\{\|7,5,9\|\}$	$\{\|[a],[a],[a]\|\}$
2	$\{\|2\|\}$	$\{\|[a]\|\}$	$\{\|7,5,3,9,4\|\}$	$\{\|[a],[a],[a],[a],[a]\|\}$

It gives *true* iff the in–degree of a node a is bigger than its out–degree.

In our sample case, the following equalities hold: $\pi_2(\sigma_{2=a}(G)) = \pi_2(\{\|[r,a]\|\})$ $= \{\|[a]\|\}$, and $\pi_1(\sigma_{1=a}(G)) = \pi_1(\{\|[a,e],[a,f]\|\}) = \{\|[a],[a]\|\}$. Thus, Equation (114) becomes $\{\|[a]\|\} - \{\|[a],[a]\|\} \neq \{\|\|\}$, which gives *false*.   □

BALG vs. relational algebra: complexity. As stated in [4], relational algebra enjoys an AC0 data complexity upper-bound and so does nested relational algebra [48]. AC0 [24] offers potential for efficient parallel evaluation. However, BALG is not in AC0: its authors state that BALG is contained in LOGSPACE.

BALG vs. relational algebra: expressiveness. A comparison of the expressive power of BALG to that of relational algebra is provided in [26]. Since BALG considers complex (nested) objects as well as flat relations, the relationship between BALG, when applied to complex objects, and nested relational algebra is also investigated.

Let *arithmetic* include the type *nat* of natural numbers, together with the operations of addition, multiplication, modified subtraction $\dot{-}$ (i.e. $n \dot{-} m = \max(0, n - m)$), and summation Σ_f. Here f is of type $T \to nat$, and Σ_f of type $\{T\} \to nat$ with the semantics $\Sigma_f(\{x_1, \ldots, x_k\}) = f(x_1) + \ldots + f(x_k)$.

A theorem states that BALG when restricted to flat bags is equivalent to relational algebra + *arithmetic*, and BALG over nested bags is equivalent to nested relational algebra + *arithmetic*. The comparison between BALG and nested relational algebra is restricted just to nested bags since it was shown [50] that nested relational algebra has no extra power with respect to relational algebra over flat relations in the sense that any nested relational algebra query from flat relations to flat relations can be defined in relational algebra.

Basically, BALG is more powerful than relational algebra and nested relational algebra since bags give a counting power. For instance, the result presented in Example 39 is quite important, as, while BALG can test the parity of relations over ordered domains, while relational algebra cannot test parity on any (ordered or unordered) domain.

However, some fundamental limitations on the expressive power of first-order logic, which are typical of relational algebra and nested relational algebra, still hold in BALG. For instance, consider the query *bag-even*, which tests the parity of the number of duplicates in a bag. Formally, given a bag B of type $\{|T|\}$ and an element c belonging to $\text{Dom}(T)$, *bag-even(B,c)* = *true* if c's multiplicity in B

Table 30. The binary relation G representing the directed graph of Fig. 4.

Arrow tail node	Arrow head node
r	a
r	b
a	d
a	e
a	f
b	f
b	g

is even, and *false* otherwise. A proposition, which is claimed to be valid for both flat and nested bags, is stated in [26], according to which the query *bag-even* is not expressible in BALG. More generally, it is stated that a property of the number of duplicates of a single constant can be tested in BALG only if such a property (or its complement) holds for a finite number of bags.

Besides, a theorem states that, for any two graphs G and G', each of which consists of a number of disconnected simple cycles having the same length, there exists a k such that, if the lengths of the cycles of both graphs are longer than k, then, for any Boolean query q, $q(G) = q(G')$. This means that the two above defined graphs cannot be distinguished based on the answers to BALG queries. From this, [26] concludes that many recursive queries are not definable in BALG over unordered domains: parity of the cardinality of a relation, transitive closure, deterministic transitive closure, testing acyclicity, testing connectivity, testing for balanced binary trees. A proof of the undefinability in BALG of parity test, transitive closure, and test for balanced binary trees is provided in [34].

Several attempts to extend the power of BALG to support the above undefinable queries have to be registered. In this respect, three further operators, introduced in [33] and listed below, have to be considered.

Powerbag. Given a bag B, the *powerbag* operator returns the bag of all subbags of B. For instance,

$$powerbag(\{|1, 1, 2|\}) = \{| \ \{||\}, \{|1|\}, \{|1|\}, \{|2|\},$$
$$\{|1, 1|\}, \{|1, 2|\}, \{|1, 2|\}, \{|1, 1, 2|\} \ \} \ . \quad (115)$$

Powerset. When applied to a bag B, the *powerset* operator returns the bag of all subbags of B, each with multiplicity 1. For instance,

$$powerset(\{|1, 1, 2|\}) = \{|\{||\}, \{|1|\}, \{|2|\}, \{|1, 1|\}, \{|1, 2|\}, \{|1, 1, 2|\}\}\} \ . \quad (116)$$

This operator is the same as the homonymous operator defined in [5]; however, here it is extended to nested bags.

Loop. Construct *loop* takes as input a function f of type $T \to T$, an object o of type T, and a bag $B = \{|o_1, \dots, o_n|\}$ of any type $\{|T'|\}$, and returns f applied n times in cascade to o (where n is the cardinality of B).

Power and complexity of extended BALG. It was shown in [33] that *powerbag* and *powerset* are interdefinable. However, even if all the queries which are not definable in BALG over unordered domains become definable in BALG+*powerset*, they are not definable efficiently. Out of efficiency BALG+*powerbag* has to be preferred. The complexity of BALG+*powerset* and BALG+*powerbag* was studied in [27].

The *loop* operator was introduced as an alternative to the structural recursion operator. The main problem with using structural recursion is that it requires some preconditions on its parameters for well-definedness. However, such preconditions are generally undecidable [9]. In [33] it was shown that nested BALG with structural recursion is equivalent to nested BALG with *loop*.

7 Conclusion

This paper has shown how the notion of a multiset affects the data model of database systems and the relevant query languages. Since most current database systems are relational, the focus is on the relational data model and on relational algebra.

The relational data model is based on the central concept of a relation. Intuitively, a relation is a set of tuples, where each tuple is, in turn, a collection of atomic values, one for each attribute of the relation. A database consists of one or several relations. Once a relational database has been created, it is possible to retrieve and modify the stored information by means of appropriate query languages. Among formal query languages for relational databases is relational algebra, which deals with the manipulation of sets. This means that each query result is expected not to include duplicates.

Even if, from the theoretical point of view of relational algebra, the result of a query is a set, from the practical point of view it is sometimes convenient to have multisets as results of queries, in particular if the information they include is intended to be processed by aggregate functions. Thus, practical query languages for relational databases, typically based on the standard query language SQL, allow the manipulation of relations in a multiset-oriented way. Based on the above considerations, relational algebra operators have been straightforwardly extended in order to manage multisets, thus obtaining some relational algebras for multisets.

Historically, the awareness has gradually grown that the relational data model is too simple to express in an intuitive way complex data, as those typically encountered in non business applications, and relationships among such data. Thus, several extensions to the relational model have been proposed as well as a number of extended relational algebras. The most noteworthy extension of the relational model, called nested relational model, was the transformation of usual (flat) relations into nested relations, where each attribute value is not necessarily atomic, rather, it can be a (nested) relation itself. The algebraic counterpart of the nested relational model is a nested relational algebra, which, although manipulating only sets, both extends relational algebra operators to nested relations and introduces new operators.

Roughly, two research streams have been focused on the extension of relational algebra. On the one hand, flat relations have been extended to flat multisets. On the other, flat relations have been extended to nested relations. The two approaches have been integrated within the notion of a complex relation, that is, a multiset of tuples where each attribute is possibly a (complex) relation. A nested relational algebra for complex relations has been defined by combining the concepts inherent to the relational algebra for multisets and the nested relational algebra.

Moreover, the paper hints at some research topics inherent to multisets in database systems. In the last decade, there has been some activity in trying to identify a standard formal query language for multisets, which may play a role similar to that of relational algebra. The first requirement of such a language is to support aggregate functions, and the second is to deal with nested data. Processing and optimization of queries in the proposed languages is an active research area. However, investigating the expressive power of multiset languages involves many difficulties. In fact, in order to cope with aggregate functions, multiset languages support built-in arithmetic and, therefore, it is hard to find a logic that captures them and that can be exploited for proving results about the expressive power. As to nested multiset languages, they involve calculi that are essentially higher-order logics, whose expressibility properties are rather unknown.

References

1. S. Abiteboul and N. Bidoit. Non-first-normal form relations to represent hierarchically organized data. In *Third ACM SIGMOD SIGACT Symposium on Principles of Database Systems*, 1984.
2. S. Abiteboul and N. Bidoit. Nonfirst normal form relations: An algebra allowing data restructuring. *Journal of Comp. and System Sc.*, 33(1):361–393, 1986.
3. S. Abiteboul, P.C. Fisher, and H.J. Schek, editors. *Nested Relations and Complex Objects in Databases.* Number 361 in LNCS. Springer-Verlag, Berlin, Germany, 1989.
4. S. Abiteboul, R. Hull, and V. Vianu. *Foundations of Databases.* Addison-Wesley, Reading, MA, 1994.
5. J. Albert. Algebraic properties of bag data types. In G. M. Lohman, A. Sernadas, and R. Camps, editors, *Seventeenth International Conference on Very Large Data Bases*, pages 211–219, Barcelona, Spain, 1991.
6. H. Arisawa, K. Moriya, and T. Miura. Operation and the properties on non-first-normal-form relational databases. In *Ninth International Conference on Very Large data Bases*, pages 197–204, Florence, Italy, 1983.
7. S. S. Bhowmick, S. K. Madria, W. K. Ng, and E.P. Lim. Web bags – are they useful in a web warehouse? In *Fifth International Conference on Foundations of Data Organization*, Kobe, Japan, 1998.
8. N. Bidoit. The Verso algebra or how to answer queries with fewer joins. *Journal of Computer and System Sciences*, 35(3):321–364, 1987.

9. V. Breazu-Tannen and R. Subrahmanyam. Logical and computational aspects of programming with sets/bags/lists. In *Eighteenth International Colloquium on Automata, Languages, and Programming*, number 510 in LNCS, Madrid, Spain, 1991.

10. F. Cacace and G. Lamperti. *Advanced Relational Programming*, volume 371 of *Mathematics and Its Applications*. Kluwer Academic Publisher, Dordrecht, The Netherlands, 1997.

11. L. Cardelli. Types for data-oriented languages. In J. W. Schmidt, S. Ceri, and M. Missikoff, editors, *International Conference on Extending Database Technology*, number 303 in LNCS, pages 1–15, Venice, Italy, 1988.

12. M. J. Carey, D. J. DeWitt, and S. L. Vandenberg. A data model and query language for EXODUS. In H. Boral and P. Larson, editors, *ACM SIGMOD International Conference of Management of Data*, pages 413–423, Chicago, IL, 1988.

13. J. Celko. *SQL for Smarties*. Morgan-Kaufmann, San Francisco, CA, 1995.

14. S. Ceri, S. Crespi-Reghizzi, G. Lamperti, L. Lavazza, and R. Zicari. Algres: An advanced database system for complex applications. *IEEE-Software*, 7(4):68–78, 1990.

15. D. D. Chamberlin, M. M. Astrahan, K. P. Eswaran, P. P. Griffiths, R. A. Lorie, J. W. Mehl, P. Reisner, and B. W. Wade. SEQUEL 2: A unified approach to data definition, manipulation, and control. *IBM Journal of Research and Development*, 20(6):560–575, 1976.

16. E.F. Codd. A relational model for large shared data banks. *Communications of the ACM*, 13(6):377–387, 1970.

17. L.S. Colby. A recursive algebra and query optimization for nested relations. In *ACM SIGMOD Conference on Management of Data*, pages 273–283, Portland, OR, 1989.

18. P. Dadam, K. Küspert, F. Andersen, H. Blanken, R. Erbe, J. Günauer, V. Lum, P. Pistor, and G. Walch. A DBMS prototype to support extended NF2 relations: An integrated view on flat tables and hierarchies. In *ACM SIGMOD Conference on Management of Data*, pages 356–366, Washington, DC, 1986.

19. C. J. Date and H. Darwen. *A Guide to the SQL Standard*. Addison-Wesley, Reading, MA, 1993.

20. U. Dayal, N. Goodman, and R. H. Katz. An extended relational algebra with control over duplicate elimination. In *ACM Symposium on Principles of Database Systems*, pages 117–123, Los Angeles, CA, 1982.

21. V. Desphande and P.Å. Larson. An algebra for nested relations. Technical Report Research Report CS-87-65, University of Waterloo, Waterloo, Ontario, December 1987.

22. P.C. Fischer and S.J. Thomas. Operators for non-first-normal-form relations. In *Seventh International Computer Software Applications Conference*, pages 464–475, Chicago, IL, 1983.

23. D. H. Fishman, D. Beech, H. P. Cate, E. C. Chow, T. Connors, J. W. Davis, N. Derrette, C. G. Hoch, W. Kent, P. Lyngbaek, B. Mahbod, M. A. Neimat, T. A. Ryan, and M.C. Shan. Iris: An object-oriented database management system. *ACM Transactions on Information Systems*, 5(1):48–69, 1987.

24. M. Furst, J. B. Saxe, and M. Sipser. Parity, circuits, and the polynomial-time hierarchy. *Mathematical System Theory*, 17:13–27, 1984.

25. T. Griffin and L. Libkin. Incremental maintenance of views with duplicates. In M. J. Carey and D. A. Schneider, editors, *ACM SIGMOD International Conference on Management of Data*, pages 328–339, San Jose, CA, 1995.

26. S. Grumbach, L. Libkin, T. Milo, and L. Wong. Query languages for bags: Expressive power and complexity. *SIGACTN: SIGACT News (ACM Special Interest Group on Automata and Computability Theory)*, 27, 1996.
27. S. Grumbach and T. Milo. Towards tractable algebras for bags. In *Twelfth ACM SIGACT-SIGMOD-SIGART Symposium on Principles of Database Systems*, pages 49–58, Washington, DC, 1993.
28. M. Hammer and D. McLeod. Database description with SDM: A semantic database model. *ACM Transaction on Database Systems*, 6(3):351–386, 1981.
29. G. Jaenschke. Recursive algebra for relations with relation valued attributes. Technical Report 85.03.002, Heidelberg Scientific Centre, IBM, Heidelberg, Germany, 1985.
30. G. Jaenschke and H.J. Schek. Remarks on the algebra for non first normal form relations. In *First ACM SIGACT SIGMOD Symposium on Principles of Database Systems*, pages 124–138, 1982.
31. A. Klausner and N. Goodman. Multirelations - semantics and languages. In A. Pirotte and Y. Vassiliou, editors, *Eleventh International Conference on Very Large Data Bases*, pages 251–258, Stockholm, Sweden, 1985.
32. A. Levy. Answering queries using views: A survey, 2000. http://www.cs.washington.edu/homes/alon/site/files/view-survey.ps.
33. L. Libkin and L. Wong. Some properties of query languages for bags. In C. Beeri, A. Ohori, and D. Shasha, editors, *Fourth International Workshop on Database Programming Languages – Object Models and Languages*, Workshops in Computing, pages 97–114, New York, NY, 1993.
34. L. Libkin and L. Wong. Query languages for bags and aggregate functions. *Journal of Computer and System Sciences*, 55(2):241–272, 1997.
35. S. K. Madria, S. S. Bhowmick, W. K. Ng, and E. P. Lim. Research issues in web data mining. In M. K. Mohania and A. Min Tjoa, editors, *Data Warehousing and Knowledge Discovery International Conference*, pages 303–312, Florence, Italy, 1999.
36. F. Manola and U. Dayal. PDM: An object-oriented data model. In K. R. Dittrich and U. Dayal, editors, *International Workshop on Object-Oriented Database Systems*, pages 18–25, Pacific Grove, CA, 1986.
37. J. Melton and A. R. Simon. *Understanding the New SQL: A Complete Guide*. Morgan-Kaufmann, San Francisco, CA, 1993.
38. I. S. Mumick, H. Pirahesh, and R. Ramakrishnan. The magic of duplicates and aggregates. In D. McLeod, R. Sacks-Davis, and H. Schek, editors, *Sixteenth International Conference on Very Large Data Bases*, pages 264–277, Brisbane, Australia, 1990.
39. G. Ozsoyoglu, Z.M. Ozsoyoglu, and V. Matos. Extending relational algebra and relational calculus with set-valued attributes and aggregate functions. *ACM Transactions on Database Systems*, 12(4):566–592, 1987.
40. M.A. Roth, H.F. Korth, and A. Silberschatz. Extended algebra and calculus for ¬1NF relational databases. *ACM Transactions on Database Systems*, 13(4):389–417, 1988.
41. H.J. Scheck and M.H. Scholl. The relational model with relation-valued attributes. *Information Systems*, 11(2):137–147, 1986.
42. H.J. Schek and P. Pistor. Data structures for an integrated database management and information retrieval systems. In *Eighth International Conference on Very Large Data Bases*, pages 197–207, Mexico City, Mexico, 1982.

43. M.H. Scholl, S. Abiteboul, F. Bancilhon, N. Bidoit, S. Gamerman, D. Plateau, P. Richard, and A. Verroust. Verso: A database machine based on nested relations. In S. Abiteboul, P.C. Fischer, and H.J. Schek, editors, *Nested relations and complex objects in databases*, number 361 in LNCS. Springer-Verlag, Heidelberg, Germany, 1989.

44. P. M. Schwarz, W. Chang, J. C. Freytag, G. M. Lohman, J. McPherson, C. Mohan, and H. Pirahesh. Extensibility in the Starburst database system. In K. R. Dittrich and U. Dayal, editors, *roceedings of the International Workshop on Object-Oriented Database Systems*, pages 85–92, Pacific Grove, CA, 1986.

45. D. W. Shipman. The functional data model and the data language DAPLEX. *ACM Transaction on Database Systems*, 6(1):140–173, 1981.

46. D. Srivastava, S. Dar, H. V. Jagadish, and A. Y. Levy. Answering queries with aggregation using views. In T. M. Vijayaraman, A. P. Buchmann, C. Mohan, and N. L. Sarda, editors, *Twenty-second International Conference on Very Large Data Bases*, pages 318–329, Mumbai, India, 1996.

47. M. Stonebraker, E. Wong, P. Kreps, and G. Held. The design and implementation of INGRES. *ACM Transaction on Database Systems*, 1(3):189–222, 1976.

48. D. Suciu and V. Tannen. A query language for NC. In *Thirteenth ACM SIGACT-SIGMOD-SIGART Symposium on Principles of Database Systems*, Minneapolis, Minnesota, 1994.

49. S. L. Vandenberg and D. J. DeWitt. Algebraic support for complex objects with arrays, identity, and inheritance. In J. Clifford and R. King, editors, *ACM SIGMOD International Conference on Management of Data*, pages 158–167, Denver, CO, 1991.

50. L. Wong. Normal form and conservative properties for query languages. In *Twelfth ACM SIGACT-SIGMOD-SIGART Symposium on Principles of Database Systems*, Washington, DC, 1993.

Tolerance Multisets

Solomon Marcus

Romanian Academy, Mathematics
Bucharest, Romania
solomon.marcus@imar.ro

Abstract. A multiset involves an equivalence relation between the copies of the same element. However, in many cases the binary relation relating an element to its copies is not exactly an equivalence one, but a weaker relation, in most cases a tolerance (i.e., reflexive and symmetric) relation, leading to a natural extension of multisets: tolerance multisets.

1   Behind Multisets Is a Relation of Sameness

Let us consider various versions of the definition of a multiset, [2,7,8,9,18,19,20]: (1) a set in which every element occurs a finite number of times; (2) a set in which every element exists in a finite number of copies; (s) given a set D and the set $\mathbf{N}$ of non-negative integers, a multiset over D is a function f from D into $\mathbf{N}$, or (4) a pair $\langle D, f \rangle$, where f is a function from D into $\mathbf{N}$; (5) a multiset over D corresponds to a (finite) string over D. For the way the multisets occur in membrane computing, see [16].

In all these versions, more or less rigorous, of the definition of a multiset, a common, unaivodable feature is to assume the existence of a binary relation of *sameness* between some objects. In version (1), the expression "finite number of times" suggests that we can order in time the objects we are considering and we can realize that some different occurrences refer to the same object. In version (2), the idea that an object is a copy of another one relates to the same relation of sameness, although the word "copy" may suggest that it refers to another object, which is the original one, to which the copy is in a relation of dependency. In version (3), we recognize the extension of the idea of a characteristic function, used to define a set; but saying that, for the object x in D, $f(x)$ is equal to n, where n is larger than one, means to assume the same relation of sameness between n different occurrences of x. In version (4), the situation is similar. In version (5), it is obviously assumed that there is a sharp distinction between the different elements of the alphabet D (as a matter of fact, this is the basic assumption for any alphabet) and that the same element of D may occur in a string several times, with no upper bound other than the length of the string.

Multisets were called sometimes *bags* [20]; this last term is both shorter and more convenient, because "multisets" may be associated with another term, used in analysis and topology, "multifunctions", denoting functions whose values are sets. Today however is too late to change the terminology in this respect, because "multiset" is quasi-generally accepted.

C.S. Calude et al. (Eds.): Multiset Processing, LNCS 2235, pp. 217–223, 2001.
© Springer-Verlag Berlin Heidelberg 2001

Version (5) considered above calls attention to the fact that multisets are only one among several types of data-structures we met in various information sciences. The terms of reference are order and repetitions. In a given arbitrary set, neither order nor repetition of the elements are relevant. In a list, both order and repetition are relevant; strings are lists. Multisets are less restrictive than lists, because only repetitions are relevant, but more restrictive than sets, which are completely free in respect to both order and repetition. It remains the possible case when only order is relevant: it is what happens with compact lists, which are lists in which contiguous occurrences of the same object are reduced to only one occurrence [2,18]. The extension we are proposing with respect to multisets can be considered for lists and compact lists too.

2 From Multisets to Tolerance Multisets: The Case of Synonymy

Examples of multisets that are not concerned by our extension are mainly those occurring in pure mathematics: the divisors of a positive integer p form a multiset where there is no reason for an extension to a tolerance multiset. The role of the set D belongs to the set of prime divisors of p. So, in the set of divisors of 72 we detect three occurrences of 2 and two occurrences of 3. Similarly, the roots of an algebraic equation such as $x^3 - x^2 - x + 1$ form a multiset in which 1 has the multiplicity equal to 2, while -1 has the multiplicity equal to 1. Here too, there is no reason for an extension to tolerance multisets. In both these examples, the relation between an element and one of its copies is clearly reflexive, symmetric, and transitive.

The situation changes as soon as we move to multisets conceived as models of some natural or social phenomena. Coins of various values in U.S.A. form a multiset and for every value we have a large number of copies. The relation between two coins of the same value is, generally speaking, an equivalence relation, but, if a coin of one dollar is too deteriorated, it may happen that some machines will not accept it. There is no sharp border between an accepted and a non-accepted coin, so what seemed to be, at a first glance, an equivalence relation is only a tolerance one.

Another significant example is the relation of synonymy in linguistics. Roughly speaking, two words or expressions are synonymous if they have the same meaning. If we remain at this representation of synonymy, then the respective relation is clearly an equivalence and we can conceive the vocabulary of a language as a multiset in which the number of copies of a word is equal to the number of its synonyms. However, things are not so simple. Exact synonyms rarely (if not never) exist. This means that it is very difficult (if not impossible) to find words x and y such that for any context $\langle u, v \rangle$ for which uxv is meaningful, uyv is meaningful too, and they both have the same meaning, while the converse is also true: for any context $\langle u, v \rangle$ for which uyv is meaningful, uxv is meaningful too and they both have the same meaning. So, we weaken the idea of synonymy in the following way: x and y are in a relation of partial synonymy if there exists at least one context $\langle u, v \rangle$ such that uxv and uyv have the same meaning. But partial synonymy is no longer transitive, so it is no longer an equivalence relation, it is only a tolerance relation. Examples in English can be obtained by means of any monolingual English dictionary. Using [3], we find that 'big' is a partial synonym of 'large', of 'important' and of 'boastful'; 'large' is a partial synonym of 'great', which, in its turn, is a partial synonym of 'elaborate' and of 'intense'. If we remain to what the dictionary tells us, we have to consider that, in respect to [3], 'big' and 'great' are not partial synonyms, although this fact cound seem strange. Any analysis of this type should be done in respect to a given dictionary, otherwise room is left to arbitrariness and subjectivity.

In French, taking as term of reference the dictionary [4], we find that 'grand' is a partial synonym of 'important', 'considerable', 'vaste', while 'important' and 'considerable' are partial synonyms of 'large', which, in its turn, is a partial synonym of 'important' and of 'considerable', both of them being, as we have seen above, partial synonyms of 'grand'. It follows that 'grand' and 'large' become partial synonyms only by mediation of 'important' and 'considerable'. This mediated synonymy could suggest to define a relation of generalized synonymy by considering the reflexive and transitive closure of the relation of partial synonymy. In this way we get again an equivalence relation.

But perhaps the most familiar example of a multiset is the bag in which, at the market, we introduce a certain number of apples, a certain number of potatoes, a certain number of oranges, etc. Beyond their common trait, to be, for instance, apples, the respective objects may be not at all similar, but there is no danger to confuse an apple with a potatoe or with an orange. Just this fact shows that in this case the extension from multisets to tolerance multisets is not necessary.

3   A Tolerance Multiset in Phonology

A more sophisticated situation appears in phonology. The basic concept of a phoneme is, in a first approximation, a class of equivalent sounds. Accepting this representation, the set of phonemes is a multiset of sounds; the phonemes are the types, while the sounds are the token. The only difficulty here is the problematic finiteness of the set of sounds associated with a given phoneme; and even in the case in which this set is really finite, its cardinal number cannot be practically evaluated. As a matter of fact, the situation here is more sophisticated and we will try in the following to describe what happens (see also [10] and [11]).

Let us consider a set S of elements called abstract sounds and let $F(S)$ be the free-monoid generated by S. Let $L(S)$ be a subset of $F(S)$; $L(S)$ is said to be a language over S. The elements of $L(S)$ are well-formed strings over S. Let r be an equivalence relation in $F(S)$. If for x, y, z in $F(S)$ and x r y we have zx r zy and xz r yz, then r is said to be a congruence relation in $F(S)$. Let R be a congruence relation in $F(S)$, such that $L(S)$ is closed with respect to R, i.e., if x is in $L(S)$ and if x R y, then y is in $L(S)$. The triad $\langle F(S), L(S), R \rangle$ is said to be a phonological system; it is a phonological system of the language

$L(S)$. The choice of $L(S)$ and of the relation R has both syntactic and semantic motivations.

Given the subsets $X(1), X(2), \ldots, X(n)$ of $F(S)$, we denote by $X(1)X(2) \ldots X(n)$ the set of all strings of the form $x(1)x(2) \ldots x(n)$, where $x(i) \in X(i)$ for all $1 \leq i \leq n$.

Proposition 1. *The free monoid $D(X)$ generated by X is identical to the union of all sets $X^n, n \geq 0$, where X^n is the cartesian product of X with itself, n times.*

Strings x and y are said to be in a relation of variation, i.e., x is a variant of y (or y is a variant of x), and we write x m y if for any strings z and w the following implications are valid: (i) if zxw is in $L(S)$, then zxw R zyw; (ii) if zyw is in $L(S)$, then zxw R zyw.

Proposition 2. *If x m y and if x is in $L(S)$ or y is in $L(S)$, then x R y.*

Proposition 3. *The relation of variation is a congruence relation in $F(S)$.*

Proposition 4. *The set $L(S)$ is closed with respect to m.*

The relation m of variation corresponds to the free variation, well known from phonemics. It is known that two sonorous strings are considered as variants (allophones) of the same entity not only when they are in free variation, but also when they are in some situations of complementarity distribution. In order to include this possibility too, we adopt an extension k of the relation m: strings x and y are in relation k, we write x k y, and we say that k is a relation of variation in the broad sense of m if for any strings z and w such that zxw is in $L(S)$ and zyw is in $L(S)$, we have also zxw R zyw. We say in this case that x (resp. y) is a variant in the broad sense of y (resp. x) and we write x v y.

We define now an equivalence relation s in $F(S)$: if x and y are in $F(S)$, we have x s y if and only if for two arbitrary strings z, w in $F(S)$ either zxw and zyw are both in $L(S)$ or zxw and zyw are both absent from $L(S)$.

Proposition 5. *If we have u s v and x s y (u, v, x and y being strings in $F(S)$), then we also have ux s vy.*

Proposition 6. *The relation s is a congruence relation in $L(S)$.*

A string x is semi-wellformed with respect to $L(S)$, if it is not wellformed, but there exist two strings y, z such that the string yxz is wellformed, i.e., yxz is in $L(S)$. A string which is neither wellformed not semi-wellformed with respect to $L(S)$ is said to be parasitic with respect to $L(S)$.

Proposition 7. *The parasitic strings with respect to $L(S)$ form a single class of s-equivalence.*

Proposition 8. *If x is semi-wellformed with respect to $L(S)$ and if y is in a relation of s-equivalence to x, then y too is semi-wellformed with respect to $L(S)$.*

Proposition 9. *If x is a variant of y, then x is a variant in the broad sense of y; the converse is not true.*

Sometimes, the phoneme is defined as a set of sounds which are either in free variation or in complementary distribution; in orther words, strings x and y belong to the same phoneme if x v y (because we never have both zxy and zyw in $L(S)$, x and y are accepted in $L(S)$ by no common context), i.e., x (resp. y) is a variant in the broad sense of y (resp. x). But, as the following theorem shows, the relation v is not an equivalence relation, so the considered definition of a phoneme is ambiguous: the same sound may belong to two different phonemes.

Theorem. *The relation of variation in the broad sense is not transitive.*

Proof. Let $S = \{a, b, c, d, e, f\}$ and $L(S) = \{ae, ce, fa, cf, dbe, dae, dce, fae\}$. Let us define R as the relation of decomposition of $F(S)$ in classes of s-equivalence. We will show that for $x = a, y = b$, and $u = c$ we have x v y and y v u, but we do not have x v u. Indeed, we have zaw in $L(S)$ and zbw in $L(S)$ iff $z = d$ and $w = e$. On the other hand, we have dbe R dae, because dbe and dae are wellformed strings whose only acceptance context is the empty one. We have proved in this way that x v y. It is also true that y v u, because zbv and zcw are in $L(S)$ iff $z = d$ and $w = e$ and, on the other hand, we have dbe R dce. It remains to observe that we do not have x v u, because if z is the empty string and if $w = e$, then the strings zxw and zuw are in $L(S)$, but we do not have zxw R zuw; indeed, zxw is accepted by the context $\langle f, \lambda \rangle$, where λ is the empty string, while zuw is not accepted by this context.   □

The above theorem clearly shows that the set of (abstract) sounds is not a multiset, but a tolerance multiset; the phonemes are the types, while the (abstract) sounds are the token. But the type-token relation is no longer an equivalence relation, it is a tolerance relation. For more details in this respect, see [5] (for free variation), [6] (for a first logical model of the phoneme), [10,11] (for the proofs of the above propositions), and [12] (for the definition of synonymy).

4   The Topological Status of the Tolerance Multisets

Let (X, t) be a tolerance space, i.e., a set X endowed with a tolerance relation t. Let A be a subset of X. Following section 4 of [14], we consider for any x in A its tolerance class $\mathsf{t}(x)$, i.e., the set of all objects y in X for which x t y, Let $T^*(A)$ be the set of all these classes, from which we select the subset $T_*(A)$ of those classes $\mathsf{t}(x)$ which are contained in A. The ordered pair $(T_*(A), T^*(A))$ will be called a *tolerance multiset*. In [14], it was called a *tolerance rough set*. $T_*(A)$ and $T^*(A)$ are the lower and the upper tolerance approximations of A by t. The set $T^*(A) - T_*(A)$ is a rough tolerance description, in the sense of [14], of the boundary of A. In the particular case when the tolerance relation t is an equivalence relation, we get the case of usual rough sets. The rough perspective in the study of data was initiated by Z. Pawlak [15].

One can extend this approach, in order to include tolerance classes which are not in $T^*(A)$. These classes are not equally situated with respect to A. Their hierarchy can be obtained by starting to put $T^0(A) = T^*(A)$. Let $T^1(A)$ be the set of those tolerance classes whose intersection with at least one tolerance class in $T^0(A)$ is non-empty. Assuming $T^n(A)$ already defined, we consider the set $T^{n+1}(A)$ of those tolerance classes whose intersection with at least one class in $T^n(A)$ is non-empty. One obtains a sequence $T^0(A), T^1(A), \ldots, T^n(A), \ldots$, where every term is contained in the next one. A similar approach can be adopted for the set $T_0(A) = T_*(A)$ and we get an increasing sequence $T_0(A)$, $T_1(A)$, ..., $T_n(A)$, ..., where $T_n(A)$ is the set of those tolerance classes which intersect at least one class in $T_{n-1}(A)$. We got in this way two infinite sequences; the former is aimed at refining our understanding of the boundary region $T^*(A) - T_*(A)$, while the latter helps us to a better understanding of what happens beyond $T^*(A)$. Obviously, if the set X is finite, then each of the considered sequences will become stationary after a finite number of steps.

The behavior of the synonymy operator, considered in a previous section, illustrates and explains, on a concrete situation, the approximation process just described.

From tolerance multisets there is only one step to Čech topology [1]. Let us first recall what happens with the topology $t(\mathsf{R})$ generated by R, when R is an equivalence relation. In this case, the approximation $\mathsf{R}_*(A)$ of the set A is just the interior of A, while the upper approximation $\mathsf{R}^*(A)$ of A is the closure of A in the topology $t(\mathsf{R})$. This topology is total, i.e., any intersection of open sets is open and any union of closed sets is closed (as a consequence of the fact that in $t(\mathsf{R})$ a set is open iff it is closed). Now let us consider the case when t is a tolerance relation and let us recall the concept of Čech topology [1]. It is defined in a set X by means of a generalized closure operator cl which associates to any subset S of X a set $cl(S)$ contained in X, such that $cl(\emptyset) = \emptyset$, $S \subseteq cl(S)$, and, if $A \subseteq B$, then $cl(A) \subseteq cl(B)$. In the particular case in which $cl(S) = cl(cl(S))$ for any subset S of X, we get the usual closure operator associated with the usual topology. The ordered pair (X, cl) is said to be a Čech topological space. Any tolerance relation t in X generates a Čech topological space, by defining $cl(A) = \{x \mid \text{there exists } y \text{ in } A \text{ such that } x \mathsf{ t } y\}$. The converse is not true; indeed, it is easy to see that the closure operator cl_t generated by the tolerance t is additive, i.e., $cl_\mathsf{t}(A \cup B) = cl_\mathsf{t}(A) \cup cl_\mathsf{t}(B)$, while the Čech closure operator does not always have this property.

Tolerance multisets can be presented in the language of Čech topology. The sequence $(T_n(A))_n$ is obtained by applying the Čech closure operator cl_t generated by the tolerance t successively to $T_*(A)$, while the sequence $T^n(A))_n$ is obtained in a similar way, but starting with $T^*(A)$, which in turn is obtained by applying the operator cl_t to A. There exists a natural relation between tolerance relations and information systems. Let us consider the correspondence f associating with each object attribute values; f determines in the set of objects a tolerance relation t: the objects x and y are in the relation t if they have at least one attribute value in common. Conversely, one can prove that any tolerance relation can be obtained in this way. For more about tolerance in linguistics, see [17]. For a relation between tolerance and learning processes, see [13].

References

1. E. Čech, *Topological Spaces*, The Publishing House of the Czech Academy, Prague, 1966.
2. A. Dovier, A. Policrity, G. Rossi, A uniform axiomatic view of lists, multisets, and sets, and the relevant unification algorithms, *Fundamenta Informaticae*, **36**, 2/3 (1998), 221–234.
3. E. Ehrlich, S. Berg Flexner, G. Garruth, J.M. Hawkins, *Oxford American Dictionary*, Avon Books, Oxford Univ. Press, New York, 1980.
4. E. Genouvrier, C. Desirat, T. Horde, *Nouveau Dictionnaire des Synonymes*, Librairie Larousse, Paris, 1980.
5. Z.S. Harris, *Structural Linguistics*, Chicago Univ. Press, Chicago, 1961.
6. S. Kanger, The notion of a phoneme, *Statistical Methods in Linguistics*, **3** (1964), 43–48.
7. D. Knuth, *The Art of Computer Programming*, vol. 2, *Seminumerical Algorithms*, Addison Wesley, Reading, Mass., 1985.
8. M. Kudlek, C. Martin-Vide, Gh. Păun, Toward FMT (Formal Macroset Theory), in *Pre-proceedings of the Workshop on Multiset Processing*, Curtea de Argeş, Romania, 2000 (C.S. Calude, M.J. Dinneen, Gh. Păun, eds.), CDMTCS Res. Report 140, Auckland Univ., 2000, 149–158.
9. Z. Manna, R. Waldinger, *The Logical Basis for Computer Programming*, vol. 1, *Deductive Reasoning*, Addison Wesley, Reading, Mass., 1985.
10. S. Marcus, Sur un ouvrage de Stig Kanger concernant le phoneme, *Statistical Methods in Linguistics*, **3** (1964), 43–48.
11. S. Marcus, *Introduction Mathématique à la Linguistique Structurale*, Dunod, Paris, 1967.
12. S. Marcus, *Mathematische Poetik*, Athenäum, Frankfurt am Main, 1973.
13. S. Marcus, Interplay of innate and acquired in some mathematical models of language learning processes, *Revue Roumaine de Linguistique*, **34** (1989), 101–116.
14. S. Marcus, Tolerance rough sets, Čech topologies, learning processes, *Bull. Polish Academy of Science. Technical Science*, **42**, 3 (1994), 471–487.
15. Z. Pawlak, *Theoretical Aspects of Reasoning About Data*, Kluwer, Dordrecht, 1991.
16. Gh. Păun, Computing with membranes, *Journal of Computer and System Sciences*, **61**, 1 (2000), 108–143, and *Turku Center for Computer Science-TUCS Report* No 208, 1998 (www.tucs.fi).
17. J. Pogonowski, *Tolerance Spaces with Applications in Linguistics*, Poznan University Press, Poznan, 1981.
18. A. Syropoulos, Mathematics of multisets, in *Pre-proceedings of the Workshop on Multiset Processing*, Curtea de Argeş, Romania, 2000 (C.S. Calude, M.J. Dinneen, Gh. Păun, eds.), CDMTCS Res. Report 140, Auckland Univ., 2000, 286–295.
19. A. Tzouvaras, The linear logic of multisets, *Logic Journal of GPL*, **6**, 6 (1998), 901–916.
20. R.R. Yager, On the theory of bags, *Intern. Journal of General Systems*, **13** (1986), 23–37.

Fuzzy Multisets and Their Generalizations

Sadaaki Miyamoto

Institute of Engineering Mechnics and Systems
University of Tsukuba, Ibaraki 305-8573, Japan
phone/fax:+81-298-53-5346
miyamoto@esys.tsukuba.ac.jp

Abstract. Fuzzy multisets with infinite membership sequences and their generalization using set-valued memberships are considered. Two metric spaces of the infinite fuzzy multisets are defined in terms of cardinality. One is the completion of fuzzy multisets of finite membership sequences; the other is derived from operations among fuzzy multisets and real-valued multisets. Theoretical properties of these infinite fuzzy multisets are discussed.

1 Introduction

Recently, many studies discuss multisets and their applications such as relational databases, information retrieval, and new computing paradigms. Multisets have sometimes been called *bags*. While the well-known book by Knuth [5] uses the term of multisets, another book by Manna and Waldinger [7] devotes a chapter to bags. The terms of multiset and bag can thus be used interchageably. The author prefers to use multisets but some readers may interpret them to be bags instead.

This paper is concerned with a generalization of multisets, that is, *fuzzy multisets* [16,3,14,4,13,15,8,9,10,6]. The ordinary nonfuzzy multisets are called crisp multisets by the common usage in fuzzy systems theory. We discuss theoretical aspects of fuzzy multisets; in particular, the focus is on infinite features of multisets. Namely, infinite sequences of memberships and a distance between fuzzy multisets are discussed. Further generalizations of fuzzy multisets are defined using a closed set in a plane as the membership range. All arguments are elementary, and hence we omit proofs of the propositions.

2 Finite Multisets and Fuzzy Multisets

Before considering fuzzy multisets, a brief review of crisp multisets is useful. In the preliminary consideration we assume finite sets and multisets for simplicity.

2.1 Crisp Multisets

Let us begin by a simple example.

Example 1. Assume that $X = \{x, y, z, w\}$ is a set of symbols and suppose we have a number of objects but they are not distinguishable except their labels of

C.S. Calude et al. (Eds.): Multiset Processing, LNCS 2235, pp. 225–235, 2001.

x, y, z, or w. For example, we have two balls with the label x and one ball with y, three with label z, but no ball with the label w. Moreover we are not allowed to put additional labels to distinguish two x's. Therefore a natural representation of the situation is that we have a collection

$$\{x, x, y, z, z, z\}.$$

We can also write

$$\{2/x, 1/y, 3/z, 0/w\}$$

to show the number for each element of the universe X, or $\{2/x, 1/y, 3/z\}$ by ignoring zero of w. We will say that there are three occurrences of x, two occurrences of y, and so on.

We proceed to general definitions. Assume $X = \{x_1, \ldots, x_n\}$ is a finite set called the universe or the basis set. A crisp multiset M of X is characterized by the function $Count_M(\cdot)$ whereby a natural number (including zero) corresponds to each $x \in X$, that is [1,5,7]:

$$Count_M \colon X \to \{0, 1, 2, \ldots\}.$$

For a crisp multiset, different expressions such as

$$M = \{k_1/x_1, \ldots, k_n/x_n\}$$

and

$$M = \{\overbrace{x_1, \ldots, x_1}^{k_1}, \ldots, \overbrace{x_n, \ldots, x_n}^{k_n}\}$$

are used. An element of X may thus appear more than once in a multiset. In the above example x_1 appears k_1 times in M, hence we have k_1 occurrences of x_1.

Consider the first example: $\{2/x, 1/y, 3/z\}$. We have

$$Count_M(x) = 2, \quad Count_M(y) = 1, \quad Count_M(z) = 3, \quad Count_M(w) = 0.$$

The following are basic relations and operations for crisp multisets.

(inclusion):
$$M \subseteq N \Leftrightarrow Count_M(x) \leq Count_N(x), \quad \forall x \in X. \tag{1}$$

(equality):
$$M = N \Leftrightarrow Count_M(x) = Count_N(x), \quad \forall x \in X. \tag{2}$$

(union):
$$Count_{M \cup N}(x) = \max\{Count_M(x), Count_N(x)\}$$
$$= Count_M(x) \vee Count_N(x). \tag{3}$$

(intersection):
$$Count_{M \cap N}(x) = \min\{Count_M(x), Count_N(x)\}$$
$$= Count_M(x) \wedge Count_N(x). \tag{4}$$

(addition):

$$Count_{M \oplus N}(x) = Count_M(x) + Count_N(x). \tag{5}$$

(subtraction):

$$Count_{M \ominus N}(x) = 0 \wedge (Count_M(x) - Count_N(x)). \tag{6}$$

The symbols $\vee$ and $\wedge$ are the infix notations of max and min, respectively. Readers should note that the operations resemble those for fuzzy sets, but the upper bound for $Count(\cdot)$ is not assumed.

Example 2. Consider the multiset M in Example 1 and

$$N = \{1/x, 4/y, 3/w\}.$$

Then,

$$M \oplus N = \{3/x, 5/y, 3/z, 3/w\},$$
$$M \cup N = \{2/x, 4/y, 3/z, 3/w\},$$
$$M \cap N = \{1/x, 1/y\},$$
$$M \ominus N = \{1/x, 3/z\}.$$

Real-valued multisets
Blizard [2] generalized multisets to real-valued multisets, from which we remark nonnegative real-valued multiplicity. Thus, the $Count$ function takes nonnegative real-values: ($Count: X \to \mathbf{R}^+ \cup \{0\}$). The basic relations and set operations use (1)–(6).

2.2 Fuzzy Multisets

Yager [16] first discussed fuzzy multisets, although he uses the term of *fuzzy bag*; an element of X may occur more than once with possibly the same or different membership values.

Example 3. Consider a fuzzy multiset

$$A = \{(x, 0.2), (x, 0.3), (y, 1), (y, 0.5), (y, 0.5)\}$$

of $X = \{x, y, z, w\}$, which means that x with the membership 0.2, x with 0.3, y with the membership 0.5, and two y's with 0.5 are contained in A. We may write

$$A = \{\{0.2, 0.3\}/x, \{1, 0.5, 0.5\}/y\}$$

in which the multisets of membership $\{0.2, 0.3\}$ and $\{1, 0.5, 0.5\}$ correspond to x and y, respectively.

$Count_A(x)$ is thus a finite multiset of the unit interval [16]. The collection of all finite fuzzy multisets of X is denoted by $\mathcal{FM}(X)$.

For $x \in X$, the membership sequence is defined to be the decreasingly-ordered sequence of the elements in $Count_A(x)$. It is denoted by

$$(\mu_A^1(x), \mu_A^2(x), \ldots, \mu_A^p(x)),$$

where $\mu_A^1(x) \geq \mu_A^2(x) \geq \cdots \geq \mu_A^p(x)$.

When we define an operation between two fuzzy multisets, say A and B, the lengths of the membership sequences $\mu_A^1(x)$, $\mu_A^2(x)$, $\ldots$, $\mu_A^p(x)$, and $\mu_B^1(x)$, $\mu_B^2(x)$, $\ldots$, $\mu_B^{p'}(x)$, should be set to be equal. We hence define the length $L(x; A)$, the length of $\mu_A^j(x)$:

$$L(x; A) = \max\{j \; : \; \mu_A^j(x) \neq 0 \},$$

and define

$$L(x; A, B) = \max\{L(x; A), L(x; B)\}.$$

When no ambiguity arises, we write $L(x) = L(x; A, B)$ for simplicity.

Example 4.

Let

$$A = \{\{0.2, 0.3\}/x, \{1, 0.5, 0.5\}/y\},$$
$$B = \{\{0.6\}/x, \{0.8, 0.6\}/y, \{0.1, 0.7\}/w\}.$$

For the representation of the membership sequence, we get

$$L(x) = 2, \quad L(y) = 3, \quad L(z) = 0, \quad L(w) = 2$$

and we have

$$A = \{(0.3, 0.2)/x, (1, 0.5, 0.5)/y, (0, 0)/w\},$$
$$B = \{(0.6, 0)/x, (0.8, 0.6, 0)/y, (0.7, 0.1)/w\}.$$

The following are basic relations and operations for fuzzy multisets [8]:

1. **[inclusion]**

$$A \subseteq B \Leftrightarrow \mu_A^j(x) \leq \mu_B^j(x), \; j = 1, \ldots, L(x), \quad \forall x \in X. \tag{7}$$

2. **[equality]**

$$A = B \Leftrightarrow \mu_A^j(x) = \mu_B^j(x), \; j = 1, \ldots, L(x), \quad \forall x \in X. \tag{8}$$

3. **[addition]**

$A \oplus B$ is defined by the addition operation in $X \times [0, 1]$ for crisp multisets [16]. Namely, if

$$A = \{(x_i, \mu_i), \ldots, (x_k, \mu_k)\}$$

and

$$B = \{(x_p, \mu_p), \ldots, (x_r, \mu_r)\}$$

are two fuzzy multisets, then

$$A \oplus B = \{(x_i, \mu_i), \ldots, (x_k, \mu_k), (x_p, \mu_p), \ldots, (x_r, \mu_r)\}. \tag{9}$$

4. **[union]**
$$\mu^j_{A \cup B}(x) = \mu^j_A(x) \vee \mu^j_B(x), \; j = 1, \ldots, L(x). \tag{10}$$

5. **[intersection]**
$$\mu^j_{A \cap B}(x) = \mu^j_A(x) \wedge \mu^j_B(x), \; j = 1, \ldots, L(x). \tag{11}$$

6. **[α-cut]**
The α-cut ($\alpha \in (0,1]$) for a fuzzy multiset A, denoted by $[A]_\alpha$, is a crisp multiset defined as follows:
$$\mu^1_A(x) < \alpha \Rightarrow Count_{[A]_\alpha}(x) = 0, \tag{12}$$
$$\mu^j_A(x) \geq \alpha, \; \mu^{j+1}_A(x) < \alpha \Rightarrow Count_{[A]_\alpha}(x) = j, \tag{13}$$
$$j = 1, \ldots, L(x). \tag{14}$$

Moreover the strong α-cut ($\alpha \in [0,1)$), denoted $]A[_\alpha$, is a crisp multiset defined as follows:
$$\mu^1_A(x) \leq \alpha \Rightarrow Count_{]A[_\alpha}(x) = 0, \tag{15}$$
$$\mu^j_A(x) > \alpha, \; \mu^{j+1}_A(x) \leq \alpha \Rightarrow Count_{]A[_\alpha}(x) = j, \tag{16}$$
$$j = 1, \ldots, L(x). \tag{17}$$

7. **[Cartesian product]**
Given two fuzzy multisets $A = \{(x, \mu)\}$ and $B = \{(y, \nu)\}$, the Cartesian product is defined by:
$$A \times B = \sum \left\{ (x, y, \mu \wedge \nu) \right\} \tag{18}$$

The combination is taken for all (x, μ) in A and (y, ν) in B.

8. **[Multirelation]**
Notice that a crisp relation R on X is a subset of $X \times X$. Given a fuzzy multiset A of X, a multirelation $\mathcal{R}$ obtained from R is a subset of $A \times A$: for all $(x, \mu), (y, \nu) \in A$,
$$(x, y, \mu \wedge \nu) \in \mathcal{R} \iff (x, y) \in R$$

When R is a fuzzy relation on X, then
$$(x, y, \mu \wedge \nu \wedge R(x, y)) \in \mathcal{R}. \tag{19}$$

(The latter includes the former as a special case.)

The following propositions are valid. The proofs are not difficult and hence omitted.

Proposition 1. Assume that A and B are fuzzy multisets of X. The necessary and sufficient condition for $A \subseteq B$ is that for all $\alpha \in (0, 1]$, $[A]_\alpha \subseteq [B]_\alpha$. Moreover, the condition for $A = B$ is that for all $\alpha \in (0, 1]$, $[A]_\alpha = [B]_\alpha$.

Proposition 2. Assume that A and B are fuzzy multisets of X. The necessary and sufficient condition for $A \subseteq B$ is that for all $\alpha \in [0, 1)$, $]A[_\alpha \subseteq]B[_\alpha$. Moreover, the condition for $A = B$ is that for all $\alpha \in [0, 1)$, $]A[_\alpha =]B[_\alpha$.

Proposition 3.   Assume that A and B are fuzzy multisets of X. Take an arbitrary $\alpha \in (0,1)$. We then have:

$$[A \cup B]_\alpha = [A]_\alpha \cup [B]_\alpha, \qquad [A \cap B]_\alpha = [A]_\alpha \cap [B]_\alpha,$$
$$[A \oplus B]_\alpha = [A]_\alpha \oplus [B]_\alpha, \qquad [A \times B]_\alpha = [A]_\alpha \times [B]_\alpha$$
$$]A \cup B[_\alpha =]A[_\alpha \cup]B[_\alpha, \qquad]A \cap B[_\alpha =]A[_\alpha \cap]B[_\alpha,$$
$$]A \oplus B[_\alpha =]A[_\alpha \oplus]B[_\alpha, \qquad]A \times B[_\alpha =]A[_\alpha \times]B[_\alpha.$$

Proposition 4.   Assume that A, B, and C are fuzzy multisets of X. The followings equalities are valid:

$$A \cup B = B \cup A,$$
$$A \cap B = B \cap A,$$
$$A \cup (B \cup C) = (A \cup B) \cup C,$$
$$A \cap (B \cap C) = (A \cap B) \cap C,$$
$$(A \cap B) \cup C = (A \cup C) \cap (B \cup C),$$
$$(A \cup B) \cap C = (A \cap C) \cup (B \cap C).$$

Therefore, the class of all fuzzy multisets of a particular universe forms a distributive lattice.

The cardinality and a metric space of fuzzy multisets
The *cardinality* of a fuzzy multiset A is given by

$$|A| = \sum_{x \in X} \sum_{j=1}^{L(x;A)} \mu_A^j(x).$$

Moreover, we define

$$|A|_x = \sum_{j=1}^{L(x;A)} \mu_A^j(x).$$

We thus have $|A| = \sum_{x \in X} |A|_x$.

We can define a metric space of fuzzy multisets. Namely, given two multisets $A, B \in \mathcal{FM}(X)$, the distance between A and B is defined by

$$d(A,B) = \sum_{x \in X} \sum_{j=1}^{L(x;A,B)} |\mu_A^j(x) - \mu_B^j(x)|. \qquad (20)$$

We can immediately have

$$d(A,B) \geq 0; \quad d(A,B) = 0 \iff A = B, \qquad (21)$$
$$d(A,B) = d(B,A), \qquad (22)$$
$$d(A,C) \leq d(A,B) + D(B,C), \qquad (23)$$

hence the set of fuzzy multisets becomes a metric space.

3   Infinite Features of Fuzzy Multisets

So far all discussions are concerned with finite multisets. It is possible to introduce the infinity $\{+\infty\}$ into the consideration of crisp multisets: $Count_M(x){:}X{\rightarrow}$ $\{0, 1, \ldots, \infty\}$. However, for the purpose of simple and effective calculations, we consider $Count_M(x)$ as being always finite. This assumption of finite crisp multisets is useful for focusing the infinite features of fuzzy multisets.

3.1   Infinite Membership Sequences

Infinite fuzzy multisets means that for $x \in X$, $Count_A(x)$ may be an infinite set of the unit interval $I = [0, 1]$. We note that α-cuts of a fuzzy multiset A gives well-defined crisp multisets $[A]_\alpha$ and $]A[_\alpha$ for $\alpha \in (0, 1)$. Since we assume a crisp multiset of finite cardinality ($Count_M(x) < \infty$), an infinite subset of I does not necessarily provide a well-defined crisp multiset.

Hence, instead of an infinite set, we use a sequence $(\nu^1, \nu^2, \ldots)$ as the $Count$ function. Namely we assume

$$Count_A(x) = (\nu^1, \nu^2, \ldots) \tag{24}$$

where $\nu^j \to 0$ as $j \to 0$.

Basic relations and operations for infinite fuzzy multisets should also be defined in terms of membership sequences using (7)–(19) with the associated sequences being infinite (e.g., $L(x) \to \infty$). We assume that, regarding fuzzy multisets, all sorting of the infinite membership sequence in (24) can be performed by an effective computational procedure.

The class of all infinite fuzzy multisets of X is denoted here by $\mathcal{FM}_0(X)$. We immediately have the next proposition.

Proposition 5. For $A, B \in \mathcal{FM}_0(X)$,

  (i) $A \oplus B \in \mathcal{FM}_0(X)$,
  (ii) $A \cup B \in \mathcal{FM}_0(X)$,
  (iii) $A \cap B \in \mathcal{FM}_0(X)$.

3.2   The Metric Space of Infinite Fuzzy Multisets

We consider the subset of $\mathcal{FM}_0(X)$ in which the cardinality is finite:

$$|A| = \sum_{x \in X} \sum_{j=1}^{\infty} \mu_A^j(x) < \infty. \tag{25}$$

This subset is denoted by $\mathcal{FM}_1(X)$:

$$\mathcal{FM}_1(X) = \{\, |A| < \infty : A \in \mathcal{FM}_0(X) \,\}.$$

For $A, B \in \mathcal{FM}_1(X)$, the distance is naturally defined by

$$d(A, B) = \sum_{x \in X} \sum_{j=1}^{\infty} |\mu_A^j(x) - \mu_B^j(x)|. \tag{26}$$

The axioms of a metric, (21)–(23), are satisfied for $d(A,B)$.

The next proposition is similar to Proposition 5.

Proposition 6. For $A, B \in \mathcal{FM}_1(X)$,

(i) $A \oplus B \in \mathcal{FM}_1(X)$,
(ii) $A \cup B \in \mathcal{FM}_1(X)$,
(iii) $A \cap B \in \mathcal{FM}_1(X)$.

The space $\mathcal{FM}_1(X)$ of infinite fuzzy multisets is justified as follows. Suppose we have an infinite Cauchy sequence of finite fuzzy multisets $A_1, A_2, \cdots \in \mathcal{FM}(X)$:

$$d(A_i, A_j) \to 0, \quad \text{as } i, j \to \infty.$$

Then, the sequence converges to a fuzzy multiset $A \in \mathcal{FM}_1(X)$. On the other hand, for an arbitrary $A \in \mathcal{FM}_1(X)$, there exists a sequence of finite fuzzy multisets $A'_1, A'_2, \cdots \in \mathcal{FM}(X)$ such that $A'_j \to A$ as $j \to \infty$.

The metric space of infinite fuzzy multisets is complete, whereas that of finite fuzzy multisets is not.

4   A Set-Valued Membership: A Generalization of Fuzzy Multisets

The membership sequence can be identified with a real-valued step function with the nonnegative real variable $y \in \mathbf{R}^+ \cup \{0\}$

$$f_A(y; x) = \begin{cases} \mu_A^1(x), & (y = 0), \\ \mu_A^j(x), & (j - 1 < y \le j, \quad j = 1, 2, \dots). \end{cases} \tag{27}$$

A crisp multiset M is represented by a simple function of

$$f_M(y; x) = 1, \quad 0 \le y \le Count_M(x), \tag{28}$$

which is a particular case of Equation (27) by putting $\mu_M^j(x) = 1$ ($j = 1, \ldots, Count_M(x)$). Notice that the last function, (28), is naturally extended to the case of positive real-valued multisets using the same definition (28) except that $Count_M(x)$ is real-valued.

We thus assume that, instead of the membership sequences, fuzzy multisets and positive-real valued multisets are characterized by these functions. Let us consider these functions in a unified framework, namely, we allow operations between $f_A(y; x)$ by (27) and $f_M(y; x)$ by (28).

The operations of union and intersection are immediately defined by using $f_A(y; x) \vee f_M(y; x)$ and $f_A(y; x) \wedge f_M(y; x)$, respectively. On the other hand, the addition, which is characterisctic to the multisets, cannot directly be defined from these functions.

For the operation of the addition, we use a closed set $\nu_A(x)$ in the (y, z)-plane naturally derived from $f_A(y; x)$. Namely, we define

$$\nu_A(x) = \{ (y, z) \in \mathbf{R}^2 : 0 \le y \le \operatorname{supp} f_A(\cdot; x), 0 \le z \le f_A(y; x) \}, \tag{29}$$

where $\operatorname{supp} f_A(\cdot; x) = \{\, y \;:\; f_A(y; x) \neq 0 \,\}$. We moreover define a function $g_A(\cdot; x)$ of the variable z:

$$g_A(z; x) = \sup \{\, y \;:\; (y, z) \in \nu_A(x) \,\}. \tag{30}$$

We can now define the addition operation by using

$$g_{A \oplus M}(z; x) = g_A(z; x) + g_M(z; x), \tag{31}$$

whereby the set $\nu_{A \oplus B}(x)$ is defined by

$$\nu_{A \oplus B}(x) = \{\, (y, z) \in \mathbf{R}^2 \;:\; 0 \le z \le \operatorname{supp} g_{A \oplus M}(\cdot; x),$$
$$0 \le y \le g_{A \oplus B}(z; x) \,\}. \tag{32}$$

We now proceed to generalize the class of functions $f_A(y; x)$. This generalization is necessary when we consider fuzzy multisets and real-valued multisets in the same framework. We easily note that the arbitrary monotone non-increasing step functions are included in this set since they are derived from operations between fuzzy multisets and real-valued multisets.

Notice also that the cardinality is extended to the norm of the function $f_A(y; x)$; thus we define

$$|f_A(\cdot; x)| = \int_0^\infty f_A(y; x)\,dy, \tag{33}$$

$$\|f_A\| = \sum_{x \in X} |f_A(\cdot; x)|. \tag{34}$$

An appropriate class of f_A is the set of monotone nonincreasing and upper-semicontinuos functions whose norms are finite. This class is denoted by

$$\mathcal{GFM}_1(X) = \{\, f_A \;:\; \|f_A\| < \infty \,\}.$$

This class is justified from the fact that for an arbitrary $f \in \mathcal{GFM}_1(X)$, we can take a sequence of monotone non-increasing step functions that converges to f. This class is moreover the completion of the class of all monotone nonincreasing step functions whose cardinality is finite.

By abuse of terminology we write $A \in \mathcal{GFM}_1(X)$. Since f_A and the associated $\nu_A(x)$ is interchangeable, we are considering a generalized fuzzy multiset A characterized by the closed set $\nu_A(x)$ for each $x \in X$.

Operations on the generalized fuzzy multisets are now summarized.

(i) inclusion

$$A \subseteq B \iff \nu_A(x) \subseteq \nu_B(x), \quad \forall x \in X.$$

(ii) equality

$$A = B \iff \nu_A(x) = \nu_B(x), \quad \forall x \in X.$$

(iii) union

$$\nu_{A \cup B}(x) = \nu_A(x) \cup \nu_B(x), \quad \forall x \in X.$$

(iv) intersection

$$\nu_{A \cap B}(x) = \nu_A(x) \cap \nu_B(x), \quad \forall x \in X.$$

(v) addition

Define $g_A(z;x)$ and $g_B(z;x)$ by (30), and calculate $g_{A \oplus B}(z;x)$ using (31), whereby $\nu_{A \oplus B}(x)$ is obtained from (32).

It is easily seen that the $\mathcal{GFM}_1$ is a distributive lattice. This class is a metric space where the metric is defined by

$$d(A,B) = \|f_A - f_B\| = \sum_{x \in X} \int_0^\infty |f_A(y;x) - f_B(y;x)| dy.$$

5   Conclusion

We have reviewed fuzzy multisets and discussed features of infiniteness in fuzzy multisets. It should be noticed that an α-cut of such a fuzzy multiset with an infinite membership sequence will produce a finite crisp multiset. Further generalization using a closed set in a plane has been considered.

We thus observe analytical features in fuzzy multisets in addition to algebraic properties from the discussion of metric spaces and analytical completion.

Future studies include further development of the theory, e.g., t-conorms in fuzzy multisets and further generalization of the multisets. Moreover, the relation between fuzzy multisets and rough sets [12] should be investigated.

Applications of fuzzy multisets include modeling of information retrieval systems [11], which we will discuss elsewhere.

References

1. W.D. Blizard, Multiset theory, *Notre Dame Journal of Formal logic*, Vol. 30, No. 1, pp. 36–66, 1989.
2. W.D. Blizard, Real-valued multisets and fuzzy sets, *Fuzzy Sets and Systems*, Vol. 33, pp. 77–97, 1989.
3. B. Li, W. Peizhang, L. Xihui, Fuzzy bags with set-valued statistics, *Comput. Math. Applic.*, Vol. 15, pp. 811–818, 1988.
4. K.S. Kim and S. Miyamoto, Application of fuzzy multisets to fuzzy database systems, *Proc. of 1996 Asian Fuzzy Systems Symposium*, Dec. 11–14, 1996, Kenting, Taiwan, R.O.C. pp. 115–120, 1996.
5. D.E. Knuth, *The Art of Computer Programming, Vol.2 / Seminumerical Algorithms*, Addison-Wesley, Reading, Massachusetts, 1969.
6. Z.-Q. Liu, S. Miyamoto (Eds.), *Soft Computing and Human-Centered Machines*, Springer, Tokyo, 2000.
7. Z. Manna and R. Waldinger, *The Logical Basis for Computer Programming, Vol. 1: Deductive Reasoning*, Addison-Wesley, Reading, Massachusetts, 1985.
8. S. Miyamoto, Fuzzy multisets with infinite collections of memberships, *Proc. of the 7th International Fuzzy Systems Association World Congress (IFSA'97)*, June 25-30, 1997, Prague, Chech, Vol.1, pp.61-66, 1997.

9. S. Miyamoto, K.S. Kim, An image of fuzzy multisets by one variable function and its application, *J. of Japan Society for Fuzzy Theory and Systems*, Vol. 10, No. 1, pp. 157–167, 1998 (in Japanese).

10. S. Miyamoto, K.S. Kim, Multiset-valued images of fuzzy sets, *Proceedings of the Third Asian Fuzzy Systems Symposium*, June 18-21, 1998, Masan, Korea, pp.543-548.

11. S. Miyamoto, Rough sets and multisets in a model of information retrieval, in F.Crestani and G.Pasi, eds., *Soft Computing in Information Retrieval: Techniques and Applications*, Springer, pp. 373–393, 2000.

12. Z. Pawlak, *Rough Sets*, Kluwer, Dordrecht, 1991.

13. A. Ramer, C.-C. Wang, Fuzzy multisets, *Proc. of 1996 Asian Fuzzy Systems Symposium*, Dec. 11-14, 1996, Kenting, Taiwan, pp. 429–434.

14. A. Rebai, Canonical fuzzy bags and bag fuzzy measures as a basis for MADM with mixed non cardinal data, *European J. of Operational Res.*, Vol. 78, pp. 34–48, 1994.

15. A. Rebai, J.-M. Martel, A fuzzy bag approach to choosing the "best" multiattributed potential actions in a multiple judgement and non cardinal data context, *Fuzzy Sets and Systems*, Vol. 87, pp. 159–166, 1997.

16. R.R. Yager, On the theory of bags, *Int. J. General Systems*, Vol. 13, pp. 23–37, 1986.

Universality Results
for Some Variants of P Systems[*]

Madhu Mutyam and Kamala Krithivasan

Department of Computer Science and Engineering
Indian Institute of Technology, Madras
Chennai-36, Tamil Nadu, India
madhu@meena.iitm.ernet.in
kamala@iitm.ernet.in

Abstract. P systems, introduced by Gh. Păun, form a new class of distributed computing models. Many variants of P systems were shown to be computationally universal. In this paper, we consider several classes of P systems with symbol-objects where we allow the catalysts to move in and out of a membrane. We prove universality results for these variants of P systems with a very small number of membranes.

1 Introduction

P systems are a class of distributed parallel computing devices of a biochemical type, introduced in [8], which can be seen as a general computing architecture where various types of objects can be processed by various operations. In the basic model one considers a membrane structure consisting of several cell-like membranes placed inside a main membrane, called the *skin membrane*. If a membrane does not contain any other membrane, then it is called an *elementary membrane*. A membrane structure is represented by means of well-formed parenthesized expressions, strings of correctly matching parentheses, placed in a unique pair of matching parentheses which represents the skin membrane. Each pair of matching parentheses corresponds to a membrane. Graphically a membrane structure is represented by a Venn diagram without intersection and with a unique superset. The membranes delimit *regions*, where we place *objects*, elements of a finite set (an alphabet). The objects evolve according to given *evolution rules*, which are associated with regions. The rules are used in a *nondeterministic maximally parallel manner*: the objects to evolve and the rules to be applied to them are chosen in a nondeterministic manner, but no object which can evolve at a given step should remain idle. An object can evolve independently of the other objects in the same region of the membrane structure, or in cooperation with other objects. In particular, *catalysts* are considered. Catalysts are objects which evolve only together with other objects, but are not modified by the evolution (they just help other objects to evolve). The evolution rules are given in the form of multiset transition rules and in their right hand

---

[*] This research work was supported by IBM-Indian Research Laboratory, India

side contain pairs $(a, here), (a, out), (a, in)$, where a is an object. The meaning is that one occurrence of the symbol a is produced and it remains in the same region, is sent out of the respective membrane, or is sent nondeterministically into one of the inner membranes (which should be reachable from the region where the rule is applied), respectively.

Another feature considered by Gh. Păun in [9] is the possibility to control the membrane thickness (by using the actions of τ and δ). This is done as follows: Initially, all membranes are considered to be of thickness 1. If a rule in a membrane of thickness 1 introduces the symbol τ, then the thickness of the membrane increases to 2. A membrane of thickness 2 does not become thicker by using further rules which introduce the symbol τ, but no object can enter or exit it. If a rule which introduces the symbol δ is used in a membrane of thickness 1, then the membrane is dissolved; if the membrane had thickness 2, then it returns to thickness 1. Whenever the skin membrane is dissolved, the whole membrane system will be destroyed. If at the same step one uses rules which introduce both δ and τ in the same membrane, then the membrane does not change its thickness. No object can be communicated through a membrane of thickness two, hence rules which introduce commands *out, in,* requesting such communications, cannot be used. However, the communication has priority over changing the thickness: if at the same step an object should be communicated and a rule introduces the action τ, then the object is communicated and after that the membrane changes the thickness.

Starting from an initial configuration (identified by the membrane structure, the objects – with multiplicities – and rules placed in its regions) and using the evolution rules, we get a *computation*. We consider a computation complete when it halts, no further rule can be applied. Two ways of assigning a result to a computation were considered: one by designating an elementary membrane as the output membrane, and the other by reading the result outside the system. We deal here with the latter variant.

In [2], with one catalyst and without moving the catalyst across membranes, universality of *P systems* was achieved with four membranes. If we relax this condition by allowing movement of catalysts across membranes, with three moving catalysts we can achieve universality with only 3 membranes, and if there is no bound on the number of catalysts, then we can achieve universality with only 2 membranes. With three moving catalysts, we can achieve universality of *uniform P systems* (open problem suggested by Gh. Păun in [11]) with 3 membranes. Similarly, with three moving catalysts, we can achieve universality of *P systems with communicating rules* (open problem suggested by Gh. Păun in [11]) with 4 membranes. Finally, in [6], with one catalyst and without moving the catalyst across membranes, universality of *P systems with membrane creation* was achieved with the degree $(1, 4)$. If we use moving catalysts, then we can achieve universality of *P systems with membrane creation* with the degree $(1, 3)$.

In the next section we give some prerequisites. In Section 3 we define P systems with moving catalysts, Section 4 discusses uniform P systems with symbol-objects, Section 5 considers P systems with communicating rules, Section 6 discusses P systems with membrane creation and the paper concludes with a brief note in Section 7.

2 Prerequisites

For an alphabet V, V^* is the free monoid generated by V, λ is the empty string, and $V^+ = V^* - \{\lambda\}$. The length of $x \in V^*$ is denoted by $|x|$, while $|x|_a$ is the number of occurrences of the symbol a in the string x. If $V = \{a_1, \cdots, a_n\}$ (the ordering is important), then $\Psi_V(x) = (|x|_{a_1}, \cdots, |x|_{a_n})$ is the *Parikh vector* of the string $x \in V^*$. For a language $L \in V^*, \Psi_V(L) = \{\Psi_V(x) \mid x \in L\}$ is the *Parikh set* of L.

A matrix grammar with appearance checking is a construct $G = (N, T, S, M, F)$, where N, T are disjoint alphabets, $S \in N$, M is a finite set of sequences of the form $(A_1 \to x_1, \cdots, A_n \to x_n), n \geq 1$, of context-free rules over $N \cup T$ (with $A_i \in N, x_i \in (N \cup T)^*$, in all cases), and F is a set of occurrences of rules in M (N is the nonterminal alphabet, T is the terminal alphabet, S is the axiom, while the elements of M are called matrices.)

For $w, z \in (N \cup T)^*$ we write $w \Rightarrow z$ if there is a matrix $(A_1 \to x_1, \cdots, A_n \to x_n)$ in M and the strings $w_i \in (N \cup T)^*, 1 \leq i \leq n + 1$, such that $w = w_1, z = w_{n+1}$, and, for all $1 < i < n$, either $(1) w_i = w_i' A_i w_i'', w_{i+1} = w_i' x_i w_i''$, for some $w_i', w_i'' \in (N \cup T)^*$, or $(2) w_i = w_{i+1}, A_i$ does not appear in w_i, and some rule $A_i \to x_i$ appears in F. (The rules of a matrix are applied in order, possibly skipping the rules in F if they cannot be applied, so we say that these rules are applied in the *appearance checking* mode.)

The language generated by G is defined by $L(G) = \{w \in T^* \mid S \Rightarrow^* w\}$. The family of languages of this form is denoted by MAT_{ac}. It is known that $MAT_{ac} = RE$.

A matrix grammar $G = (N, T, S, M, F)$ is said to be in the binary normal form if $N = N_1 \cup N_2 \cup \{S, \dagger\}$, with these three sets mutually disjoint, and the matrices in M are in one of the following forms:

1. $(S \to XA)$, with $X \in N_1, A \in N_2$;
2. $(X \to Y, A \to x)$, with $X, Y \in N_1, A \in N_2, x \in (N_2 \cup T)^*$;
3. $(X \to Y, A \to \dagger)$, with $X, Y \in N_1, A \in N_2$;
4. $(X \to \lambda, A \to x)$, with $X \in N_1, A \in N_2, x \in T^*$;

Moreover, there is only one matrix of type 1 and F consists exactly of all rules $A \to \dagger$ appearing in matrices of type 3; $\dagger$ is called a trap symbol, because once introduced, it is never removed. A matrix of type 4 is used only once, in the last step of a derivation.

According to [1], for each matrix grammar there is an equivalent matrix grammar in the binary normal form.

For an arbitrary matrix grammar $G = (N, T, S, M, F)$, let us denote by $ac(G)$, the cardinality of the set $\{A \in N \mid A \to \alpha \in F\}$. From the construction in the proof of Lemma 1.3.7 in [1] one can see that if we start from a matrix grammar G and we get the grammar G' in the binary normal form, then $ac(G') = ac(G)$.

Improving the result from [7] (six nonterminals, all of them used in the appearance checking mode, suffice in order to characterize RE with matrix grammars), in [3] it was proved that four nonterminals are sufficient in order to characterize RE with matrix grammars and out of them only three are used in appearance checking rules. Of interest here is another result from [3]; if the total number of nonterminals is not restricted, then each recursively enumerable language can be generated by a matrix grammar G such that $ac(G) \leq 2$.

Consequently, to the properties of a grammar G in the binary normal form we can add the fact that $ac(G) \leq 2$. This is called the *strong binary normal form* for matrix grammars.

A P system (of degree $m \geq 1$) is a construct

$$\Pi = (V, T, C, \mu, w_1, \cdots, w_m, R_1, \cdots, R_m),$$

where:

1. V is an alphabet; its elements are called *objects*;
2. $T \subseteq V$, is the output alphabet;
3. $C \subseteq V, C \cap T = \phi$, is the set of catalysts;
4. μ is a membrane structure consisting of m membranes, with the membranes and the regions labeled in a one-to-one manner with elements in a given set; in this section we use the labels $1, \cdots, m$;
5. $w_i, 1 \leq i \leq m$, are strings representing multisets over V associated with the regions $1, \cdots, m$ of μ;
6. $R_i, 1 \leq i \leq m$, are finite set of *evolution rules* over V associated with the regions $1, \cdots, m$ of μ;

An evolution rule is of the form $u \rightarrow v$, where u is a string over V and $v = v'$ or $v = v'\delta$ or $v = v'\tau$, where v' is a string over $\{(a, here), (a, out), (a, in) \mid a \in (V - C)\}$, and δ and τ are special symbols not in V. The length of u is called *the radius* of the rule $u \rightarrow v$.

When presenting the evolution rules, the indication *here* is in general omitted.

If Π contains rules of radius greater than one, then we say that Π is a system *with cooperation*. Otherwise, it is a *non-cooperative* system. A particular class of cooperating systems is that of *catalytic* systems: the only rules of radius greater than one are of the form $ca \rightarrow cv$, where $c \in C, a \in (V - C)$, and v contains no catalyst; moreover, no other evolution rules contain catalysts (there is no rule of the form $c \rightarrow v$ or $a \rightarrow v_1 c v_2$, for $c \in C$).

The membrane structure and the multisets in Π constitute the *initial configuration* of the system. We can pass from a configuration to another one by using the evolution rules. This is done in parallel: all objects, from all membranes, which can be the subject of local evolution rules, should evolve simultaneously. A rule can be used only if there are objects which are *free* at the moment when we check its applicability.

The application of a rule $ca \rightarrow cv$ in a region containing a multiset w means to remove a copy of the object a in the presence of a copy of c, providing that such copies exist, then follow the prescriptions given by v: If an object appears in v in the form $(b, here)$, then it remains in the same region; if it appears in the form (b, out), then a copy of the object b will be introduced in the region of the membrane placed outside the region of the rule $ca \rightarrow cv$; if it appears in the form (b, in), then a copy of b is introduced into one of the inner membranes, nondeterministically chosen.

A sequence of transitions between configurations of a given P system Π is called a *computation* with respect to Π. A computation is *successful* if and only if it halts, that is, there is no rule applicable to the objects present in the last configuration. The result of a successful computation consists of the vector describing the multiplicity of all objects from T which are sent out of the system during the computation. The set of all vectors of natural numbers produced in this way by a system Π is denoted by $N(\Pi)$. The family of all such sets $N(\Pi)$ of vectors of natural numbers generated by a P system with catalysts and the actions of both δ, τ, and of degree at most $m, m \geq 1$, using target indications of the form $here, out, in$, is denoted by $NP_m(Cat, i/o, \delta, \tau)$; when one of the features $\alpha \in \{Cat, \delta, \tau\}$ is not present, we replace it with $n\alpha$. When the number of membranes is not bounded, we replace the subscript m with $*$.

By $PsRE$, we denote the family of recursively enumerable sets of vectors of natural numbers, i.e., the sets which are Parikh images of recursively enumerable languages. If we can compute $PsRE$ with a class of P systems, we say that universality can be achieved in that class of P systems.

3   P Systems with Moving Catalysts

So far in all the variants of P systems with symbol-objects, the evolution rules with a catalyst were of the form $ca \rightarrow cv$, where $c \in C$, $v = v'$ or $v = v'\delta$ or $v = v'\tau$, where v' is a string over $\{(a, here), (a, out), (a, in) \mid a \in (V - C)\}$. So, a catalyst will never go out of the region where it was initially placed.

Here we remove this restriction and allow a catalyst to move from one membrane to another one. At the same time we stick to the point that a catalyst evolves only together with other objects, but is not modified by the evolution (they just help other objects to evolve). In this new framework, an evolution rule with a catalyst will be of the form $ca \rightarrow (c, tar)v$, where $c \in C$, $tar \in \{here, out, in\}$, $v = v'$ or $v = v'\delta$ or $v = v'\tau$, where v' is a string over $\{(a, here), (a, out), (a, in) \mid a \in (V - C)\}$.

The family of sets of vectors of natural numbers $N(\Pi)$ generated by P systems with moving catalysts, and the actions δ, τ, and of degree at most $m \geq 1$, using target indications of the form $here, out, in$, using at most n catalysts, is denoted by $NP_m(MCat_n, i/o, \delta, \tau)$. In [2], authors proved

$$PsRE = NP_4(Cat, i/o, \delta, \tau).$$

Here we improve this result by achieving universality with only 3 membranes with 3 moving catalysts. If there is no bound on the number of catalysts, then we can achieve universality with only 2 membranes.

Theorem 1. $PsRE = NP_3(MCat_3, i/o, \delta, \tau)$.

Proof. We prove only the inclusion $PsRE \subseteq NP_3(MCat_3, i/o, \delta, \tau)$, the reverse inclusion can be proved in a straightforward manner. Let us consider a matrix grammar with appearance checking, $G = (N, T, S, M, F)$, in the strong binary normal form, that is with $N = N_1 \cup N_2 \cup \{S, \dagger\}$, with $ac(G) \leq 2$, and with the rules of the four forms mentioned in Section 2.

Assume that we are in the worst case, with $ac(G) = 2$, and let $B^{(1)}$ and $B^{(2)}$ be the two symbols in N_2 for which we have rules $B^{(j)} \rightarrow \dagger$ in matrices of M. Let us assume that we have h matrices of the form $m'_i : (X \rightarrow Y, B^{(j)} \rightarrow \dagger), X, Y \in N_1, j \in \{1, 2\}, 1 \leq i \leq h$, and k matrices of the form $m_i : (X \rightarrow \alpha, A \rightarrow x), X \in N_1, A \in N_2, \alpha \in N_1 \cup \{\lambda\}$, and $x \in (N_2 \cup T)^*, 1 \leq i \leq k$. The matrices of the form $(X \rightarrow Y, B^{(j)} \rightarrow \dagger), X, Y \in N_1$ are labeled by m'_i, with $i \in lab_j$, for $j \in \{1, 2\}$, such that lab_1, lab_2 and $lab_0 = \{1, 2, \cdots, k\}$ are mutually disjoint sets.

We construct a P system with symbol-objects of degree 3 as

$$\Pi = (V, T, C, \mu, w_1, w_2, w_3, R_1, R_2, R_3),$$

with the following components:

$$V = N_1 \cup N_2 \cup \{X_i, A_i \mid X \in N_1, A \in N_2, 1 \leq i \leq 3k\}$$
$$\cup \{X'_{3i}, A'_{3i} \mid X \in N_1, 1 \leq i \leq k\} \cup \{X^1_i, X^2_i \mid X \in N_1, 1 \leq i \leq h\}$$
$$\cup \{Z, E, E_1, \dagger\} \cup \{c_0, c_1, c_2\};$$
$$C = \{c_0, c_1, c_2\};$$
$$\mu = [_1[_2[_3]_3]_2]_1;$$
$$w_1 = \phi;$$
$$w_2 = \{c_0, c_1, c_2, X, A\};$$
$$w_3 = \{E\};$$

R_1 contains the following rules:

1. $X_{3i-2} \rightarrow X_{3i-1}Z\tau, 1 \leq i \leq k;$
2. $A_{3i-2} \rightarrow A_{3i-1}\delta, 1 \leq i \leq k;$
3. $X_{3i-1} \rightarrow X_{3i}Z\tau, 1 \leq i \leq k;$
4. $A_{3i-1} \rightarrow A_{3i}\delta, 1 \leq i \leq k;$
5. $X_{3i} \rightarrow X'_{3i}Z\tau, 1 \leq i \leq k - 1;$
6. $A_{3i} \rightarrow A'_{3i}\delta, 1 \leq i \leq k - 1;$
7. $X'_{3i} \rightarrow X_{3i+1}Z\tau, 1 \leq i \leq k - 1;$
8. $A'_{3i} \rightarrow A_{3i+1}\delta, 1 \leq i \leq k - 1;$
9. $Z \rightarrow (\lambda, out);$
10. $Z \rightarrow \dagger;$
11. $c_0X_{3i} \rightarrow (c_0Y, in)$, for $m_i : (X \rightarrow Y, A \rightarrow x), 1 \leq i \leq k;$
12. $X_{3i} \rightarrow \lambda$, for $m_i : (X \rightarrow \lambda, A \rightarrow x), 1 \leq i \leq k;$
13. $A_{3i} \rightarrow (x, in)$, for $m_i : (X \rightarrow \alpha, A \rightarrow x), 1 \leq i \leq k, \alpha \in N_1 \cup \{\lambda\};$
14. $X_{3k} \rightarrow \dagger;$

15. $A_{3k} \to \dagger$;
16. $a \to (a, out), a \in T$;
17. $\dagger \to \dagger$;

R_2 contains the following rules:

1. $X \to (X_1, out)\tau$;
2. $c_0 A \to (c_0 A_1, out)\delta$;
3. $c_j X \to (c_j X_i^j, in)$, for $m_i' : (X \to Y, B^{(j)} \to \dagger), j \in \{1, 2\}$;
4. $A \to (A, in), A \in N_2$;
5. $a \to (a, out), a \in T$;

R_3 contains the following rules:

1. $E \to E_1 \tau$;
2. $E_1 \to E\delta$;
3. $c_j X_i^j \to (c_j Y, out)$, for $m_i' : (X \to Y, B^{(j)} \to \dagger), j \in \{1, 2\}$;
4. $A \to (A, out)$;
5. $c_j B^{(j)} \to c_j \dagger, j \in \{1, 2\}$;
6. $\dagger \to \dagger$;

The system works as follows.

Initially, membrane 2 contains the objects c_0, c_1, c_2, X, and A. If we want to simulate a matrix m_i of type 2, then the object X will be replaced with X_1 and sent out when also increasing the thickness of membrane 2. At the same time, the object A with the help of the catalyst c_0 will be replaced with A_1 and sent out when also decreasing the thickness of membrane 2. Since we apply these two rules simultaneously, the thickness of membrane 2 will not be changed. On the other hand, if we apply only $X \to (X_1, out)\tau$, then a trap symbol will be introduced in the skin membrane by using the rule $Z \to \dagger$ (since we can not apply the rule $Z \to (\lambda, out)$). Similarly, if we apply only $c_0 A \to (c_0 A_1, out)\delta$, then the whole system will be destroyed by using a rule $A_1 \to A_2 \delta$ in the skin membrane. So, after the simultaneous application of the rules $X \to (X_1, out)\tau$ and $c_0 A \to (c_0 A_1, out)\delta$, the skin membrane gets X_1 and $c_0 A_1$. The subscripts of X and A are incremented step by step. At any time we can apply the rules $c_0 X_{3i} \to (c_0 Y, in)$ and $A_{3i} \to (x, in)$. If we apply these two rules in different steps, then either a trap symbol will be introduced or the whole system will be destroyed. Thus, we can simulate a matrix m_i of type 2.

If we simulate a matrix m_i of type 4, then the procedure is the same as above, excepting the last step, where we apply the rule $X_{3i} \to \lambda$ instead of $c_0 X_{3i} \to (c_0 Y, in)$. After simulating a matrix m_i of type 4, if there is any symbol $A \in N_2$ present in membrane 2, then it will keep on evolving by using the rules $A \to (A, in)$ in membrane 2 and $A \to (A, out)$ in membrane 3, so that the computation never halts.

If we simulate a matrix m_i' of type 3, then we use the rules $c_j X \to (c_j X_i^j, in)$, for $m_i' : (X \to Y, B^{(j)} \to \dagger), j \in \{1, 2\}$, and $A \to (A, in), A \in N_2$. If we use the rule $c_0 A \to (c_0 A_1, out)\delta$ simultaneously with the above set of rules, then the whole system will be destroyed by using the rule $A_1 \to A_2 \delta$. Once the objects

c_j, X_i^j and $A, A \in N_2$ enter membrane 3, its thickness will be increased to 2 (due to the application of the rule $E \to E_1\tau$). As the thickness of the membrane 3 is two, we cannot apply the rules $c_j X_i^j \to (c_j Y, out)$, for $m_i' : (X \to Y, B^{(j)} \to$ $\dagger), j \in \{1, 2\}$ and $A \to (A, out)$. At this step, if there is a nonterminal $B^{(j)}$ in membrane 3, then we can apply the rule $c_j B^{(j)} \to c_j\dagger, j \in \{1, 2\}$, so that the computation never halts. Otherwise, after reducing the thickness to 1 (by using the rule $E_1 \to E\delta$), we can apply $c_j X_i^j \to (c_j Y, out)$, for $m_i' : (X \to Y, B^{(j)} \to$ $\dagger), j \in \{1, 2\}$ and $A \to (A, out)$. In this way, we can simulate a matrix m_i' of type 3. Thus, the equality $\Psi_T(L(G)) = N(\Pi)$ follows.   □

Theorem 2. $PsRE = NP_2(MCat_*, i/o, \delta, \tau)$.

Proof. We proceed as in the previous proof, by considering a matrix grammar in the strong binary normal form, $G = (N, T, S, M, F)$. Let us assume that we have h matrices of the form $m_i' : (X \to Y, B^{(j)} \to \dagger), X, Y \in N_1, j \in \{1, 2\}, 1 \le i \le h$, and k matrices of the form $m_i : (X \to Y, A \to x), X \in N_1, A \in N_2, Y \in N_1 \cup \{\lambda\}$, and $x \in (N_2 \cup T)^*, 1 \le i \le k$. Each matrix of the form $(X \to \lambda, A \to x), X \in N_1, A \in N_2, x \in T^*$, is replaced by $(X \to f, A \to x)$, where f is a new symbol. We continue to label the obtained matrix in the same way as the original one. Here we consider the catalyst set $C = \{c_1, \cdots, c_k, d_1, d_2\}$ where each c_i is for $m_i : (X \to Y, A \to x), 1 \le i \le k$ and each d_j is for $m_i' : (X \to Y, B^{(j)} \to \dagger), j \in \{1, 2\}$. We construct a P system with symbol-objects of degree 2 as

$$\Pi = (V, T, C, \mu, w_1, w_2, R_1, R_2),$$

with the following components:

$V = N_1 \cup N_2 \cup \{X', X^I, X_1 \mid X \in N_1\} \cup \{f, f_1, f', \dagger\} \cup \{c_1, c_2, \cdots, c_k, d_1, d_2\};$

$C = \{c_1, c_2, \cdots, c_k, d_1, d_2\};$

$\mu = [_1[_2]_2]_1;$

$w_1 = \{c_1, c_2, \cdots, c_k, d_1, d_2, X\},$ for $(S \to XA)$ of G;

$w_2 = \{A\},$ for $(S \to XA)$ of G;

R_1 contains the following rules:

1. $c_i X \to (c_i Y', in)$, for $m_i : (X \to Y, A \to x), Y \in N_1 \cup \{f\}$, of type 2 and 4;
2. $d_j X \to (d_j Y^I, in)$, for $m_i' : (X \to Y, B^{(j)} \to \dagger), j \in \{1, 2\}$, of type 3;
3. $f \to \lambda;$
4. $a \to (a, out), a \in T;$
5. $Y_1 \to \dagger, Y \in N_1;$
6. $A \to \dagger, A \in N_2;$
7. $\dagger \to \dagger;$

R_2 contains the following rules:

1. $Y' \to Y_1\tau, Y \in N_1 \cup \{f\};$
2. $c_i A \to c_i x\delta$, for $m_i : (X \to Y, A \to x), Y \in N_1 \cup \{f\}$, of type 2 and 4;

3. $c_i Y_1 \to (c_i Y, out), Y \in N_1$;
4. $Y_1 \to Y_1, Y \in N_1$;
5. $f_1 \to f\delta$;
6. $f \to f$;
7. $Y^I \to Y$;
8. $d_j B^{(j)} \to d_j\dagger$, for $m_i : (X \to Y, B^{(j)} \to \dagger), j \in \{1,2\}$, of type 3;
9. $d_j Y \to (d_j Y, out), j \in \{1,2\}$;
10. $a \to (a, out), a \in T$;
11. $\dagger \to \dagger$;

The system works as follows.

The skin membrane contains $c_1, \cdots, c_k, d_1, d_2$, and X and membrane 2 contains A. (The symbols X and A are for the matrix $(S \to XA)$.)

To simulate a matrix m_i of type 2 or 4, with the help of a catalyst c_i, X will be replaced with $Y', Y \in N_1 \cup \{f\}$, and move into the inner membrane 2 along with c_i. In membrane 2, Y' will be replaced with Y_1 and the thickness of the membrane will be incremented. At the same time if any symbol $A \in N_2$ is present (for a matrix $m_i : (X \to Y, A \to x)$), then with the help of the catalyst c_i, it will be replaced with x and the thickness will be decremented. Otherwise, the thickness is 2 so that we cannot apply the rule $c_i Y_1 \to (c_i Y, out)$ and the computation never halts due to the application of the rule $Y_1 \to Y_1, Y \in N_1 \cup \{f\}$. On the other hand, if there are multiple copies of $A \in N_2$ for $m_i : (X \to Y, A \to x)$, and instead of applying the rule $c_i Y_1 \to (c_i Y, out)$, we apply again the same rule $c_i A \to c_i x \delta$, then a trap symbol will be introduced in the skin membrane by using the rule $Y_1 \to \dagger, Y \in N_1$. If we simulate a matrix of type 4, then the symbol f will be erased in the skin membrane. After simulating a matrix of type 4, if there is any symbol $A \in N_2$ is present in the skin membrane, then a trap symbol will be introduced using the rule $A \to \dagger$.

To simulate a matrix m_i' of type 3, with the help of a catalyst $d_j, j \in \{1,2\}$, X will be replaced with Y^I and move into the inner membrane 2 together with d_j. In membrane 2, Y^I will be replaced with Y. At the same time, if any symbol $B^{(j)} \in N_2$ (for $m_i : (X \to Y, B^{(j)} \to \dagger), j \in \{1,2\}$) is present in the membrane 2, then a trap symbol will be introduced by using the rule $d_j B^{(j)} \to d_j\dagger$ so that the computation never halts. Otherwise, with the help of the same catalyst d_j, Y will be sent out (using the rule $d_j Y \to (d_j Y, out)$) from membrane 2. In this way, we can simulate a matrix of type 3. Thus, the equality $\Psi_T(L(G)) = N(\Pi)$ follows. □

4   Uniform P Systems with Symbol-Objects

In [5], the author considered a variant of P systems: simple or uniform P systems based on rewriting and splicing. Here we consider another variant, i.e., uniform P systems with symbol-objects. The idea of having such a system and to study its properties was suggested as an open problem in [11]. Uniform P systems are the systems for which we have a single set of rules for all the membranes.

Formally, a *uniform P system with symbol-objects of degree* $m, m \geq 1$ is a construct

$$\Pi = (V, T, C, \mu, w_1, \cdots, w_m, R),$$

with the usual notation as given in Section 2, and where R is a finite set of *evolution rules* over V.

As usual, an evolution rule is of the form $u \to v$, where u is a string over V and $v = v'$ or $v = v'\delta$ or $v = v'\tau$, where v' is a string over $\{(a, here), (a, out), (a, in) \mid a \in (V - C)\}$, and δ and τ are special symbols not in V. The computations in Π are defined as in a system having R as present in all its regions. The family of sets of vectors of natural numbers $N(\Pi)$ generated by uniform P systems with symbol objects with at most n moving catalysts, and the actions δ, τ, and of degree at most $m \geq 1$, using target indications of the form $here, out, in$, is denoted by $NUP_m(MCat_n, i/o, \delta, \tau)$.

Theorem 3. $PsRE = NUP_3(MCat_3, i/o, \delta, \tau)$.

Proof. We proceed as in the previous proofs: starting from a matrix grammar in the strong binary normal form, $G = (N, T, S, M, F)$, we construct a uniform P system with symbol-objects of degree 3 as

$$\Pi = (V, T, C, \mu, w_1, w_2, w_3, R),$$

with the following components :

$$V = N_1 \cup N_2 \cup \{X_i, A_i, A' \mid X \in N_1, A \in N_2, 1 \leq i \leq 3k\}$$
$$\cup \{X'_{3i} \mid X \in N_1, 1 \leq i \leq k\} \cup \{X_i^1, X_i^2 \mid X \in N_1, 1 \leq i \leq h\}$$
$$\cup \{Z, E, E_1, \dagger\} \cup \{c_0, c_1, c_2\};$$
$$C = \{c_0, c_1, c_2\};$$
$$\mu = [_1[_2[_3]_3]_2]_1;$$
$$w_1 = \phi;$$
$$w_2 = \{c_0, c_1, c_2, X, A\};$$
$$w_3 = \{E\};$$

R contains the following rules:

1. $X \to (X_1, out)\tau$;
2. $c_0 A \to (c_0 A_1, out)\delta$;
3. $X_{3i-2} \to X_{3i-1}Z\tau, 1 \leq i \leq k$;
4. $A_{3i-2} \to A_{3i-1}\delta, 1 \leq i \leq k$;
5. $X_{3i-1} \to X_{3i}Z\tau, 1 \leq i \leq k$;
6. $A_{3i-1} \to A_{3i}\delta, 1 \leq i \leq k$;
7. $X_{3i} \to X'_{3i}Z\tau, 1 \leq i \leq k - 1$;
8. $A_{3i} \to A'_{3i}\delta, 1 \leq i \leq k - 1$;
9. $X'_{3i} \to X_{3i+1}Z\tau, 1 \leq i \leq k - 1$;
10. $A'_{3i} \to A_{3i+1}\delta, 1 \leq i \leq k - 1$;

11. $Z \to (\lambda, out)$;
12. $Z \to \dagger$;
13. $c_0 X_{3i} \to (c_0 Y, in)$, for $m_i : (X \to Y, A \to x), 1 \le i \le k$;
14. $X_{3i} \to \lambda$, for $m_i : (X \to \lambda, A \to x), 1 \le i \le k$;
15. $A_{3i} \to (x, in)$, for $m_i : (X \to \alpha, A \to x), 1 \le i \le k, \alpha \in N_1 \cup \{\lambda\}$;
16. $X_{3k} \to \dagger$;
17. $A_{3k} \to \dagger$;
18. $c_j X \to (c_j X_i^j, in)$, for $m_i' : (X \to Y, B^{(j)} \to \dagger), j \in \{1, 2\}$;
19. $A \to (A', in), A \in N_2$;
20. $E \to E_1 \tau$;
21. $E_1 \to E\delta$;
22. $c_j X_i^j \to (c_j Y, out)$, for $m_i' : (X \to Y, B^{(j)} \to \dagger), j \in \{1, 2\}$;
23. $A' \to (A, out), A \in N_2$;
24. $c_j B^{(j)'} \to c_j \dagger, j \in \{1, 2\}$;
25. $a \to (a, out), a \in T$;
26. $\dagger \to \dagger$;

The proof of the equality $\Psi_T(L(G)) = N(\Pi)$ is similar to Theorem 1. The main difference is in rules 19 and 23, where a symbol object $A \in V$ becomes A' while it enters membrane 3, and becomes A while comes back to membrane 2. $\square$

5   P Systems with Communicating Rules

We consider now another open problem from [11], about *P systems with no object evolving rules*, i.e., no rule of the form $a \to bc$. So, such a system consists of only rules used for object communication from one region to another region. Because no object can be produced inside the system, we have to consider rules for bringing objects from a specified source, for example, from the outer region. This can be done by considering a new command associated with objects, besides, *here, in, out*, for instance, *come*; rules introducing symbols $(b, come)$ will be used in the skin membrane only, with the obvious meaning, of bringing a copy of b from outside of the system. Note that even simple rules of the form $a \to b$ are not allowed, here we have either to allow changing of objects when passing through a membrane, or to simulate them, for instance, by sending the symbol a outside the system and, at the same time, bringing b from the outer region.

Formally a *P system with communicating rules* of degree $m \ge 1$, is a construct

$$\Pi = (V, T, C, \mu, w_1, \cdots, w_m, R_1, \cdots, R_m),$$

with the usual notation as given in Section 2 and R_i sets of rules over V associated with the regions of μ.

The rules are of the following types:

1. If a is an object, then a rule of the form $a \to (v, come)$, where v is a string of objects, can be used in the skin membrane only;

2. If a is an object and c is a catalyst, then a rule of the form $ca \rightarrow (c, tar)$ $(v, come), tar \in \{here, in\}$, where v is a string of objects, can be used in the skin membrane only;

3. If a and b are single objects, then a rule of the form $a \rightarrow (b, tar), tar \in \{out, in\}$, can be used in any membrane;

4. If a, b are single objects and c is a catalyst, then a rule of the form $ca \rightarrow (c, tar_1)(b, tar_2), tar_1 \in \{here, in, out\}$ and $tar_2 \in \{out, in\}$, can be used in any membrane;

5. If a is an object, then a rule of the form $a \rightarrow (a, tar), tar \in \{here, out, in\}$, can be used in any membrane;

6. If a is an object and c is a catalyst, then a rule of the form $ca \rightarrow (c, tar)(a, tar)$, $tar \in \{here, out, in\}$, can be used in any membrane;

The application of a rule $ca \rightarrow (c, tar)(v, come)$ in the skin membrane containing a multiset w means to remove a copy of the object a in the presence of the catalyst c, provided that such copies exist, and adding the objects specified v to it. At the same time, the catalyst c will follow the prescription given by tar: if tar is $here$, then the catalyst c remains in the same region; if tar is in, then the catalyst c will be introduced in one of the inner membranes. The main idea behind the application of this rule is to send one copy of a, in presence of the catalyst c, out of the skin membrane and at the same time to bring the objects specified by v into the skin membrane. The application of a rule $ca \rightarrow (c, tar)(a, tar)$ in a region containing a multiset w means to remove a copy of the object a in presence of the catalyst c, provided that such copies exist, then to follow the prescriptions given by tar. The application of a rule $ca \rightarrow c(b, tar), tar \in \{out, in\}$, is similar. Similarly, a single object can move either into a membrane or out of the membrane by transforming into another object by rules of the form $a \rightarrow (b, tar), tar \in \{out, in\}$. The use of τ and δ for varying thickness of the membrane are as discussed in Section 2. Computations are defined in the usual manner. The family of sets of vectors of natural numbers $N(\Pi)$ generated by a P system with communicating rules with at most n moving catalysts, the actions δ, τ, and of the degree at most $m \geq 1$, using target indications of the form $here, out, in$, is denoted by $NPC_m(MCat_n, i/o, \delta, \tau)$.

Theorem 4. $PsRE = NPC_4(MCat_3, i/o, \tau, \delta)$

Proof. We proceed as in the previous proofs: starting from a matrix grammar in the strong binary normal form, $G = (N, T, S, M, F)$, we construct a *P system with communicating rules* of degree 4 as

$$\Pi = (V, T, C, \mu, w_1, w_2, w_3, w_4, R_1, R_2, R_3, R_4),$$

where

$$V = N_1 \cup N_2 \cup \{X', X_i, A_i \mid X \in N_1, A \in N_2, 1 \leq i \leq 2k\} \cup \{X_i^1, X_i^2 \mid X \in N_1, 1 \leq i \leq h\}$$
$$\cup \{X_{2i}', X_{2i}'', A_{2i}', A_{2i}'' \mid X \in N_1, A \in N_2, 1 \leq i \leq k-1\}$$
$$\cup \{Z, E_1, E_2, \dagger\} \cup \{c_0, c_1, c_2\};$$

$C = \{c_0, c_1, c_2\}$;

$\mu = [_1[_2[_3[_4]_4]_3]_2]_1$;

$w_2 = \{c_0, c_1, c_2, X, A\}$;

$w_1 = w_4 = \phi$;

$w_3 = \{E_1\}$;

R_1 contains the following rules:

1. $X_{2i-1} \to (X_{2i}Z, come)\tau, 1 \le i \le k$;
2. $A_{2i-1} \to (A_{2i}, come)\delta, 1 \le i \le k$;
3. $X_{2i} \to (X'_{2i}Z, come)\tau, 1 \le i \le k-1$;
4. $A_{2i} \to (A'_{2i}, come)\delta, 1 \le i \le k-1$;
5. $X'_{2i} \to (X''_{2i}Z, come)\tau, 1 \le i \le k-1$;
6. $A'_{2i} \to (A''_{2i}, come)\delta, 1 \le i \le k-1$;
7. $X''_{2i} \to (X_{2i+1}Z, come)\tau, 1 \le i \le k-1$;
8. $A''_{2i} \to (A_{2i+1}, come)\delta, 1 \le i \le k-1$;
9. $X_{2i} \to (Y'Z, come)\tau$, for $m_i : (X \to Y, A \to x), 1 \le i \le k$;
10. $X_{2i} \to (Z\lambda, come)\tau$, for $m_i : (X \to \lambda, A \to x), 1 \le i \le k$;
11. $A_{2i} \to (x, come)\delta, 1 \le i \le k$;
12. $c_0Y' \to (c_0Y, in)$;
13. $Z \to (\lambda, out)$;
14. $Z \to (\dagger, in)$;
15. $A \to (A, in), A \in N_2$;
16. $X_{2k} \to (\dagger, in)$;
17. $A_{2k} \to (\dagger, in)$;
18. $a \to (a, out), a \in T$;

R_2 contains the following rules:

1. $X \to (X_1, out)\tau$;
2. $c_0A \to (c_0A_1, out)\delta$;
3. $c_jX \to (c_jX_i^j, in)$, for $m'_i : (X \to Y, B^{(j)} \to \dagger), j \in \{1,2\}$;
4. $A \to (A, in), A \in N_2$;
5. $\dagger \to \dagger$;

R_3 contains the following rules:

1. $E_1 \to (E_2, in)\tau$;
2. $E_2 \to (E_1, in)\delta$;
3. $c_jX_i^j \to (c_jY, out)$, for $m'_i : (X \to Y, B^{(j)} \to \dagger), 1 \le i \le h, j \in \{1,2\}$;
4. $c_jB^{(j)} \to c_j(\dagger, in), j \in \{1,2\}$;
5. $A \to (A, out), A \in N_2$;

R_4 contains the following rules:

1. $E_1 \to (E_1, out)$;
2. $E_2 \to (E_2, out)$;
3. $\dagger \to \dagger$;

The system works as follows.

Initially, membrane 2 contains the objects c_0, c_1, c_2, X, and A. If we want to simulate a matrix m_i of type 2, then the object X will be replaced with X_1 and sent out while increasing the thickness of membrane 2. At the same time, the object A with the help of the catalyst c_0 will be replaced with A_1 and sent out while decreasing the thickness of membrane 2. Since we apply these two rules simultaneously, the thickness of membrane 2 will not be changed. On the other hand, if we apply only $X \to (X_1, out)\tau$, then a trap symbol will be introduced by using the rule $Z \to (\dagger, in)$ (since we can not apply the rule $Z \to (\lambda, out)$). Similarly, if we apply only $c_0 A \to (c_0 A_1, out)\delta$, then the whole system will be destroyed by using a rule $A_1 \to (A_2, come)\delta$ in the skin membrane. So, after the simultaneous application of the rules $X \to (X_1, out)\tau$ and $c_0 A \to (c_0 A_1, out)\delta$, the skin membrane gets X_1 and $c_0 A_1$. The subscripts of X and A are incremented step by step. At any time we can apply the rules $X_{2i} \to (Y'Z, come)\tau$ and $A_{2i} \to (x, come)\delta$. If we apply these two rules in different steps, then either a trap symbol will be introduced or the whole system will be destroyed. In the next step, Y' will be replaced with Y and sent into membrane 2 and all $A \in N_2$ will also be sent to membrane 2. Thus, we can simulate a matrix m_i of type 2.

If we simulate a matrix m_i of type 4, then the procedure is the same as above, except in the last step, where we apply the rule $X_{2i} \to (Z\lambda, come)\tau$ instead of $X_{2i} \to (Y'Z, come)\tau$. After simulating a matrix m_i of type 4, if there is any symbol $A \in N_2$ is present in membrane 2, then it will keep on evolving by using the rules $A \to (A, in)$ in membrane 2 and $A \to (A, out)$ in membrane 3, so that the computation never halts.

If we simulate a matrix m'_i of type 3, then we use the rules $c_j X \to (c_j X_i^j, in)$, for $m'_i : (X \to Y, B^{(j)} \to \dagger), j \in \{1, 2\}$ and $A \to (A, in), A \in N_2$. If we use the rule $c_0 A \to (c_0 A_1, out)\delta$ simultaneously with above set of rules, then the whole system will be destroyed by using the rule $A_1 \to (A_2, come)\delta$. Once the objects c_j, X_i^j and $A(\in N_2)$ enters membrane 3, its thickness will be increased to 2 (due to simultaneous application of the rule $E_1 \to (E_2, in)\tau$). As the thickness of membrane 3 is two, we can not apply the rules $c_j X_i^j \to (c_j Y, out)$, for $m'_i : (X \to Y, B^{(j)} \to \dagger), j \in \{1, 2\}$ and $A \to (A, out)$. At this step, if there is a nonterminal $B^{(j)}$ in the membrane 3 then we can apply the rule $c_j B^{(j)} \to c_j(\dagger, in), j \in \{1, 2\}$, so that the computation never halts. Otherwise, after reducing the thickness to 1 (by using the rule $E_2 \to (E_1, in)\delta$) we can apply $c_j X_i^j \to (c_j Y, out)$, for $m'_i : (X \to Y, B^{(j)} \to \dagger), j \in \{1, 2\}$ and $A \to (A, out)$. In this way we can simulate a matrix m'_i of type 3. Thus, the equality $\Psi_T(L(G)) = N(\Pi)$ follows.

$\square$

6   P Systems with Membrane Creation

P systems with membrane creation [4] constitute a variant of P systems where each membrane contains both productive and non-productive objects. A productive object is an object which can create a new membrane and transforms into another object. A non-productive object only transforms into another object without creating a new membrane.

Formally a *P system with membrane creation* of degree (m, n), $n \geq m \geq 1$, is a construct

$$\Pi = (V, T, C, \mu, w_1, \cdots, w_m, R_1, \cdots, R_n),$$

with the usual notation as given in Section 2 and R_i, $1 \leq i \leq n$, are finite set of evolution rules over V. The evolution rules are of two types:

1. If a is a non-productive object, then the evolution rule is in the form $a \rightarrow v$ or $ca \rightarrow (c, tar)v$, where $c \in C$, $tar \in \{in, out, here\}$, $a \in (V - C)$, $v = v'$ or $v = v'\delta$ or $v = v'\tau$, where v' is a multiset of objects over $((V - C) \times \{here, in, out\})$ and δ, τ are special symbols not in V.
2. If a is a productive object, then the evolution rule is in the form $a \rightarrow [_i v]_i$ or $ca \rightarrow (c, tar)[_i v]_i$, where $c \in C$, $tar \in \{in, out, here\}$, $a \in (V - C)$, v is a multiset of objects over $(V - C)$. A rule of the form $a \rightarrow [_i v]_i$ means that the object identified by a is transformed into the objects identified by v, surrounded by a new membrane having the label i.

The family of all sets of vectors of natural numbers $N(\Pi)$ generated by P systems with membrane creation with at most r moving catalysts and the actions of both δ, τ, and of degree $(m, n), n \geq m \geq 1$ is denoted by

$$NPMC_{(m,n)}(MCat_r, i/o, \delta, \tau).$$

In [6], it is proved that

$$PsRE = NPMC_{(1,4)}(Cat, tar, \tau, \delta).$$

Here we improve this result by achieving $PsRE$ with the degree $(1,3)$, but making use of moving catalysts.

Theorem 5. $PsRE = NPMC_{(1,3)}(MCat_3, i/o, n\tau, \delta)$.

Proof. We proceed as in the previous proofs: starting from a matrix grammar in the strong binary normal form, $G = (N, T, S, M, F)$, we construct a *P system with membrane creation* as

$$\Pi = (V, T, C, \mu, w_1, R_1, R_2, R_3),$$

with the following components :

$$\begin{aligned}
V = \, & N_1 \cup N_2 \cup \{X', X_i \mid X \in N_1, 1 \leq i \leq k\} \\
& \cup \{X_i^1, X_i^2 \mid X \in N_1, 1 \leq i \leq h\} \\
& \cup \{A_i \mid A \in N_2, 1 \leq i \leq k\} \\
& \cup \{E, E_1, f, Z, \dagger\} \cup \{c_0, c_1, c_2\} \\
C = \, & \{c_0, c_1, c_2\}; \\
\mu = \, & [_1]_1; \\
w_1 = \, & \{c_0, c_1, c_2, X, A\}, \text{ for } (S \rightarrow XA) \text{ being the initial matrix of } G;
\end{aligned}$$

and the sets R_1, R_2 and R_3 contains the following rules:

- R_1 :
  1. $X \rightarrow [_2 X']_2$;
  2. $X \rightarrow [_3 E X_i^j]_3$, for $m_i' : (X \rightarrow Y, B^{(j)} \rightarrow \dagger)$;
  3. $c_0 A \rightarrow (c_0 A_1, in), A \in N_2 - \{B^{(1)}, B^{(2)}\}$;
  4. $c_j B^{(j)} \rightarrow (c_j B^{(j)}, in), j \in \{1, 2\}$;
  5. $X_i^{(j)} \rightarrow Y$, for $m_i' : (X \rightarrow Y, B^{(j)} \rightarrow \dagger)$;
  6. $A_i \rightarrow \dagger, 1 \leq i \leq k$;
  7. $f \rightarrow [_3 \lambda]_3$;
  8. $Z \rightarrow \lambda$;
  9. $a \rightarrow (a, out)$;
  10. $\dagger \rightarrow \dagger$;
- R_2 :
  1. $X' \rightarrow X_1$;
  2. $X_i \rightarrow X_{i+1}, 1 \leq i \leq (k-1)$;
  3. $A_i \rightarrow A_{i+1}, 1 \leq i \leq (k-1)$;
  4. $X_i \rightarrow Y\delta$, for $m_i : (X \rightarrow Y, A \rightarrow x), X \in N_1, Y \in N_1 \cup \{f\}$;
  5. $A_i \rightarrow xZ$, for $m_i : (X \rightarrow Y, A \rightarrow x), X \in N_1, Y \in N_1 \cup \{f\}$;
  6. $Z \rightarrow \dagger$;
  7. $X_k \rightarrow \dagger$;
  8. $A_k \rightarrow \dagger$;
  9. $\dagger \rightarrow \dagger$;
- R_3 :
  1. $E \rightarrow E_1$;
  2. $c_j X_i^j \rightarrow c_j \dagger, j \in \{1, 2\}$;
  3. $E_1 \rightarrow \lambda\delta$;
  4. $A_1 \rightarrow A, A \in N_2 - \{B^{(1)}, B^{(2)}\}$;
  5. $A \rightarrow A, A \in N_2$;

The system works as follows.

Initially, the skin membrane contains the objects X, A and the catalysts c_0, c_1, c_2. Here X is a productive object and creates a membrane. In order to simulate a matrix of type 2 or 4, the object X will create a membrane with label 2 and transforms into X'. Whenever a membrane with index 2 appears, with the help of catalyst c_0, a non-productive object $A \in N_2$ will move into membrane 2 as A_1. At the same time, X' will transform into X_1. Once we have X_1 and A_1 in membrane 2, the subscripts of X and A are increased step by step. At any time we can apply the rules $X_i \rightarrow Y\delta$ and $A_i \rightarrow Zx$. But if we apply one of these rules first, then a trap symbol will be introduced. So, in order to get the correct simulation, we have to apply these two rules at the same step. In this way we can complete the simulation of a matrix m_i of type 2 or 4. If the symbol A in membrane 2 is not the one in the matrix $m_i : (X \rightarrow Y, A \rightarrow x)$, then we cannot apply the rule $A_i \rightarrow Zx$ and a trap symbol will be introduced in the skin membrane by the rule $A_i \rightarrow \dagger$. After simulating a matrix $m_i : (X \rightarrow f, A \rightarrow x)$ of type 4, f will create a membrane with label 3 so that if there is any nonterminal $A \in N_2$ present in the skin membrane, then the computation never halts.

For simulating a matrix of type 3, X will create a membrane with label 3 and transform into $EX^{(j)}$. If there is a symbol $B^{(j)}$ present in the skin membrane, then with the help of a catalyst c_j, it will move into the membrane 3. In membrane 3, the object E will be replaced with E_1 and at the same time the object $X^{(j)}$, with the help of a catalyst c_j (if exists in membrane 3), will be replaced with a trap symbol ($\dagger$) and the computation never halts. Otherwise, in the next step, the object E_1 will be erased and thus dissolve membrane 3. In this way we can complete the simulation of a matrix m_i'. Thus, the equality $\Psi_T(L(G)) = N(\Pi)$ follows. (It should be noted that here even though we are using 3 membranes, the depth of the membrane system is always 2 only.)   □

7   Conclusion

In this paper we considered several variants of P systems with symbol-objects where the catalysts were allowed to move in and out of a membrane. We have proved the universality results for these variants of P systems with a very small number (2, 3 or 4 depending upon the case) of membranes. We feel that this is the optimum, i.e., to achieve universality we require this many membranes in each case. It still remains to prove the optimality rigorously.

References

1. J. Dassow, Gh. Păun, *Regulated Rewriting in Formal Language Theory*, Springer-Verlag, Berlin, 1989.
2. R. Freund, C. Martin-Vide, Gh. Păun, Computing with membranes: Three more collapsing hierarchies, submitted, 2000.
3. R. Freund, Gh. Păun, On the number of nonterminals in graph-controlled, programmed, and matrix grammars, *Proc. of MCU Conf.*. Chişinău, Moldova, 2001.
4. M. Ito, C. Martin-Vide, Gh. Păun, A characterization of Parikh sets of ET0L languages in terms of P systems, in vol. *Words, Semigroups, Transducers* (M. Ito, Gh. Păun, S. Yu, eds.), World Scientific, Singapore, 2001, in press.
5. S.N. Krishna, Computing with simple P systems, Workshop on Multiset Processing, Curtea de Argeş, Romania, 2000.
6. M. Madhu, K. Krithivasan, P systems with membrane creation: Universality and efficiency, *Proc. of MCU Conf.*. Chişinău, Moldova, 2001.
7. Gh. Păun, Six nonterminals are enough for generating each r.e. language by a matrix grammar, *Intern. J. Computer Math.*, 15 (1984), 23–37.
8. Gh. Păun, Computing with membranes, *Journal of Computer and System Sciences*, 61, 1 (2000), 108–143.
9. Gh. Păun, Computing with membranes. A variant: P systems with polarized membranes, *Intern. J. of Foundations of Computer Science*, 11, 1 (2000), 167–182.
10. Gh. Păun, Computing with membranes: Attacking NP-complete problems, *Unconventional Models of Computation-UMC2K* (I. Antoniou, C. S. Calude, M. J. Dinneen, eds.), Springer-Verlag, London, 2000, 94–115.
11. Gh. Păun, Computing with membranes (P systems): Twenty six research topics, *Auckland University, CDMTCS Report* No 119, 2000. (www.cs.auckland.ac.nz/CDMTCS).

Multiset and K-Subset Transforming Systems

Taishin Yasunobu Nishida

Faculty of Engineering, Toyama Prefectural University
Kosugi-machi, Toyama 939-0398, Japan
nishida@pu-toyama.ac.jp

Abstract. We introduce K-subset transforming systems as a generalization of multiset transformation. A K-subset, which is a generalization of a multiset where "multiplicities" take values in a semiring, is considered by S. Eilenberg. We construct an example of K-subset transforming system which models a chaotic discrete dynamical system. We show that for every basic reaction of multiset transformation we can construct a K-subset transforming system which expresses the multiset transformation. We also show that for every phrase structure grammar there is a K-subset transforming system such that the system simulates derivations of the grammar.

1   Introduction

Recently a number of new computing models are proposed, such as quantum computing [14], DNA computing [12], membrane computing or P systems [3,9,10,11], and so on. These new models and the "traditional" models, such as Turing machines, phrase structure grammars, term rewriting systems, cellular automata, etc, have a quite different appearance, but we can find a common feature of them: they all obey a discrete time development. We can say that computational science is a science of discrete dynamical systems; by contrast, the physical science have been described by continuous differential equations.

In this paper we try to build a general framework of discrete dynamical systems. We adopt K-subset [5] to express objects in discrete dynamical systems. A K-subset is a generalization of a multiset as it is a multiset where "multiplicities" take values in a semiring. Among many varieties of multiset theory, K-subset has a firm theoretical background [4]. Our model has a set of rules which are pairs of a condition and an action over K-subsets. Adding an initial K-subset, we obtain a K-subset transforming system.

A multiset transformation is a good model of a chemical reaction. However, in such a situation when a molecule changes its conformation according to a concentration of other molecules, such as pH or calcium ion, rational or real "multiplicities" will be useful. Similarly, in quantum computing, the "multiplicities" of quantum states are complex numbers. This is why we introduce non-integral multiplicities.

After preliminaries describing semirings and K-subsets, we define K-subset transforming systems and give an example which models a chaotic discrete dynamical system (Section 3). Then we mention the relation between K-subset

C.S. Calude et al. (Eds.): Multiset Processing, LNCS 2235, pp. 255–265, 2001.

transforming systems and multiset transformation (Section 4). Finally, in Section 5, we show that K-subset transforming systems encompass string rewriting systems: phrase structure grammars and L systems. Although some results in this paper are obtained by only simulating existing models with K-subset transforming systems, we believe that K-subset transforming systems open up new research vistas in computer science. Example 2 and Theorem 5 suggest the wide possibilities of K-subset transforming systems.

2   Preliminaries

First we introduce the notion of semiring from [5]. A set K is said to be a *semiring* if K has two operations, addition and multiplication, K is a commutative monoid with respect to addition, it is a monoid with respect to multiplication, and addition and multiplication are connected by the following equations

$$x(y + z) = xy + xz,$$
$$(x + y)z = xz + yz,$$
$$x0 = 0 = 0x,$$

for every $x, y, z \in K$, where 0 is the unit element with respect to addition, $+$ stands for addition, and omitted $\cdot$ stands for multiplication. The unit element with respect to multiplication is denoted by 1. Thus for every element $x, y, z \in K$ we have

$$x + y = y + x,$$
$$x + (y + z) = (x + y) + z,$$
$$x + 0 = x,$$
$$x(yz) = (xy)z,$$
$$1x = x = x1.$$

Clearly, any ring is a semiring. We give a few examples of semirings which will appear in this paper.

Example 1. The following sets are all semirings.

1. $\mathcal{B} = \{0, 1\}$ with the operations:

$$0 + 0 = 0, \ 0 + 1 = 1 + 0 = 1, \ 1 + 1 = 1,$$
$$00 = 01 = 10 = 0, \ 11 = 1.$$

   So 0 is the unit element for addition and 1 is the unit element for multiplication. Notice that $\mathcal{B}$ is different from $GF(2)$[1].
2. The set of all nonnegative integers $\mathbb{N}$.
3. The set of all real numbers $\mathbb{R}$.

---

[1] $GF(2)$ is the only field of two elements and satisfies $1 + 1 = 0$.

A semiring K is called *commutative* if for every $x, y \in K$ we have $xy = yx$.

Then we define K-subsets [5]. We assume that K is a nontrivial semiring, i.e., $0 \neq 1$, or equivalently K has at least two elements. We also assume that K is commutative. Let X be a set. A K-*subset* A of X is a function

$$A : X \rightarrow K.$$

For every $x \in X$ the element $A(x)$ of K is called the *multiplicity with which x belongs to A* or *multiplicity of x in A*. The union $\cup$ and the intersection $\cap$ of two K-subsets A and B of X are defined by

$$(A \cup B)(x) = A(x) + B(x) \text{ and}$$
$$(A \cap B)(x) = A(x)B(x).$$

For every $k \in K$ the operation kA is defined by

$$(kA)(x) = kA(x).$$

We note that every $\mathcal{B}$-subset B of X corresponds a normal subset S of X by the next equation.

$$S = \{x \in X \,|\, B(x) = 1\}.$$

A "normal" multiset is a collection in which elements may be duplicated, for example,

$$\{a, a, a, b, c, c\}.$$

The number of times an element occurs in a multiset is called its multiplicity. In this paper we denote a multiset in the form

$$\{\text{multiplicity of } a \cdot a, \text{multiplicity of } b \cdot b, \ldots\},$$

so the multiset above is expressed as

$$\{3 \cdot a, 1 \cdot b, 2 \cdot c\}.$$

Then we have the next proposition.

Proposition 1. *For a set X, there exist functions μ and ν such that μ maps every $\mathbb{N}$-subset A of X to a multiset $\mu(A)$ over X, ν maps every multiset M over X to an $\mathbb{N}$-subset $\nu(M)$ of X, and $\mu\nu$ and $\nu\mu$ are identities.*

Proof. The functions μ and ν are defined by

$$\mu(A) = \bigcup_{x \in X \wedge A(x) > 0} \{A(x) \cdot x\}$$

and

$$\nu(M)(x) = \begin{cases} \text{the multiplicity of } x \text{ in } M & \text{if } x \in M \\ 0 & \text{otherwise} \end{cases}.$$

Then the conclusions are obvious. $\qquad\qquad\square$

For every $x, y \in \mathbb{N}$ we define $x \mathbin{\dot{-}} y$ by

$$x \mathbin{\dot{-}} y = \begin{cases} x - y & \text{if } x \geq y \\ 0 & \text{otherwise} \end{cases}.$$

A semiring has some relations over its elements: every semiring has equality $=$ and inequality $\neq$ and if a semiring is a ordered set, then it has two more binary relations, $<$ and $\leq$. Let $A : X \to K$ be a K-subset. A *term* over the K-subset A is defined by:

1. An element $a \in K$ is a term.
2. For every $x \in X$, $A(x)$ is a term.
3. For a variable letter z, z and $A(z)$ are terms.
4. For terms t_1 and t_2, $t_1 + t_2$, $t_1 t_2$, and (t_1) are terms.
5. Nothing else is a term.

A *predicate* over the K-subset A is defined by:

1. $R(t_1, \ldots, t_n)$ is a predicate where R is an n-ary relation over K and $t_1, \ldots, t_n$ are terms over A.
2. For two predicates P_1 and P_2, $P_1 \vee P_2$, $P_1 \wedge P_2$, and $\neg P_1$ are predicates.
3. For a predicate P and a variable letter z, $(\exists z \in X)(P)$, $(\forall z \in X)(P)$, $(\exists z \in K)(P)$, and $(\forall z \in K)(P)$ are predicates.
4. Nothing else is a predicate.

A predicate is called *closed* if every variable is in the scope of a quantifier. A closed predicate is *satisfiable* on a K-subset A if the following conditions hold.

1. For $(\exists z \in X)(P)$, there exists $x \in X$ such that $P|_{A(z) \to A(x)}$ is true where $P|_{A(z) \to A(x)}$ stands for the predicate in which every occurrence of $A(z)$ in P is replaced with $A(x)$.
2. For $(\forall z \in X)(P)$, $P|_{A(z) \to A(x)}$ is true for every $x \in X$.
3. $P_1 \vee P_2$ is satisfiable if and only if P_1 or P_2 is satisfiable, $P_1 \wedge P_2$ is satisfiable if and only if P_1 and P_2 are satisfiable, and $\neg P$ is satisfiable if and only if P is not satisfiable.

Unless otherwise stated, for every predicate P, we assume that $\forall i\,(P)$ and $\exists i\,(P)$ stand for $(\forall i \in \mathbb{N})(P)$ and $(\exists i \in \mathbb{N})(P)$, respectively.

3   Definitions of K-Subset Transforming Systems

Here we give the definition of a K-subset transforming system.

Definition 1. *A K-subset transforming system is a 4-tuple $G = \langle X, K, R, A_0 \rangle$ where X is a set, K is a semiring, R is a set of rules of the form*

$$\langle \text{condition} \rangle : \langle \text{action} \rangle$$

in which condition *is a closed predicate over K-subset $X \to K$ and* action *consists a set of formulas that give a new K-subset from the current K-subset, in* *which the current K-subset and the new K-subset are denoted by A and A', respectively. So an action consists of equations of the form*

$$A'(x_1, \ldots, x_n) = \text{(a term over } A)$$

where $x_i \in X$ or x_i is a variable letter for $1 \leq i \leq n$. If there are variable letters in an action which are quantified in the condition of the action, then the variable letters take values which make the condition satisfiable. We omit, from actions, the definition of multiplicities of elements of X which take the same multiplicities in the current K-subset. And $A_0 : X \to K$ is the initial K-subset.

Let $G = \langle X, K, R, A_0 \rangle$ be a K-subset transforming system and let $B : X \to K$ be a K-subset. Two rules r_1 and r_2 in R are said to be *competing in the conditions* on B if there are an element $x \in X$ and two variables z_1 and z_2 occurring r_1 and r_2, respectively, such that the conditions of r_1 and r_2 are satisfiable for z_1 and z_2 to be assigned to x. Two rules r_1 and r_2 are called *overlapping in the actions* if there is an element $x \in X$ such that r_1 has an equation

$$A'(x) = \text{(a term)}$$

and r_2 has an equation

$$A'(x) = \text{(another term)}.$$

We note that it is generally undecidable whether two rules are competing in the conditions or not. Now we define derivations by a K-subset transforming system.

Definition 2. *Let $G = \langle X, K, R, A_0 \rangle$ be a K-subset transforming system and let B and $B' : X \to K$ be K-subsets. Then B' is said to be directly derived from B by G if there are rules $r_1, \ldots, r_n$ such that their conditions are satisfiable on B and B' is obtained from B by the actions of the rules $r_1, \ldots, r_n$. If there are competing rules in the conditions on B, then one rule is selected nondeterministicly. If there are overlapping rules in the actions, then one rule is selected nondeterministicly. A sequence of K-subsets $(A_0, A_1, \ldots)$ is generated by G if and only if A_n is derived from A_{n-1} by G for $n = 1, 2, \ldots$. If there is no rule whose condition is satisfiable for a K-subset A_n, then G derives no K-subset from A_n and the sequence is terminated at A_n.*

Now we give a simple but illustrative example.

Example 2. Let $G = \langle \{y\}, \mathbf{R}, R, A_0 \rangle$ be an $\mathbf{R}$-subset transforming system where R consists of

$$A(y) \geq 0 \; : \; A'(y) = -2A(y) + 1,$$
$$A(y) < 0 \; : \; A'(y) = 2A(y) + 1,$$

and $A_0(y) = x_0$ for some $x_0 \in [-1, 1]$. We note that the derivation by G is deterministic because the two rules will never compete in the conditions. Then the multiplicities of y in the sequence $(A_0, \ldots, A_n, \ldots)$ generated by G give the trajectory of the discrete dynamical system

$$x_{n+1} = -2|x_n| + 1,$$

i.e.,

$$x_n = A_n(y).$$

This is an example of chaotic dynamical systems (Example 6.2.1 of [8]).

4   Multiset Transformation and $\mathbb{N}$-Subset Transforming System

The $\mathbb{N}$-subset transforming systems are able to carry out multiset transformations. For example, the sort program written in GAMMA [1,2,6] looks like:

Example 3. Let $G = \langle X, \mathbb{N}, R, A_0 \rangle$ be an $\mathbb{N}$-subset transforming system where

$$X = \{1, \ldots, n\} \times \{x_1, \ldots, x_n\}, \ 1, \ldots, n \in \mathbb{N}, \ x_i \in \mathbb{R}$$

A_0 is an $\mathbb{N}$-subset of X, and R consists of the rule

$$\exists i \exists j \, (\exists x, y \in \mathbb{R})(A((i, x))A((j, y)) > 0 \wedge i < j \wedge x > y) :$$
$$A'((i, x)) = A((i, x)) - 1, \ A'((j, y)) = A((j, y)) - 1,$$
$$A'((i, y)) = A((i, y)) + 1, \ A'((j, x)) = A((j, x)) + 1.$$

Now obviously G generates a finite sequence $(A_0, \ldots, A_k)$ such that

$$i \leq j \text{ and } x \leq y \text{ if } A_k((i, x))A_k((j, y)) > 0. \tag{1}$$

Since there may be more than one combination of pairs (i, x) and (j, y) which make the condition of the rule satisfiable, G may generate more than one sequences $(A_0, \ldots, A_k)$. It is easily proved that (1) holds for every A_k in every sequence.

Example 3 is generalized to the basic reaction of GAMMA program.

Theorem 1. *Let X be a set and let*

$$G : x_1, \ldots, x_n \rightarrow A(x_1, \ldots, x_n) \Leftarrow R(x_1, \ldots, x_n)$$

be a basic reaction of multiset over X where $x_1, \ldots, x_n$ are variables and R and A are of arity n [6]. Then there is an $\mathbb{N}$-subset transforming system H such that G transform a multiset M to M' if and only if H generates $\mathbb{N}$-subset M' from M.

Proof. Let $H = \langle X, \mathbb{N}, P, A_0 \rangle$ be an $\mathbb{N}$-subset transforming system where P consists of

$$(\exists x_1, \ldots, x_n \in X) R(x_1, \ldots, x_n) : B'(x_i) = B(x_i) + \nu(A(x_1, \ldots, x_n))(x_i) \dot{-} 1,$$
$$i = 1, \ldots, n$$
$$B'(y) = B(y) + \nu(A(x_1, \ldots, x_n))(y)$$
$$\text{for } y \notin \{x_1, \ldots, x_n\}$$

and $A_0 = \nu(M_0)$ where M_0 is the initial multiset for G and ν is the function defined in Proposition 1. Then the conclusion follows immediately. □

We do not treat sequential and parallel composition operators of GAMMA [6]. But by Theorem 3 in Section 5, K-subset transforming systems have computational universality.

5   Phrase Structure Grammars, L Systems, and K-Subset Transforming Systems

In this section we consider the relation between string rewriting systems and K-subset transforming systems. We assume the reader is familiar with basics of phrase structure grammars (see [13]) and L systems (see [7]).

Let Σ be a finite alphabet. A $\mathcal{B}$-subset A of $\mathbb{N} \times \Sigma$ is said to be linearizable if A satisfies

1. For every $i \in \mathbb{N}$ and $a, b \in \Sigma$ such that $a \neq b$ we have $A((i, a))A((i, b)) = 0$.
2. For every $i < j < k$ and every $a, b, c \in \Sigma$ we have $\neg(A((i, a)) = 1 \wedge A((j, b)) = 0 \wedge A((k, c)) = 1)$.

For a linearizable $\mathcal{B}$ subset A, a mapping $\phi : A \to \Sigma^*$ or Σ^ω is defined by

$$\phi(A) = \begin{cases} a_i \cdots a_j \in \Sigma^* & \text{if } A((i, a_i)) = \cdots = A((j, a_j)) = 1 \wedge \\ & A((k, b)) = 0 \text{ for } k < i, k > j \\ a_i \cdots a_j \cdots \in \Sigma^\omega & \text{if } A((j, a_j)) = 1 \text{ for some } i \in \mathbb{N} \text{ and every } j \geq i \end{cases}.$$

Then the next theorem describes the fact that $\mathcal{B}$-subset transforming systems cover the phrase structure grammars.

Theorem 2. *Let $G = \langle V, \Sigma, P, S \rangle$ be a grammar. Then there exists a $\mathcal{B}$-subset transforming system H such that for every sentential form w generated by G there is a $\mathcal{B}$-subset A of $\mathbb{N} \times (V \cup \Sigma)$ generated by H which satisfies*

$$w = \phi(A).$$

Proof. Let $H = \langle \mathbb{N} \times (V \cup \Sigma), \mathcal{B}, R, A_0 \rangle$ where $A_0((1, S)) = 1$, $A_0((i, x)) = 0$ for $i \neq 1$ and $x \in (V \cup \Sigma)$, and R consists of the following rules: For every $a_1 \cdots a_k \to b_1 \cdots b_l \in P$ and $l \geq 1$

$$\exists i \, (A((i, a_1)) = \cdots = A((i + k, a_k)) = 1 \wedge a_1 \cdots a_k \to b_1 \cdots b_l) :$$

$$A'((i, b_i)) = \cdots = A'((i + l, b_l)) = 1,$$
$$A'((i, a_1)) = \cdots = A'((i + k, a_k)) = 0,$$
$$A'((i + l + j, c_{i+k+j})) = A((i + k + j, c_{i+k+j})) \text{ for } j > l, \text{ and}$$
$$A'((j, c_j)) = A((j, c_j)) \text{ for } j < i.$$

For every $a_1 \cdots a_k \to \varepsilon \in P$,

$$\exists i \, (A(i, a_1)) = \cdots = A((i + k, a_k)) = 1 \wedge a_1 \cdots a_k \to \varepsilon) :$$
$$A'((i, a_i)) = \cdots = A'((i + k, a_k)) = 0,$$
$$A'((i + j - 1, c_{i+k+j})) = A((i + k + j, c_{i+k+j})) \text{ for } j > 0, \text{ and}$$
$$A'((j, c_j)) = A((j, c_j)) \text{ for } j < i.$$

First we observe that every $\mathcal{B}$-subset A of $\mathbb{N} \times (V \cup \Sigma)$ generated by H is linearizable. Then the definition of ϕ leads the conclusion.   □

Since the above theorem says that K-subset transforming systems can simulate type 0 grammars, we have the following theorem.

Theorem 3. *The K-subset transforming systems generate all recursively enumerable languages. There is a K-subset transforming system generating a sequence of K-subsets which is not recursively enumerable.*

Proof. The first assertion is a corollary of Theorem 2. Since a chaotic dynamical system shows quite different behaviour by any infinitesimal change in the initial value, the different initial $\mathbb{R}$-subset in Example 2 gives a different $\mathbb{R}$-subset transforming system. So the cardinality of possible $\mathbb{R}$-subset transforming systems in Example 2 is the cardinality of continuum. Then the second assertion is true.   □

We note that, by Church's hypothesis, the class of effectively computable K-subset transforming systems must coincide with the class of Turing machines.

Next we consider L systems.

Theorem 4. *Let $G = \langle \Sigma, P, \#, w \rangle$ be a $(1, 1)L$ system where $\#$ is the environmental marker not in Σ. Then there is a $\mathcal{B}$-subset transforming system H such that for every $u \in \Sigma^+$ derived by G, H generates the linearizable $\mathcal{B}$-subset A of $\mathbb{N} \times (\Sigma \cup \{\#\})$ satisfying*

$$u\# = \phi(A).$$

Proof. Let $H = \langle X, \mathcal{B}, R, A_0 \rangle$ be the $\mathcal{B}$-subset transforming system where

$$X = \{-1\} \cup \mathbb{N} \cup \{\$\} \times \mathbb{N} \cup \mathbb{N} \times (\Sigma \cup \{\#\}) \cup (\{0, 1\} \times \mathbb{N}) \times (\Sigma \cup \{\#\}),$$
$$A_0((i, a_i)) = 1, \, A_0((l + 1, \#)) = 1, \text{ and } A_0(x) = 0 \text{ for other } x \in X,$$

where $w = a_0 \cdots a_l$, and R has the following rules:

1. $\exists i (\exists a \in \Sigma \cup \{\#\})(A((i, a)) = 1) : A'(((0, i), a)) = 1, \, A'((i, a)) = 0.$

2. $\forall i (\forall a \in \Sigma \cup \{\#\})(A((i,a)) = 0) \wedge$
$(\exists a, x \in \Sigma)(A(((0,0),a))A(((0,1),x)) = 1 \wedge (\#,a,x) \to b_1 \cdots b_k \in P) :$
$A'(((1,0),b_1)) = \cdots = A'(((1,k-1),b_k)) = 1,$
$A'(((0,0),a)) = 0, A'(1) = 1, A'(($\$$,k-1)) = 1.$

2'. $\forall i (\forall a \in \Sigma \cup \{\#\})(A((i,a)) = 0) \wedge$
$(\exists a \in \Sigma)(A(((0,0),a))A(((0,1),\#)) = 1 \wedge (\#,a,\#) \to b_1 \cdots b_k \in P) :$
$A'(((1,0),b_1)) = \cdots = A'(((1,k-1),b_k)) = 1,$
$A'(((0,0),a)) = 0, A'(-1) = 1, A'(((1,k),\#)) = 1.$

3. $\exists j\, (A(j) = 1) \wedge \exists l\, (A(($\$$,l)) = 1) \wedge$
$(\exists a, x, y \in \Sigma)(A(((0,j-1),x))A(((0,j),a))A(((0,j+1),y)) = 1 \wedge$
$(x,a,y) \to b_1 \cdots b_k \in P) :$
$A'(((1,l+1),b_1)) = \cdots = A'(((1,l+k),b_k)) = 1, A'(((0,j),a)) = 0,$
$A'(j+1) = 1, A'(j) = 0, A'(($\$$,l)) = 0, A'(($\$$,l+k)) = 1.$

4. $\exists j\, (A(j) = 1) \wedge \exists l\, (A(($\$$,l)) = 1) \wedge$
$(\exists a, x \in \Sigma)(A(((0,j-1),x))A(((0,j),a))A(((0,j+1),\#)) = 1 \wedge$
$(x,a,\#) \to b_1 \cdots b_k \in P) :$
$A'(((1,l+1),b_1)) = \cdots = A'(((1,l+k),b_k)) = A'(((1,l+k+1),\#)) = 1,$
$A'(((0,j),a)) = 0, A'(j) = 0, A'(($\$$,l)) = 0, A'(-1) = 1.$

5. $A(-1) = 1 \wedge \exists i (\exists a \in \Sigma \cup \{\#\})(A(((1,i),a)) = 1) :$
$A'((i,a)) = 1, A'(((1,i),a)) = 0.$

6. $A(-1) = 1 \wedge \forall i (\forall a \in \Sigma \cup \{\#\})(A(((1,i),a)) = 0) : A'(-1) = 0.$

Now we show that for every $\mathcal{B}$-subset A of $\mathbb{N} \times (\Sigma \cup \{\#\})$ satisfying $u\# = \phi(A)$ for some $u \in \Sigma^+$, $u \Rightarrow_G v$ if and only if there exists a $\mathcal{B}$-subset B which is generated from A by H and $\phi(B) = v\#$. Let $u = a_0 \cdots a_{n-1}$ where $a_i \in \Sigma$, $i = 0, \ldots, n-1$, let $A((0,a_0)) = \cdots = A((n-1,a_{n-1})) = A((n,\#)) = 1$, and let $A((j,a)) = 0$ for $j < 0$ or $j > n$. Then by iterating the rule 1 $n+1$ times, we have $\mathcal{B}$-subset A_1 such that

$$A_1(((0,0),a_0)) = \cdots = A_1(((0,n-1),a_{n-1})) = A_1(((0,n),\#)) = 1.$$

We note that rule 2 cannot be iterated until rule 1 is iterated $n+1$ times by the first condition of rule 2:

$$\forall i (\forall a \in \Sigma \cup \{\#\})(A((i,a)) = 0).$$

Next rules 2, 3, and 4 (or 2') simulate the derivation of G from left to right. After the rule 4 (or 2') is iterated, we have $\mathcal{B}$-subset A_2 satisfying

$$A_2(((1,0),b_0)) = \cdots = A_2(((1,m),b_m)) = A_2(((1,m+1),\#)) = 1,$$
$$A_2(-1), \text{ and } b_0 \cdots b_m = v.$$

Finally rules 5 and 6 generate the desired $\mathcal{B}$-subset B. Then it is proved that H generates $\mathcal{B}$-subset A of $\mathbb{N} \times (\Sigma \cup \{\#\})$ if only if G generates $u \in \Sigma^+$ such that

$$u\# = \phi(A).$$

If G generates ε, then H generates $\mathcal{B}$-subset A_ε and A_ε derives nothing where

$$A_\varepsilon(((0,0),\#)) = 1, \ A_\varepsilon(x) = 0 \text{ for other } x \in X.$$

Since all L systems generate ε from ε, this is also correct.   □

The $\mathcal{B}$-subset transforming system constructed in the above proof is quite unefficient. It simulates one step derivation of an L system by means of many steps. We should find K-subset transforming systems which can generate strings in parallel.

But by considering an $\mathbb{N}$-subset of $\Sigma^* \times \mathbb{N}$, we can measure the multiplicity of a word, that is, the total number of different derivations of a word in an L system.

Theorem 5. *Let* $G = \langle \Sigma, P, \#, w \rangle$ *be an L system. Then there is an $\mathbb{N}$-subset transforming system H such that for every word u derived by G in i steps H generates an $\mathbb{N}$-subset A of $\Sigma^* \times \mathbb{N}$ and that $A((u,i))$ gives the multiplicity of u.*

Proof. Let $H = \langle \Sigma^* \times \mathbb{N}, \mathbb{N}, R, A_0 \rangle$ where $A_0((w,0)) = 1$, $A_0((x,i)) = 0$ for $x \neq w$ or $i \neq 0$ and R consists of the following rule

$$\exists i (\exists u \in \Sigma^*)(A((u,i)) > 0 \wedge (\exists v \in \Sigma^*)(u \Rightarrow_G v)) :$$

$$A'((v, i+1)) = A((v, i+1)) + A((u,i)), A'((u,i)) = 0.$$

Then obviously $A((u,i))$ gives the multiplicity of u derived by G in i steps.   □

References

1. J. Banâtre, A. Coutant, and D. Le Metayer, A parallel machine for multiset transformation and its programming style, *Future Generations Computer Systems*, **4** (1988), 133–144.
2. J. Banâtre and D. Le Métayer, Programming by multiset transformation, *Communications of the ACM*, **36** (1993), 98–111.
3. J. Dassow and G. Păun, On the power of membrane computing, *Journal of Universal Computer Science*, **5** (1999), 33–49.
4. W. D. Blizard, The development of multiset theory, *Modern Logic*, **1** (1991), 319–352[2].
5. S. Eilenberg, *Automata, Languages, and Machines Volume A*, Academic Press, New York, 1974.
6. C. Hankin, D. Le Métayer, and D. Sands, Refining multiset transformers, *Theoretical Computer Science*, **192** (1998), 233–258.
7. G. T. Herman and G. Rozenberg, *Developmental Systems and Languages*, North-Holland, Amsterdam, 1975.
8. M. Martelli, *Discrete Dynamical Systems and Chaos*, Longman Scientific & Technical, Harlow, 1992.
9. G. Păun, Computing with membranes, *Journal of Computer and System Sciences*, **61** (2000), 108–143, (and *Turku Centre for Computer Science-TUCS Report* No 208, 1998 (http://www.tucs.fi)).

---

[2] There is a correction to this paper, but you need not look at it. The correction is "Item [8] on p. 349 of this paper should have read as follows:
[8] Blizard, W., Dedekind Multisets and Function Shells, *Theoretical Computer Science* **110** (1993) 79–98."

10. G. Păun, Computing with membranes. An introduction, *Bulletin of the EATCS* **67** (1999) 139–152.
11. G. Păun, P systems: an early survey, *The Third International Colloquium on Words, Languages and Combinatorics* March 2000, Kyoto (Proceedings will be published by World Scientific, Singapore).
12. G. Păun, G. Rozenberg, and A. Salomaa, *DNA Computing*, Springer-Verlag, Berlin, 1998.
13. A. Salomaa, *Formal Languages*, Academic Press, New York, 1973.
14. P. W. Shor, Algorithm for quantum computation: discrete log and factoring, in: *Proceedings of the 35th Annual IEEE Symposium on Foundations of Computer Science*, 1994.

On P Systems with Active Membranes Solving the Integer Factorization Problem in a Polynomial Time

Adam Obtułowicz

Institute of Mathematics, Polish Academy of Sciences
Śniadeckich 8, P.O. Box 137
00-950 Warsaw, Poland
adamo@impan.gov.pl

Abstract. There are presented deterministic P systems with active membranes which are able to solve the Integer Factorization Problem in a polynomial time, which is the main result of the paper. There is introduced a class of programs for the description and correct implementation of algorithms of elementary number theory in nonstandard computing systems, especially in P systems with active membranes. By using some of these programs there is achieved the main result.

Introduction

Recently in the area of membrane computing, a branch of molecular computing, the known NP-complete SAT problem was successfully attacked by membrane systems (also called P systems) with active membranes, which ae able to solve this problem in a linear time by massively parallel computations. These P systems with active membranes are described in the pioneering paper Păun (2000a). Thus the old Integer Factorization Problem belonging to the class of NP-problems can be also solved in a polynomial time by P systems with active membranes. One can get this kind of a solution by the reduction of Integer Factorization Problem to SAT problem in order to apply those P systems which solve SAT problem in a linear time.

We recall that the Integer Factorization Problem, formulated as a problem of finding a non-trivial divisor of a given natural number, is an important problem of number theory and public-key cryptography, cf. Menezes et al. (1996) and Koblitz (1998).

In the present paper we present deterministic P systems with active membranes which solve Integer Factorization Problem in the time $O(\ln^2 n)$[1] for a given natural number n to be factored. We do not apply the cumbersome and time consuming reduction of Integer Factorization Problem to SAT problem (see the proof of Cook's Theorem in Papadimitriou (1994), where there is given a general method of reduction of NP-problems to SAT problem). There is only used

---

[1] For the O notation see Papidimitriou (1994) or Koblitz (1998).

C.S. Calude et al. (Eds.): Multiset Processing, LNCS 2235, pp. 267–285, 2001.

a "school" algorithm for computing the remainder of the division of two binary presented natural numbers. We achieve the result by writing this algorithm in a form of a program belonging to a class of simple programs called liberal loop programs and related to loop programs defined in Meyer and Ritchie (1967). The liberal loop programs, introduced in the present paper, are aimed to describe algorithms of elementary number theory in terms of bit operations (cf. Koblitz (1998)) for correct and immediate implementation of these algorithms in non-standard computing systems, especially in P systems with active membranes.

The methods presented in the paper can be also applied to define P systems with active membranes which solve Discrete Logarithm Problems (cf. Koblitz (1998)) in a polynomial time.

In Section 1 of the paper we introduce liberal loop programs. Then in Section 2 we describe P systems with active membranes, called briefly P systems in the paper, and give a precise definition of processes generated by P systems. The processes generated by P systems are counterparts of processes of computation. Deterministic P systems solving Integer Factorization Problem are described in Section 3. In the Appendix we present the rules for an implementation of liberal loop programs in P systems.

For all unexplained terms of complexity of computation we refer the reader to Papadimitriou (1994).

Acknowledgements

I would like to thank Peter Hegarty for discussions and his comments concerning probabilistic algorithms for solving Integer Factorization Problem in a subexponential time. Furthermore, I thank Jerzy Urbanowicz for his encouragement.

1   Liberal Loop Programs

Liberal loop programs, briefly called l-programs and related to loop programs defined in Meyer and Ritchie (1967), form a restricted class of programs for manipulating binary strings[2] stored in registers. There is no limit on the size of registers and every register contains exactly one binary string.

An l-program is a finite sequence of instructions of five types:

(1) $X := Y$,
(2) $X := \Lambda$
(3) $X := \mathrm{op}(\ldots, X, Y, Z, \ldots)$ for $\mathrm{op} \in \mathrm{BOP}$,
(4) $\mathrm{LOOP}\,|X|$,
(5) END,

where $\ldots, X, Y, Z, \ldots$ are names of registers, and the symbols $\mathrm{op} \in \mathrm{BOP}$[3] are symbols of bit operations (cf. Koblitz (1998)). From now on, names of registers

---

[2] A binary string is a finite sequence of digits **0** and **1**.

[3] The set BOP may be specified separately for every l-program according to the practical need of particular bit operations to be used.

are denoted by capital Latin letters. The occurrences of instructions LOOP $|X|$ and END in an l-program are matching like occurrences of parentheses "(" and ")", respectively in a well formed propositional formula, e.g., as in formula

$$((A \wedge B) \vee \neg(C)) \wedge (B \wedge C).$$

An l-program $\mathcal{P}$ is *realized* by the execution of instructions in the sequential order in which they appear in $\mathcal{P}$. The instructions are executed as follows:

- according to $X := Y$ the current content of register Y is transferred to register X,
- $X := \Lambda$ means to clear register X or equivalently to put the empty string Λ in register X,
- $X := \mathrm{op}(\ldots, X, Y, Z, \ldots)$ means to put the result (value) of a bit operation op determined by the current contents of registers $\ldots, X, Y, Z, \ldots$ in register X,
- for an l-program of the form

  (∗)   LOOP $|X|\mathcal{P}$ END

  if the first occurrence of LOOP $|X|$ and the last occurrence of END in this program are matching, then the first occurrence of LOOP $|X|$ means to realize $|x|$-times the piece $\mathcal{P}$, where $|x|$ is the length of string contained in register X upon entering the loop,
- if an l-program $\mathcal{P}'$ of the form (∗) appears in another l-program $\mathcal{P}''$ and the first occurrence of LOOP $|X|$ and the last occurrence of END in $\mathcal{P}'$ are matching, then the corresponding occurrence of LOOP $|X|$ in $\mathcal{P}''$ is executed analogously as above.

All basic conventions concerning l-programs are the same as the conventions for loop programs which are adopted in Meyer and Ritchie (1967), except the execution of instructions $X := \mathrm{op}(\ldots, X, Y, Z, \ldots)$ and LOOP $|X|$. Thus, for more explanations we refer the reader to Meyer and Ritchie (1967) or Brainerd and Landweber (1974).

For an l-program $\mathcal{P}$ we write $\mathrm{RGR}(\mathcal{P})$ to denote the set of names of registers which occur in $\mathcal{P}$. By *input data* $\mathcal{D}$ for a program $\mathcal{P}$ we mean a family of binary strings s_X $(X \in I)$ for a set $I \subseteq \mathrm{RGR}(\mathcal{P})$ of input registers such that s_X is contained in register X for every $X \in I$, respectively, when realization of $\mathcal{P}$ starts. We assume that when realization of $\mathcal{P}$ starts, the non-input registers, i.e. belonging to $\mathrm{RGR}(\mathcal{P}) - I$, contain the empty string.

By the *time of realization of a program $\mathcal{P}$ for input data $\mathcal{D}$*, written $\mathrm{TIME}(\mathcal{P}, \mathcal{D})$, we mean the total number of executions of single instructions occurring in $\mathcal{P}$ during the realization of $\mathcal{P}$ for input $\mathcal{D}$. By the *space of realization* of a *program $\mathcal{P}$ for input data $\mathcal{D}$*, written $\mathrm{SPACE}(\mathcal{P}, \mathcal{D})$, we mean the length of the longest string contained in registers of $\mathrm{RGR}(\mathcal{P})$ during the realization of $\mathcal{P}$ for input $\mathcal{D}$.

For an l-program $\mathcal{P}$ we denote by $|\mathcal{P}|$ the length of $\mathcal{P}$ as a sequence of instructions.

We present now some l-programs describing certain simple algorithms of elementary number theory. We identify natural numbers with their binary presentations if there is no risk of confusion. A finite string of digit **0** may appear on the left in a binary presentation of a natural number. We say that *meaningless zeros do not occur* in a binary presentation of a natural number given by a binary string s if digit **0** occurs as the first symbol of s whenever s is of length 1 and s presents number 0.

The l-program

$$
\begin{aligned}
&W := \Lambda \\
&Z := \Lambda \\
&\text{LOOP } |X| \\
&\quad W := \sharp(X, Y, Z) \\
&\quad Z := \oplus(X, Y, Z) \\
&\quad X := \downarrow(X) \\
&\quad Y := \downarrow(Y) \\
&\text{END} \\
&W := W \,!\, Z
\end{aligned}
$$

computes the sum of two binary presented natural numbers m and n contained in the input registers X and Y respectively, where $|m| \geq |n|$ and $|m|, |n|$ are the lengths of binary strings presenting m and n, respectively. The symbols $\sharp$, $\oplus$, $\downarrow$, ! are elements of BOP. The binary presented sum $m + n$ is contained in the output register W after the realization of the above program. The realization of the above program coincides with the "school" computation of the sum $m + n$ of binary presented natural numbers m and n, where register Z is aimed to store digit **1** carried during the computation and where execution of new instructions is defined in the following way. According to instruction $W := \sharp(X, Y, Z)$ one writes digit **1** on the left end of the current content of register W if all registers X, Y, Z currently contain those strings whose last symbol is digit **1** or exactly one of the registers X, Y, Z contains that string whose last symbol is digit **1**, otherwise one writes digit **0** on the left and of the current content of W. The instruction $Z := \oplus(X, Y, Z)$ is executed in such a way that one puts one element string "1" in register Z if at least two of the registers X, Y, Z contain those strings whose last symbol is **1**, otherwise one puts empty string in register Z. According to the instructions $X := \downarrow(X)$, $Y := \downarrow(Y)$ one erases the last symbol of the strings currently contained in registers X and Y. When registers X, Y contain the empty string, the execution of these instructions leaves the empty string contained in them. According to instruction $W := W \,!\, Z$ one writes on the left end of the current content of register W the last symbol of that string which is currently contained in register Z. When register Z contains the empty string, the content of register W remains unchanged after the execution of $W := W \,!\, Z$. We use expression $\mathcal{P}[W := X + Y]$ as an abbreviation of the program written above. The realization of the l-program $\mathcal{P}[W := X + Y]$ does not save the input data contained in the registers X and Y because of the execution of instructions $X := \downarrow(X)$ and $Y := \downarrow(Y)$. One avoids this defect by writing a new l-program which is juxtaposition of instructions $P := X$, $Q := Y$, and $\mathcal{P}[W := P + Q]$ given by

$$P := X$$
$$Q := Y$$
$$\mathcal{P}[W := P + Q]$$

where $\mathcal{P}[W := P + Q]$ is the l-program resulting from the replacement of occurrences of X, Y in $\mathcal{P}[W := X + Y]$ by occurrences of P and Q, respectively. We denote the improved program by $\mathcal{P}[P := X, Q := Y, W := P + Q]$.

The l-program

$$W := \Lambda$$
$$Z := \Lambda$$
$$\text{LOOP } |X|$$
$$\quad W := \sharp (X, Y, Z)$$
$$\quad Z := \ominus (X, Y, Z)$$
$$\quad X := \downarrow(X)$$
$$\quad Y := \downarrow(Y)$$
$$\text{END}$$
$$\text{LOOP } |Z|$$
$$\quad W := \Lambda$$
$$\text{END}$$

computes a function described by

$$m \div n = \begin{cases} \text{the binary presented subtraction } m - n, & \text{if } m \geq n, \\ \text{the empty string } \Lambda, & \text{otherwise} \end{cases}$$

for two binary presented natural numbers m and n contained in the input registers X and Y, respectively, where $|m| \geq |n|$, and $|m|$, $|n|$ are the lengths of binary presentations of m and n, respectively. The symbols $\sharp$, $\ominus$, $\downarrow$ are elements of BOP, where the meaning of $\sharp$ and $\downarrow$ has been already explained in the case of the program for computing the sum $m + n$.

The value $m \div n$ is contained after realization in the output register W. The realization of the above program coincides with the "school" computation of subtraction $m - n$ of binary presented natural numbers m and n, where register Z is aimed to store digit 1 borrowed during the computation. The execution of instruction $Z := \ominus(X, Y, Z)$ provides correct storing of borrowed 1 in register Z and is similar to the execution of instruction $Z := \oplus(X, Y, Z)$. We use the expression $\mathcal{P}[W := X \div Y]$ as an abbreviation of the above program. We improve the program $\mathcal{P}[W := X \div Y]$ by writing new l-program

$$J := X$$
$$K := Y$$
$$\mathcal{P}[W := J \div K]$$

whose realization saves the input data contained in registers X and Y. The improved program is denoted by $\mathcal{P}[J := X, K := Y, W := I \div K]$.

We use the following auxiliary l-program

$$M := Y$$
$$\text{LOOP } |R|$$
$$\quad M := \downarrow(M)$$
$$\text{END}$$
$$T := M$$
$$T := \Lambda \ominus T$$

which verifies whether $|y| \leq |r|$ for two binary strings y and r contained in the input registers Y and R, respectively, where $|y|$ and $|r|$ denote the lengths of y and r, respectively. The output register T contains after realization one-element string "**0**" if $|y| \leq |r|$, otherwise T contains the empty string. The execution of instruction $T := \Lambda \ominus T$ means to put one-element string "**0**" in register T if register T contains currently the empty string, otherwise to clear register T.

We write $\mathcal{P}[T := |Y| \leq |R|]$ to denote the above auxiliary program. We write also $\mathcal{P}[T := \Lambda \ominus M]$ to denote the two-instruction program

$$T := M$$
$$T := \Lambda \ominus T$$

which verifies whether register M contains the empty string, where the execution of instruction $T := \Lambda \ominus T$ is defined as above.

We describe now some useful constructions of new l-programs from old l-programs. If $\mathcal{P}$ is an l-program, then the juxtaposition of instruction $\text{LOOP } |X|$, the program $\mathcal{P}$, and instruction END, written

$$\text{LOOP } |X|$$
$$\mathcal{P}$$
$$\text{END}$$

is an l-program. For two programs $\mathcal{P}$ and $\mathcal{P}'$ their juxtaposition, written

$$\mathcal{P}$$
$$\mathcal{P}'$$

is in an l-program.

Hence if the realization of an l-program $\mathcal{P}$ verifies a property $\Phi(\mathcal{D})$ of input data $\mathcal{D}$ such that after the realization the output register T contains a string of length 1 when $\Phi(\mathcal{D})$ holds, otherwise T contains empty string, then for two l-programs $\mathcal{P}'$ and $\mathcal{P}''$ the juxtaposition given by

$$\mathcal{P}$$
$$\text{LOOP } |T|$$
$$\quad \mathcal{P}'$$
$$\text{END}$$
$$T := \Lambda \ominus T$$
$$\text{LOOP } |T|$$
$$\quad \mathcal{P}''$$
$$\text{END}$$

is an l-program whose realization means the computation according to the flow diagram

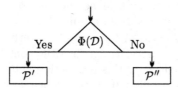

where the name T does not appear on the left hand side of the symbol $:=$ in any instruction occurring in $\mathcal{P}'$. The execution of $T := \Lambda \ominus T$ has been already defined.

By using these constructions we can introduce more complex programs.

The l-program written in Fig. 1 computes the remainder $\mathrm{rem}(m/n)$ of the division of m by n of two binary presented natural numbers m and n contained in the input registers X and Y respectively, where $m > n > 0$ and meaningless zeros do not occur in the binary presentations of m and n. The result $\mathrm{rem}(m/n)$ written as a binary string is contained in the output register R after the realiza-

$$
\begin{aligned}
&R := \Lambda\\
&\mathrm{LOOP}\,|Y|\\
&\quad R := R \,\&\, X\\
&\quad X := \uparrow(X)\\
&\mathrm{END}\\
&U := X\\
&U := U \,\&\, \mathbf{0}\\
&\mathrm{LOOP}\,|U|\\
&\quad \mathcal{P}[T := |Y| \leq |R|]\\
&\quad \mathrm{LOOP}\,|T|\\
&\qquad \mathcal{P}[J := R, K := Y, W := J \div K]\\
&\qquad \mathcal{P}[E := \Lambda \ominus W]\\
&\qquad E := \Lambda \ominus E\\
&\qquad \mathrm{LOOP}\,|E|\\
&\qquad\quad R := W\\
&\qquad \mathrm{END}\\
&\quad \mathrm{END}\\
&\quad \mathcal{P}[S := \Lambda \ominus X]\\
&\quad S := \Lambda \ominus S\\
&\quad \mathrm{LOOP}\,|S|\\
&\qquad \mathrm{LOOP}\,|R|\\
&\qquad\quad R := \mathbf{0} \uparrow R\\
&\qquad \mathrm{END}\\
&\qquad R := R \,\&\, X\\
&\qquad X := \uparrow(X)\\
&\quad \mathrm{END}\\
&\mathrm{END}
\end{aligned}
$$

Fig. 1.

tion of the program. The realization of the program in Fig. 1 coincides with the "school" computation of remainder of division of two natural numbers, where the execution of new instructions is defined in the following way. According to instruction $R := R\&X$ one writes the first symbol of the string currently contained in register X on the right and of the string currently contained in register R. If register X contains the empty string, then the content of register R remains unchanged after the execution of this instruction. The execution of instruction $U := U \& \mathbf{0}$ means to write digit $\mathbf{0}$ on the right end of the current context of register U.

According to instruction $X := \uparrow(X)$ one erases the first symbol from the current content of register X. If X contains the empty string, then its content remains unchanged after the execution of this instruction.

According to instruction $R := \mathbf{0} \uparrow R$ one erases the first symbol from the current content of register R if this symbol is digit $\mathbf{0}$, otherwise the content of register remains unchanged, also when R contains the empty string. The above new instructions are described in terms of bit operations. The program written in Fig. 1 is denoted by $\mathcal{P}[R := \mathrm{rem}(X/Y)]$.

The program written in Fig. 2 checks whether a binary string s presents a non-trivial divisor of a binary presented natural number m for m and s contained in the input registers X and Y, respectively, where $1 < |s| < |m|$ and meaningless zeros do not occur in m. If s does not present a non-trivial divisor of m, then the output register A contains the empty string after the realization of the program.

$$
\begin{aligned}
&\text{LOOP } |Y| \\
&\quad Y := \mathbf{0} \uparrow Y \\
&\text{END} \\
&B := Y \\
&B := |B| \leq 1 \\
&\text{LOOP } |B| \\
&\quad A := \Lambda \\
&\text{END} \\
&B := \Lambda \ominus B \\
&\text{LOOP } |B| \\
&\quad \mathcal{P}[R := \mathrm{rem}(X/Y)] \\
&\quad \text{LOOP } |R| \\
&\quad\quad R := \mathbf{0} \uparrow R \\
&\quad \text{END} \\
&\quad R := \Lambda \ominus R \\
&\quad \text{LOOP } |R| \\
&\quad\quad A := Y \\
&\quad \text{END} \\
&\quad R := \Lambda \ominus R \\
&\quad \text{LOOP } |R| \\
&\quad\quad A := \Lambda \\
&\quad \text{END} \\
&\text{END}
\end{aligned}
$$

Fig. 2.

If s presents a non-trivial divisor of m, then after the realization of the program the output register A contains that non-empty binary string which is the result of erasing of meaningless zeros in s.

New instruction $B := |B| \leq 1$ means to put one element string "**0**" in register B if B contains currently a string of length no greater than 1, otherwise to clear register B.

The program written in Fig. 2 is denoted by $\mathcal{P}_{\text{fct}}$.

Proposition. *Let input data $\mathcal{D}$ for program $\mathcal{P}[R := \text{rem}(X/Y)]$ be given by binary strings s_X and s_Y which present natural numbers m and n, respectively such that $m > n > 0$ and meaningless zeros do not occur in both s_X and s_Y. Then we have that*

$$\text{TIME}(\mathcal{P}[R := \text{rem}(X/Y)], \mathcal{D}) \leq C \cdot |s_X| \cdot |s_Y|$$
$$and \quad \text{SPACE}(\mathcal{P}[R := \text{rem}(X/Y)], \mathcal{D}) = |s_X|$$

for some constant natural number $C > 1$, where $|s_X|$ and $|s_Y|$ denote the lengths of strings s_X and s_Y, respectively.

Proof. One proves this proposition by a careful inspection of the realization of the piece of $\mathcal{P}[R := \text{rem}(X/Y)]$ between the occurrence of instruction LOOP $|U|$ and the occurrence of END matching with the occurrence of LOOP $|U|$.

Corollary. *Let input data $\mathcal{D}$ for program $\mathcal{P}_{\text{fct}}$ be given by binary strings s_X and s_Y such that $1 < |s_Y| < |s_X|$ and s_X presents a natural number such that meaningless zeros do not occur in s_X. Then we have that*

$$\text{TIME}(\mathcal{P}_{\text{fct}}, \mathcal{D}) \leq C \cdot |s_X| \cdot |s_Y| \quad and \quad \text{SPACE}(\mathcal{P}_{\text{fct}}, \mathcal{D}) = |s_X|.$$

for some constant natural number $C > 1$, where $|s_X|$ and $|s_Y|$ denote the lengths of s_X and s_Y, respectively.

Proof. This is an immediate consequence of the above proposition.

2   P Systems

We recall that for a set A a *multiset* of elements of A is a function $f : A \to N$ defined on A and valued in the set N of natural numbers with 0. For two multisets $f, g : A \to N$ we define their sum $f + g$ and difference $f \doteq g$ to be the multisets of elements of A which are given by

$$(f + g)(a) = f(a) + g(a), \quad (f \doteq g)(a) = f(a) \doteq g(a) \text{ for all } a \in A,$$

where $m \doteq n = m - n$ for $m \geq n$, otherwise $m \doteq n = 0$.

For an element a of A we define a multiset $\langle a \rangle : A \to N$ by

$$\langle a \rangle(a') = \begin{cases} 1 & \text{if } a' = a, \\ 0 & \text{otherwise.} \end{cases}$$

If w is a finite sequence (string) $a_1 \ldots a_n$ of elements of A, we define the multiset $\langle w \rangle$ or $\langle a_1 \ldots a_n \rangle$ by

$$\langle w \rangle = \sum_{i=1}^{n} \langle a_i \rangle = \langle a_1 \rangle + \ldots + \langle a_n \rangle.$$

We use the symbol $\mathbb{O}$ to denote the *empty multiset*, i.e. that constant function defined on A whose value is equal to 0. For two multisets f, g we write $f \leq g$ if $f(a) \leq g(a)$ for all $a \in A$.

We recall according to Păun (2000a), (2000b) that a *membrane system* $\mathcal{S}$ is given by the following data:

- a finite non-empty set $\mathbb{B}$ of balls of finite diameters > 0 (in Euclidean space E^n for $n \geq 1$) such that the frontiers[4] of the balls contained in $\mathbb{B}$ are pairwise disjoint sets, and there exists the greatest ball b_0 in $\mathbb{B}$ with respect to the inclusion relation $\subseteq$; the set $\mathbb{B}$ is called the *underlying set* of balls of $\mathcal{S}$.
- three labelling functions $l : \mathbb{B} \to L$, $e : \mathbb{B} \to \{-, 0, +\}$, $M : \mathbb{B} \to N^O$, where L is the *set of labels* of $\mathcal{S}$, the function e assigns the electric charge $e(b)$ to every ball $b \in \mathbb{B}$, O is the *set of objects* of $\mathcal{S}$ and M is a function valued in the set N^O of multisets of elements of O.

For a ball $b \in \mathbb{B}$ the multiset $M(b)$ describes the numbers of copies of individual elements of O contained in b in the space (region) between the frontier of b and the frontiers of those balls in $\mathbb{B}$ which are immediate subsets of b.

For two balls b_1, b_2 of the underlying set $\mathbb{B}$ of a membrane system we write $b_1 \prec b_2$ iff b_1 is an *immediate subset* of b_2, i.e., $b_1 \subsetneq b_2$, and there is no any other ball $b' \in \mathbb{B}$ such that $b_1 \subsetneq b' \subsetneq b_2$.

The membrane systems are transformed into new membrane systems by applications of the evolution rules, described in Păun (2000a), (2000b), to the balls and objects of systems. We consider in the paper the following five types of evolution rules which are written according to the conventions adopted in Păun (2000a):

(1) $[_h u \to w]_h^\alpha$,

(2) $[_h a]_h^0 \to [_h c]_h^- [_h d]_h^+$,

(3) $[_h a]_h^\alpha \to c$,

(4) $[_{h_0} [_{h_1}]_{h_1}^- [_{h_2}]_{h_2}^+]_{h_0}^0 \to [_{h_0} [_{h_1}]_{h_1}^0]_{h_0}^0 [_{h_0} [_{h_2}]_{h_2}^0]_{h_0}^0$,

(5) $[_h a]_h^\alpha \to [_h]_h^{\alpha'} c$,

where u, w are finite sequences (strings) of objects of a given membrane system, $\alpha, \alpha' \in \{-, 0, +\}$, a, c, d are objects of a given membrane system, and h, h_0, h_1, h_2 are elements of the set of labels of a given membrane system.

The applications of the evolution rules of the above five types are described in the following way for a membrane system $\mathcal{S}$ given by $\mathbb{B}$, l, e, M.

---

[4] The frontier of a ball $\{x \in E^n \mid d(s, x) \leq r\}$ is the sphere $\{x \in E^n \mid d(s, x) = r\}$, where $d(s, x)$ is the distance between the center s and x.

Ad (1). A rule of this type applies to those balls $b \in \mathbb{B}$ for which $l(b) = h$, $e(b) = \alpha$, and $\langle u \rangle \leq M(b)$. According to this rule there is only changed the content $M(b)$ of b into a new $M'(b)$ given by

$$M'(b) = (M(b) \dot{-} \langle u \rangle) + \langle w \rangle.$$

Ad (2). A rule of this type applies to those innermost[5] balls $b \in \mathbb{B}$ for which $l(b) = h$, $e(b) = 0$, and $\langle a \rangle \leq M(b)$. According to this rule a ball b is divided (or replaced by) into new innermost disjoint balls b^-, b^+ of opposite electric charges $e(b^-) = -$, $e(b^+) = +$, and $l(b^-) = l(b^+) = h$, $M(b^-) = (M(b) \dot{-} \langle a \rangle) + \langle c \rangle$, $M(b^+) = (M(b) \dot{-} \langle a \rangle) + \langle d \rangle$. Thus the old ball b is deleted from the system. The new balls b^-, b^+ are immediate subsets of that ball $b' \in \mathbb{B}$ for which $b \prec b'$ before division of b.

Ad (3). A rule of this type applies to those innermost balls $b \in \mathbb{B}$ for which $l(b) = h$, $e(b) = \alpha$, and $\langle a \rangle \leq M(b)$. According to this rule a ball b is deleted from the system. There is also changed the content $M(b')$ of that ball $b' \in \mathbb{B}$ for which $b \prec b'$ before deletion of b. A new content $M'(b')$ is given by

$$M'(b') = M(b') + ((M(b) \dot{-} \langle a \rangle) + \langle c \rangle).$$

Ad (4). A rule of this type applies to those three different balls $b_1, b_2, b_3 \in \mathbb{B}$ for which we have that

- b_1 is an immediate subset of the greatest ball $b_0 \in \mathbb{B}$, $b_2 \prec b_1$, $b_3 \prec b_1$ and there is no any other ball $b \in \mathbb{B}$ such that $b \prec b_1$,
- $e(b_1) = 0$, $e(b_2) = -$, $e(b_3) = +$, $l(b_1) = h_0$, $l(b_2) = h_1$, $l(b_3) = h_2$.

According to this rule the ball b_1 is divided (or replaced by) into new disjoint balls b', b'' such that

- $b_2 \prec b' \prec b_0$ and $b_3 \prec b'' \prec b_0$,
- $l(b') = l(b'') = h_0$, $e(b') = e(b'') = 0$, and $M(b') = M(b'') = M(b_1)$.

Thus the old ball b_1 is deleted from the system but the other balls of $\mathbb{B}$, including the balls $b \subsetneq b_1$, remain unchanged except the electric charges of the balls b_2, b_3 are changed into new charges $e'(b_3) = e'(b_2) = 0$. The frontiers of balls in $(\mathbb{B} - \{b_1\}) \cup \{b', b''\}$ are pairwise disjoint. The rules of type (4) are formulated in a more general way in Păun (2000a).

Ad (5). A rule of this type applies to those balls $b \in \mathbb{B}$ for which $l(b) = h$, $e(b) = \alpha$, and $\langle a \rangle \leq M(b)$. According to this rule one copy of the object a is deleted from a ball b whose electric charge is changed into α'. One copy of the object c is placed in that ball b' for which $b \prec b'$. If b is the greatest ball in $\mathbb{B}$, then one copy of c is sent out of the system.

By a *place of application* of an evolution rule to a membrane system S we mean an ordered pair $p = (r_p, x_p)$, where r_p is an expression written in one of the forms given in (1)–(5) and x_p is described in the following way:

- if r_p is in one of the forms in (1)–(3),(5) then x_p is a ball of S for which r_p can be applied as it has been already described in Ad (1)–Ad (3), Ad (5), respectively,

---

[5] Innermost is meant as minimal with respect to $\subseteq$.

– if r_p is in the form in (4), then x_p is an ordered triple (b_1, b_2, b_3) of balls of $\mathcal{S}$ for which r_p can be applied as it has been described in Ad (4).

A set $\mathcal{P}$ of places of application of evolution rules to a membrane system $\mathcal{S}$ is called *unambiguous* if for all places $p, p' \in \mathcal{P}$ the simultaneous applications of r_p and $r_{p'}$ to x_p and $x_{p'}$, respectively do not lead to a case when an object in a given ball or a ball itself is a subject of two different changes resulting from applications of the rules. The simultaneous applications of the rules of type (1) and of type (4) are always unambiguous.

A P *system* is defined in Păun (2000a) to be an ordered pair $\Pi = (\mathcal{S}, \mathcal{R})$, where $\mathcal{S}$ is a membrane system and $\mathcal{R}$ is a set of expression like those given in (1)–(5) which present the evolution rules. The P systems generate processes of evolution. We consider in the paper only finite processes generated by these systems. A *finite process of evolution generated by* a P system $\Pi = (\mathcal{S}, \mathcal{R})$ is a pair of sequences $(\mathcal{S}_1)_{i=0}^n$ and $(\mathcal{P}_i)_{i=1}^n$ of membrane systems and non-empty sets of places of application, respectively, written

$$\mathcal{S}_0 \stackrel{\mathcal{P}_1}{\Longrightarrow} \mathcal{S}_1 \stackrel{\mathcal{P}_2}{\Longrightarrow} \mathcal{S}_2 \ldots \mathcal{S}_{n-1} \stackrel{\mathcal{P}_n}{\Longrightarrow} \mathcal{S}_n,$$

such that the following conditions hold:

(P_1) $\mathcal{P}_i$ is a maximal (with respect to $\subseteq$) unambiguous set of places of application of evolution rules to $\mathcal{S}_{i-1}$ such that $\{r_p \mid p \in \mathcal{P}_i\} \subseteq \mathcal{R}$ for $1 \le i \le n$,

(P_2) $\mathcal{S}_0$ is $\mathcal{S}$ and for $1 \le i \le n$ the system $\mathcal{S}_i$ is the result of simultaneous applications of r_p to x_p in $\mathcal{S}_{i-1}$ for all $p \in \mathcal{P}_i$, respectively,

(P_3) there is no any rule in $\mathcal{R}$ which can be applied to $\mathcal{S}_n$ which means that *process of evolution stops after n steps*.

A P system $\Pi = (\mathcal{S}, \mathcal{R})$ is called *deterministic* if it generates a process given by $(\mathcal{S}_i)_{i=0}^n$, $(\mathcal{P}_i)_{i=1}^n$ such that the sets $\mathcal{P}_i$ of places of application are the greatest (with respect to $\subseteq$) unambiguous sets of places of application of evolution rules to $\mathcal{S}_{i-1}$ such that $\{r_p \mid p \in \mathcal{P}_i\} \subseteq \mathcal{R}$.

One can also define a *weak deterministic* P system to be that system which generates only finite processes and these processes stop after the same number of steps reaching the same (up to isomorphism) membrane system in the last step.

The P systems used in Păun (2000a) to solve SAT-problem are weak deterministic systems.

3   P Systems for Solving the Integer Factorization Problem

We introduce in this section P systems which solve the Integer Factorization Problem in time $O(\ln^2 n)$ for a natural number $n > 2$ to be factored. We apply an l-program $\mathcal{P}_{\mathrm{fct}}$ to define these P systems.

For a natural number $n > 2$ presented by a binary string $\sigma_1 \ldots \sigma_k$ of length $k = 1 + [\log_2 n]$ (meaningless zeros do not occur in this string) we define a P system $\Pi^{(n)} = (\mathcal{S}^{(n)}, \mathcal{R}^{(n)})$ such that

- the membrane system $\mathcal{S}^{(n)}$ is such that its underlying set $\mathbb{B}$ of balls contains exactly three balls $b_2 \subsetneq b_1 \subsetneq b_0$, the set L of labels of $\mathcal{S}^{(n)}$ is the set $\{0, 1, 2\}$, and $l(b_i) = i$, $e(b_i) = 0$ for all $i \in \{0, 1, 2\}$,
- the set O of objects of $\mathcal{S}^{(n)}$ is the set of ordered triples (x, y, z), written $(x)_z^y$, such that
  - $x \in \{0, 1, \triangleright\} \cup \{0, 1, \ldots, |\mathcal{P}_{\text{fct}}| + 1\}$, one can verify that $|\mathcal{P}_{\text{fct}}| = 68$,
  - $y \in \text{RGR}(\mathcal{P}_{\text{fct}}) \cup \{-@, @, +@, \perp\}$,
  - $z \in \{0, 1, \ldots, k+1\} \cup \{\perp, @\}$,
  where digits $0, 1$, and auxiliary symbols $-@, @, +@, \perp, \triangleright$ are different from natural numbers and names of registers in $\text{RGR}(\mathcal{P}_{\text{fct}})$,
- the labelling function $M : \mathbb{B} \to N^O$ valued in the set N^O of multisets is given by

$$M(b_0) = M(b_1) = \mathbb{O},$$

$$M(b_2) = \left(\sum_{i=1}^{k} \langle (\sigma_i)_i^X \rangle \right) + \langle (\triangleright)_{k+1}^X \rangle + \left(\sum_{j=0}^{k-2} \langle (\triangleright)_j^{\perp} \rangle \right)$$
$$+ \langle (\triangleright)_{k-1}^@ \rangle + \sum_{Z \in I^-} \langle (\triangleright)_1^Z \rangle,$$

where the binary string $\sigma_1 \ldots \sigma_k$ presents natural number n and $I^- = \text{RGR}(\mathcal{P}_{\text{fct}}) - \{X, Y\}$ for input registers X and Y of l-program $\mathcal{P}_{\text{fct}}$,

- the set $\mathcal{R}^{(n)}$ of rules contains the rules given by the following schemes:

(S_1) $[_2 (\triangleright)_i^@]_2^0 \to [_2 (\triangleright)_i^{-@}]_2^- [_2 (\triangleright)_i^{+@}]_2^+$ for $1 \leq i \leq k - 1$,

(S_2) $[_2 (\triangleright)_{i-1}^{\perp} (\triangleright)_i^{-@} \to (\triangleright)_{i-1}^@ (0)_i^Y]_2^-$, for $1 \leq i \leq k - 1$ where 0 is digit,

(S_3) $[_2 (\triangleright)_{i-1}^{\perp} (\triangleright)_i^{+@} \to (\triangleright)_{i-1}^@ (1)_i^Y]_2^+$, for $1 \leq i \leq k - 1$, where 1 is digit,

(S_4) $[_1 [_2]_2^- [_2]_2^+]_1^0 \to [_1 [_2]_2^0]_1^0 [_1 [_2]_2^0]_1^0$,

(S_5) $[_2 (\triangleright)_0^@]_2^0 \to (\triangleright)_k^Y (1)_{@}^{\perp}]_2^0$, where 1 is number,

(P) the set of rules determined by l-program $\mathcal{P}_{\text{fct}}$ and described in Appendix by the clauses (1)–(16),

(F_1) $[_2 (|\mathcal{P}_{\text{fct}}| + 1)_{@}^{\perp} (\triangleright)_j^A \to (1)_{\perp}^{\perp} (\triangleright)_j^A]_2^0$ for $2 < j < k$, where A is output register of $\mathcal{P}_{\text{fct}}$,

(F_2) $[_2 (1)_{\perp}^{\perp}]_2^0 \to (1)_{\perp}^{\perp}$,

(F_3) $[_i (1)_{\perp}^{\perp}]_i^0 \to [_i]_i^+ (1)_{\perp}^{\perp}$ for $i \in \{0, 1\}$.

There is no any other rule in $\mathcal{R}^{(n)}$ than that described by the above schemes.

Theorem. *For every natural number $n > 2$ the P system $\Pi^{(n)} = (\mathcal{S}^{(n)}, \mathcal{R}^{(n)})$ is a deterministic system and the process of evolution generated by it stops after the number of steps no greater than $C \cdot \ln^2 n$ for some constant C. Moreover, the process generated by $\Pi^{(n)}$ reaches in the last step that membrane system $\mathcal{S}$ whose innermost balls distinguished by electric charge $+$ assigned to them contain all non-trivial divisors of n if they exist and also greatest ball of $\mathcal{S}$ has electric charge $+$ in this case, otherwise its charge is 0.*

Proof. We outline a proof of the theorem by describing a process of evolution generated by $\Pi^{(n)} = (\mathcal{S}^{(n)}, \mathcal{R}^{(n)})$. We write $\mathcal{S}_i$ to denote that membrane system which results from transformation of the system $\mathcal{S}^{(n)}$ after i steps of evolution realized by simultaneous applications of the rules contained in $\mathcal{R}^{(n)}$. We write $\mathbb{B}_i$ and M_i to denote the underlying set of balls of $\mathcal{S}_i$ and the labelling functions of $\mathcal{S}_i$ valued in the set N^O of multisets, respectively.

After $2 \cdot (k-1)$ steps the membrane system $\mathcal{S}^{(n)}$ is transformed into membrane system $\mathcal{S}_{2 \cdot (k-1)}$ whose underlying set $\mathbb{B}_{2 \cdot (k-1)}$ of balls contains $2^k + 1$ balls $b_0 \supsetneq b'_j \supsetneq b''_j$ $(1 \le j \le 2^{k-1})$. The transformation of $\mathcal{S}^{(n)}$ into $\mathcal{S}_{2 \cdot (k-1)}$ is realized by applications of the rules given by schemes (S_1)–(S_4) such that in an odd step some rule in the form (S_1) is applied and in an even step the rules in the forms (S_2)–(S_4) are applied simultaneously; see figure below

In all steps $i > 2 \cdot (k-1)$ the underlying sets $\mathbb{B}_i$ of balls of $\mathcal{S}_i$ remain unchanged (i.e., $\mathbb{B}_i = \mathbb{B}_{2 \cdot (k-1)}$ for $i > 2 \cdot (k-1)$) until the object $(1)^{\perp}_{\perp}$ appears in some innermost balls.

There is a one to one correspondence between 2^{k-1} binary strings of length $k-1$ and 2^{k-1} innermost balls in $\mathbb{B}_{2 \cdot k - 1}$ which is determined in consequence of applying (S_5) by the inequality:

$$(*) \qquad \langle (\sigma'_1)^Y_1 \ldots (\sigma'_{k-1})^Y_{k-1} (\triangleright)^Y_k \rangle \le M_{2 \cdot k - 1}(b''_j)$$

for an innermost ball b''_j and a binary string $\sigma'_1 \ldots \sigma'_{k-1}$ of length $k-1$. We have also that

$$M_{2 \cdot k - 1}(b_0) = M_{2 \cdot k - 1}(b'_j) = \mathbb{O} \quad \text{for every } 1 \le j \le 2^{k-1}.$$

In each step $i > 2 \cdot k - 1$ for every innermost ball b''_j the objects contained in b''_j evolve according the rules described in (P) such that the process of evolution simulates the realization of the l-program $\mathcal{P}_{\text{fct}}$ for the input data given by strings

$$(**) \qquad s_X = \sigma_1 \ldots \sigma_k \quad \text{and} \quad s_Y = \sigma'_1 \ldots \sigma'_{k-1},$$

where s_X presents the number n and inequality $(*)$ holds for s_Y. The contents of registers Z in $\text{RGR}(\mathcal{P}_{\text{fct}})$ processed during the realization of $\mathcal{P}_{\text{fct}}$ are represented by multisets

$$\langle (\sigma''_1)^Z \ldots (\sigma''_m)^Z_m (\triangleright)_{m+1} \rangle \le M_i(b''_j)$$

for binary strings $\sigma''_1 \ldots \sigma''_m$ contained in Z.

The objects $(j)^{\perp}_{@}$ for $1 \le j \le |\mathcal{P}_{\text{fct}}|$ are used to control the simulation of the execution of single instructions in the order in which they appear in $\mathcal{P}_{\text{fct}}$.

The multisets $\langle (m)^{\perp}_1 \ldots (m)^{\perp}_q (m)^@_{q+1} \rangle \le M_i(b''_j)$ are aimed to control the simulation of the execution of the instruction $\text{LOOP } |Z|$ appearing as an m-th element of $\mathcal{P}_{\text{fct}}$ (see clause (6) in the Appendix) for a register $Z \in \text{RGR}(\mathcal{P}_{\text{fct}})$ and $1 \le m \le |\mathcal{P}_{\text{fct}}| - 1$.

Thus in each step $i > 2 \cdot k - 1$ the process of evolution simultaneously simulates 2^{k-1} realizations of l-program $\mathcal{P}_{\text{fct}}$ for 2^{k-1} different input data given by strings s_X and s_Y described in $(**)$, where s_X is a constant string presenting the number n and s_Y is an arbitrary binary string of length $k - 1$.

Hence the process of evolution stops after a number of steps no greater than $3C \cdot \ln^2 n$, see the Corollary from Section 1.

The rules given by schemes (F_1), (F_2), (F_3) are used to assign the electric charge $+$ to those balls which contain nontrivial divisors of number n if they exist. By using rules given by scheme (F_3) there is assigned the electric charge $+$ to the greatest ball if there exist nontrivial divisors of n, otherwise the electric charge of this ball remains 0.

Concluding Remarks. Reading the Theorem and its proof one finds that the Integer Factorization Problem can be solved in a polynomial time by using P systems but with an exponential growth of the number of processors, where processors are meant here as balls of membrane systems. We conjecture that by combining the methods outlined in Lenstra and Lenstra (1993) and an idea of probabilistic P systems one could obtain probabilistic P systems which can solve the Integer Factorization Problem in a polynomial time and with a subexponential growth of the number of processors.

One can modify P systems $\Pi^{(n)}$ $(n > 2)$ to obtain new P systems which besides computing non-trivial divisors of n select also those divisors which are prime numbers. A modification can be done by using some new rules related to the rules of type (4) described in Section 2.

4   Appendix

Let $\mathcal{P}$ be an l-program written by using only instructions introduced in Section 1 and let $\mathcal{D}$ be input data for $\mathcal{P}$ such that $\text{SPACE}(\mathcal{P}, \mathcal{D}) \le k$ for a natural number $k > 1$. We define the *set $O_{\mathcal{P},\mathcal{D}}$ of objects determined by $\mathcal{P}$ and $\mathcal{D}$* in the same way as the set of objects of the membrane system $\mathcal{S}^{(n)}$ in Section 3, except $\mathcal{P}_{\text{fct}}$ is replaced by $\mathcal{P}$. The *set of rules determined by occurrences of instructions in $\mathcal{P}$ and by $\mathcal{D}$*, written $\mathcal{R}_{\mathcal{P},\mathcal{D}}$, is defined by the following clauses by using $O_{\mathcal{P},\mathcal{D}}$. We use here capital Latin letters as variables ranging over $\text{RGR}(\mathcal{P})$.

(1) if $1 \le j \le |\mathcal{P}|, X, Y \in \text{RGR}(\mathcal{P})$, and instruction $X := Y$ is j-th element of $\mathcal{P}$, then the rules given by the following schemes belong to $\mathcal{R}_{\mathcal{P},\mathcal{D}}$:

$(s_{1,1})$ $[_2(j)_{\circledcirc}^{\perp}(\triangleright)_{m+1}^X(\triangleright)_{n+1}^Y \to (\triangleright)_{n+1}^X(\triangleright)_{n+1}^Y u_m w_n]_2^0$ for $0 \le m \le k, 0 \le n \le k$, where u_m and u_n are strings of objects belonging to $O_{\mathcal{P},\mathcal{D}}$ such that

$$u_m = \begin{cases} (j)_0^X & \text{if } m = 0, \\ (j)_0^X \dots (j)_m^X & \text{if } m > 0, \end{cases}$$

$$w_n = \begin{cases} \text{empty string} & \text{if } n = 0, \\ (j)_1^Y \dots (j)_n^Y & \text{if } n > 0, \end{cases}$$

$(s_{1,2})$ $[_2(j)_i^X(\sigma)_i^X \to (0)_{\perp}^{\perp}]_2^0$ for $1 \leq i \leq k$ and for a digit $\sigma \in \{0,1\}$,

$(s_{1,3})$ $[_2(j)_i^Y(\sigma)_i^Y \to (\sigma)_i^X(\sigma)_i^Y]_2^0$ for $1 \leq i \leq k$ and for a digit $\sigma \in \{0,1\}$,

$(s_{1,4})$ $[_2(j)_0^X \to (j+1)_{@}^{\perp}]_2^0$

(2) if $1 \leq j \leq |\mathcal{P}|, X \in \mathrm{RGR}(\mathcal{P})$, and instruction $X := \Lambda$ is j-th element of $\mathcal{P}$, then the rules given by the following schemes belong to $\mathcal{R}_{\mathcal{P},\mathcal{D}}$:

$(s_{2,1})$ $[_2(j)_{@}^{\perp}(\triangleright)_{m+1}^X \to (\triangleright)_1^X v_m]_2^0$ for $0 \leq m \leq k$, where v_m is a string of objects on $O_{\mathcal{P},\mathcal{D}}$ such that

$$v_m = \begin{cases} (j+1)_{@}^{\perp} & \text{if } m = 0, \\ (j)_0^X \ldots (j)_m^X & \text{if } m > 0, \end{cases}$$

$(s_{2,2})$ the scheme $(s_{1,2})$,

$(s_{2,3})$ the scheme $(s_{1,4})$,

(3) if $1 \leq j \leq |\mathcal{P}|, X \in \mathrm{RGR}(\mathcal{P})$, and instruction $X := 0$ is j-th element of $\mathcal{P}$, then the rules given by the following schemes belong to $\mathcal{R}_{\mathcal{P},\mathcal{D}}$:

$(s_{3,1})$ $[_2(j)_{@}^{\perp}(\triangleright)_{m+1}^X \to (0)_1^X(\triangleright)_2^X v_m]_2^0$ for $0 \leq m \leq k$ where v_m is a string of objects in $O_{\mathcal{P},\mathcal{D}}$ as in $(s_{2,1})$,

$(s_{3,2})$ the scheme $(s_{1,2})$,

$(s_{3,3})$ the scheme $(s_{1,4})$,

(4) if $1 \leq j \leq |\mathcal{P}|, X \in \mathrm{RGR}(\mathcal{P})$, and instruction $X := \Lambda \ominus X$ is j-th element of $\mathcal{P}$, then the rules given by the following schemes belong to $\mathcal{R}_{\mathcal{P},\mathcal{D}}$:

$(s_{4,1})$ $[_2(j)_{@}^{\perp}(\triangleright)_{m+1}^X \to (\triangleright)_{m+1}^X u_m]_2^0$ for $0 \leq m \leq k$, where u_m is a string of objects in $O_{\mathcal{P},\mathcal{D}}$ as in $(s_{1,1})$,

$(s_{4,2})$ $[_2(j)_0^X(\triangleright)_{m+1}^X \to (j+1)_{@}^{\perp}(\triangleright)_1^X]_2^0$ for $1 \leq m \leq k$,

$(s_{4,3})$ $[_2(j)_0^X(\triangleright)_1^X \to (j+1)_{@}^{\perp}(0)_1^X(\triangleright)_2^X]_2^0$, where 0 is a digit,

$(s_{4,4})$ the scheme $(s_{1,2})$,

(5) if $1 \leq j \leq |\mathcal{P}|, X \in \mathrm{RGR}(\mathcal{P})$, and instruction $X := |X| \leq 1$ is j-th element of $\mathcal{P}$, then the rules given by the following schemes belong to $\mathcal{R}_{\mathcal{P},\mathcal{D}}$:

$(s_{5,1})$ $[_2(j)_{@}^{\perp}(\triangleright)_m^X \to t_m]_2^0$ for $1 \leq m \leq k$, where t_m is a string of objects belonging to $O_{\mathcal{P},\mathcal{D}}$ such that

$$t_m = \begin{cases} (j+1)_{@}^{\perp}(0)_1^X(\triangleright)_2^X & \text{if } m = 1 \text{ (here } 0 \text{ is digit)}, \\ (j)_0^X(j)_1^X(\triangleright)_2^X & \text{if } m = 2, \\ (j)_0^X \ldots (j)_{m-1}^X(\triangleright)_m^X & \text{if } m > 2, \end{cases}$$

$(s_{5,2})$ $[_2(j)_0^X(\triangleright)_m^X \to (j+1)_{@}^{\perp}(\triangleright)_1^X]_2^0$ for $2 < m \leq k+1$,

$(s_{5,3})$ $[_2(j)_0^X(\triangleright)_2^X \to (j+1)_{@}^{\perp}(0)_1^X(\triangleright)_2^X]_2^0$, where 0 is a digit,

$(s_{5,4})$ the scheme $(s_{1,2})$,

(6) if $1 \leq j \leq |\mathcal{P}|, 1 \leq q \leq |\mathcal{P}|, X \in \mathrm{RGR}(\mathcal{P})$, and both the occurrence of $\mathrm{LOOP}\,|X|$ as j-th element of $\mathcal{P}$ and the occurrence of END as q-th element of $\mathcal{P}$ are matching, then the rules given by the following schemes belong to $\mathcal{R}_{\mathcal{P},\mathcal{D}}$:

$(s_{6,1})$ $[_2(j)_{@}^{\perp}(\triangleright)_1^X \to (q+1)_{@}^{\perp}(\triangleright)_1^X]_2^0$,

$(s_{6,2})$ $[_2(j)_{@}^{\perp}(\triangleright)_{m+1}^X \to (q)_{@}^{\perp}(j)_1^{\perp} \ldots (j)_m^{\perp}(j)_{m+1}^{@}(\triangleright)_{m+1}^X]_2^0$ for $1 \leq m \leq k$,

$(s_{6,3})$ $[_2(q)_{@}^{\perp}(j)_i^{\perp}(j)_{i+1}^{@} \to (j+1)_{@}^{\perp}(j)_i^{@}]_2^0$ for $1 \leq i \leq k$,

$(s_{6,4})$ $[_2(q)\frac{1}{@}(j)_1^@ \to (q+1)\frac{1}{@}]_2^0$.

(7) if $1 \leq j \leq |\mathcal{P}|, X \in \mathrm{RGR}(\mathcal{P})$, and instruction $X := \uparrow(X)$ is j-th element of $\mathcal{P}$, then the rules given by the following schemes belong to $\mathcal{R}_{\mathcal{P},\mathcal{D}}$:

$(s_{7,1})$ $[_2(j)\frac{1}{@}(\triangleright)_{m+1}^X \to u_m(\triangleright)_{m+1}^X]_2^0$ for $0 \leq m \leq k$, where u_m is a string of objects as in $(s_{1,1})$,

$(s_{7,2})$ $[_2(j)_0^X(j)_1^X(\sigma)_1^X(\triangleright)_{m+1}^X \to (j+1)\frac{1}{@}(\triangleright)_m^X]_2^0$ for $1 \leq m \leq k$ and for a digit $\sigma \in \{0,1\}$,

$(s_{7,3})$ $[_2(j)_0^X(\triangleright)_1^X \to (j+1)\frac{1}{@}(\triangleright)_1^X]_2^0$

$(s_{7,4})$ $[_2(j)_{i+1}^X(\sigma)_{i+1}^X \to (\sigma)_i^X]_2^0$ for $1 \leq i \leq k$ and a digit $\sigma \in \{0,1\}$,

(8) if $1 \leq j \leq |\mathcal{P}|, X \in \mathrm{RGR}(\mathcal{P})$, and instruction $X := 0 \uparrow X$ is j-th element of $\mathcal{P}$, then the rules given by the following schemes belong to $\mathcal{R}_{\mathcal{P},\mathcal{D}}$:

$(s_{8,1})$ $[_2(j)\frac{1}{@}(\triangleright)_1^X \to (j+1)\frac{1}{@}(\triangleright)_1^X]_2^0$,

$(s_{8,2})$ $[_2(j)\frac{1}{2}(\sigma)_1^X(\triangleright)_{m+1}^X \to u_m^\sigma(\triangleright)_{m+1}^X]_2^0$ for $1 \leq m \leq k$ and a digit $\sigma \in \{0,1\}$, where u_m^σ is a string of objects in $O_{\mathcal{P},\mathcal{D}}$ such that

$$u_m^\sigma = \begin{cases} (j+1)\frac{1}{@}(1)_1^X & \text{if } \sigma \text{ is digit } \mathbf{1}, \\ (j)_0^X \ldots (j)_m^X & \text{if } \sigma \text{ is digit } \mathbf{0}, \end{cases}$$

$(s_{8,3})$ $[_2(j)_0^X(j)_1^X(0)_1^X(\triangleright)_{m+1}^X \to (j+1)\frac{1}{@}(\triangleright)_m^X]_2^0$ for $1 \leq m \leq k$,

$(s_{8,4})$ the scheme $(s_{7,4})$,

(9) if $1 \leq j \leq |\mathcal{P}|, X \in \mathrm{RGR}(\mathcal{P})$, and instruction $X := \downarrow(X)$ is j-th element of $\mathcal{P}$, then the rules given by the following schemes belong to $\mathcal{R}_{\mathcal{P},\mathcal{D}}$:

$(s_{9,1})$ $[_2(j)\frac{1}{@}(\triangleright)_1^X \to (j+1)\frac{1}{@}(\triangleright)_1^X]_2^0$

$(s_{9,2})$ $[_2(j)\frac{1}{@}(\sigma)_m^X(\triangleright)_{m+1}^X \to (j+1)\frac{1}{@}(\triangleright)_m^X]_2^0$ for $1 \leq m \leq k$ and a digit $\sigma \in \{0,1\}$,

(10) if $1 \leq j \leq |\mathcal{P}|, X,Y \in \mathrm{RGR}(\mathcal{P})$, and instruction $X := X \& Y$ is j-th element of $\mathcal{P}$, then the rules given by the following schemes belong to $\mathcal{R}_{\mathcal{P},\mathcal{D}}$:

$(s_{10,1})$ $[_2(j)\frac{1}{@}(\triangleright)_1^Y \to (j+1)\frac{1}{@}(\triangleright)_1^Y]_2^0$,

$(s_{10,2})$ $[_2(j)\frac{1}{@}(\sigma)_1^Y(\triangleright)_{m+1}^X \to (j+1)\frac{1}{@}(\sigma)_1^Y(\sigma)_{m+1}^X(\triangleright)_{m+2}^X]_2^0$ for $0 \leq m < k$ and for a digit $\sigma \in \{0,1\}$,

(11) if $1 \leq j \leq |\mathcal{P}|, X \in \mathrm{RGR}(\mathcal{P})$, and instruction $X := X \& 0$ is j-th element of $\mathcal{P}$, then the rules given by the following scheme belong to $\mathcal{R}_{\mathcal{P},\mathcal{D}}$:

$(s_{11,1})$ $[_2(j)\frac{1}{@}(\triangleright)_{m+1}^X \to (j+1)\frac{1}{@}(0)_{m+1}^X(\triangleright)_{m+2}^X]_2^0$, for $0 \leq m < k$,

(12) if $1 \leq j \leq |\mathcal{P}|, X,Y \in \mathrm{RGR}(\mathcal{P})$ and instruction $X := X \,!\, Y$ is j-th element of $\mathcal{P}$, then the rules given by the following schemes belong to $\mathcal{R}_{\mathcal{P},\mathcal{D}}$:

$(s_{12,1})$ $[_2(j)\frac{1}{@}(\triangleright)_{m+1}^X(\triangleright)_{n+1}^Y \to s_m^n(\triangleright)_{m+1}^X(\triangleright)_{n+1}^Y]_2^0$ for $0 \leq m \leq k, 0 \leq n \leq k$, where s_m^n is a string of objects in $O_{\mathcal{P},\mathcal{D}}$ such that

$$s_m^n = \begin{cases} (j)_0^X & \text{if } m = 0 \text{ or } n = 0, \\ (j)_0^X \ldots (j)_m^X & \text{if } m > 0 \text{ and } n > 0, \end{cases}$$

$(s_{12,2})$ $[_2(j)_0^X(\triangleright)_{m+1}^X(\sigma)_n^Y(\triangleright)_{n+1}^Y \to (j+1)\frac{1}{@}(\sigma)_1^X(\triangleright)_{m+2}^X(\sigma)_n^Y(\triangleright)_{n+1}^Y]_2^0$ for $0 \leq m < k, 0 \leq n \leq k$, and for a digit $\sigma \in \{0,1\}$,

$(s_{12,3})$ $[_2(j)_0^X(\triangleright)_{m+1}^X(\triangleright)_1^Y \to (j+1)\frac{1}{@}(\triangleright)_{m+1}^X(\triangleright)_1^Y]_0^0$ for $0 \leq m \leq k$,

$(s_{12,4})$ $[_2(j)_i^X(\sigma)_i^X \to (\sigma)_{i+1}^X]_2^0$ for $1 \leq i < k$ and a digit $\sigma \in \{0,1\}$,

(13) if $1 \leq j \leq |\mathcal{P}|, X, Y, Z \in \mathrm{RGR}(\mathcal{P})$, and instruction $Z := \oplus(X, Y, Z)$ is j-th element of $\mathcal{P}$, then the rules given by the following schemes belong to $\mathcal{R}_{\mathcal{P}, \mathcal{D}}$:

$(s_{13,1})$ $[_2(j)_{\circledash}^{\perp}(\triangleright)_m^Z \to (j+1)_{\circledash}^{\perp}(\triangleright)_m^Z]_2^0$ for $2 < m \leq k+1$,

$(s_{13,2})$ $[_2(j)_{\circledash}^{\perp}(\mathbf{1})_1^Z(\triangleright)_2^Z u \to (j+1)_{\circledash}^{\perp}(\mathbf{1})_1^Z(\triangleright)_2^Z u]_2^0$ for a string u of objects in $O_{\mathcal{P}, \mathcal{D}}$ in one of the following forms:

(f_1) $(\mathbf{1})_i^X(\triangleright)_{i+1}^X(\mathbf{1})_q^Y(\triangleright)_{q+1}^Y$,

(f_2) $(\mathbf{1})_i^X(\triangleright)_{i+1}^X(\mathbf{0})_q^Y(\triangleright)_{q+1}^Y$,

(f_3) $(\mathbf{0})_i^X(\triangleright)_{i+1}^X(\mathbf{1})_q^Y(\triangleright)_{q+1}^Y$,

(f_4) $(\mathbf{1})_i^X(\triangleright)_{i+1}^X(\triangleright)_1^Y$,

for $1 \leq i \leq k$ and $1 \leq q \leq k$

$(s_{13,3})$ $[_2(j)_{\circledash}^{\perp}(\triangleright)_1^Z u \to (j+1)_{\circledash}^{\perp}(\mathbf{1})_1^Z(\triangleright)_2^Z u]_2^0$ for u of the form (f_1)

$(s_{13,4})$ $[_2(j)_{\circledash}^{\perp}(\mathbf{1})_1^Z(\triangleright)_2^Z u \to (j+1)_{\circledash}^{\perp}(\triangleright)_1^Z u]_2^0$ for u in one of the following forms:

(f_5) $(\mathbf{0})_i^X(\triangleright)_{i+1}^X(\mathbf{0})_q^Y(\triangleright)_{q+1}^Y$,

(f_6) $(\mathbf{0})_i^X(\triangleright)_{i+1}^X(\triangleright)_1^Y$,

for $1 \leq i \leq k$ and $1 \leq q \leq k$,

$(s_{13,5})$ $[_2(j)_{\circledash}^{\perp}(\triangleright)_1^Z u \to (j+1)_{\circledash}^{\perp}(\triangleright)_1^Z u]_2^0$ for u in one of the forms (f_2)–(f_6),

(14) if $1 \leq j \leq |\mathcal{P}|, X, Y, Z \in \mathrm{RGR}(\mathcal{P})$, and instruction $Z := \ominus(X, Y, Z)$ is j-th element of $\mathcal{P}$, then the rules given by the following schemes belong to $\mathcal{R}_{\mathcal{P}, \mathcal{D}}$:

$(s_{14,1})$ the same as $(s_{13,1})$,

$(s_{14,2})$ $[_2(j)_{\circledash}^{\perp}(\triangleright)_1^Z u \to (j+1)_{\circledash}^{\perp}(\mathbf{1})_1^Z(\triangleright)_2^Z u]_2^0$ for u of the form (f_3),

$(s_{14,3})$ $[_2(j)_{\circledash}^{\perp}(\mathbf{1})_1^Z(\triangleright)_2^Z u \to (j+1)_{\circledash}^{\perp}(\mathbf{1})_1^Z(\triangleright)_2^Z u]_2^0$ for u in one of the forms (f_1), (f_3), (f_5), (f_6),

$(s_{14,4})$ $[_2(j)_{\circledash}^{\perp}(\triangleright)_1^Z u \to (j+1)_{\circledash}^{\perp}(\triangleright)_1^Z u]_2^0$ for u in one of the forms (f_2), (f_4), (f_5), (f_6),

$(s_{14,5})$ $[_2(j)_{\circledash}^{\perp}(\mathbf{1})_1^Z(\triangleright)_2^Z u \to (j+1)_{\circledash}^{\perp}(\triangleright)_1^Z u]_2^0$ for u in one of the forms (f_2), (f_4),

(15) if $1 \leq j \leq |\mathcal{P}|, W, X, Y, Z \in \mathrm{RGR}(\mathcal{P})$, and instruction $W := \sharp(X, Y, Z)$ is j-th element of $\mathcal{P}$, then the rules given by the following schemes belong to $\mathcal{R}_{\mathcal{P}, \mathcal{D}}$:

$(s_{15,1})$ $[_2(j)_{\circledash}^{\perp}(\triangleright)_{m+1}^W \to p_m]_2^0$ for $0 \leq m \leq k$ and for a string p_m of objects in $O_{\mathcal{P}, \mathcal{D}}$ such that

$$p_m = \begin{cases} (j)_0^W & \text{if } m = 0, \\ (j)_0^W \ldots (j)_{m+1}^W & \text{if } m > 0, \end{cases}$$

$(s_{15,2})$ $[_2(j)_i^W(\sigma)_i^W \to (\sigma)_{i+1}^W]_2^0$ for $1 \leq i \leq k$ and $\sigma \in \{\mathbf{0}, \mathbf{1}, \triangleright\}$,

$(s_{15,3})$ $[_2(j)_0^W u \to (\sigma)_1^W u(j+1)_{\circledash}^{\perp}]_2^0$ for a string u of objects in $O_{\mathcal{P}, \mathcal{D}}$ in one of the following forms

$$(\sigma_1)_i^X(\triangleright)_{i+1}^X(\sigma_2)_m^Y(\triangleright)_{m+1}^Y(\sigma_3)_q^Z(\triangleright)_{q+1}^Z,$$
$$(\sigma_1)_i^X(\triangleright)_{i+1}^X(\triangleright)_1^Y(\triangleright)_1^Z,$$

$$(\sigma_1)_i^X (\triangleright)_{i+1}^X (\sigma_2)_m^Y (\triangleright)_{m+1}^Y (\triangleright)_1^Z ,$$
$$(\sigma_1)_i^X (\triangleright)_{i+1}^X (\triangleright)_1^Y (\sigma_3)_q^Z (\triangleright)_{q+1}^Z$$

(for $1 \leq i \leq k$, $1 \leq m \leq k$, $1 \leq q \leq k$, and for digits $\sigma_1, \sigma_2, \sigma_3 \in \{0, 1\}$) and for a digit σ depending on u in such a way that σ is **1** if digit **1** appears three times in u or **1** appears one time in u (in the places where $\sigma_1, \sigma_2, \sigma_3$ occur), otherwise σ is digit **0**,

$(s_{15,4})$ $[_2(j)_0^W (\triangleright)_1^X \to (\mathbf{0})_1^W (\triangleright)_1^X (j+1)_{@}^{\perp}]_2^0,$

(16) there is no any other rule in $\mathcal{R}_{\mathcal{P},\mathcal{D}}$ than those given by the above clauses (1)–(15).

References

[1974] Brainerd, W. S., Landweber, L. H. (1974): *Theory of Computation.* New York

[1998] Koblitz, N. (1998): *Algebraic Aspects of Cryptography.* Berlin

[1993] Lenstra, A. K., Lenstra, H. W., Jr. (1993): The Development of the Number Field Sieve. *Lecture Notes in Mathematics,* **1554**, Berlin

[1996] Menezes, A. J., van Oorschot, P. C., Vanstone, S. A. (1996): *Handbook of Applied Cryptography.* CRC Press, Boca Raton

[1967] Meyer, A. R., Ritchie, D. M. (1967): The complexity of loop programs. *Proceedings of the ACM National Meeting,* ACM Pub. P-67, 465–469

[1994] Papadimitriou, Ch. P. (1994): *Computational Complexity.* Reading, Massachusetts

[2000a] Păun, Gh. (2000a): P-Systems with Active Membranes: Attacking NP Complete Problems. *Journal of Automata, Languages and Combinatorics,* **6** (2000), 75–90

[2000b] Păun, Gh. (2000b): Computing with Membranes. *Journal of Computer and System Sciences,* **61** (2000) 108–143

The Linear Theory
of Multiset Based Dynamic Systems

Wolfgang Reisig

Humboldt-Universität zu Berlin, Institut für Informatik,
Rudower Chausse 25, D-12489 Berlin

1 Multisets in Unconventional Models of Computation

All operational models of computation define two kinds of elementary concepts:
Data and discrete change of data. Data are usually aggregated in *data locations*.
Data are changed by *actions*. An action usually affects some, but not all data
locations.

In conventional models, data locations are composed of variables. Actions
update variables, i.e. replace a variable's present value by a new value.

In unconventional models, data locations carry frequently sets of data. Occurence of an action coincidently updates some of those locations (i.e. adds or
removes elementary data). Actions may concurrently access data locations: Two
actions may add and/or remove data to or from a location, without interfering
with each other. In particular, they may add two instances of the same data
item, or an action may add an instance of a data item of which another item
is already present in that location. Hence, ordinary sets are not adequate to
describe concurrent access to data locations; instead, multisets (bags) are the
adequate choice. In fact, many unconventional system models are based on multisets, including Petri Nets [4] and the Chemical Abstract Machine [1]. Milner's
unifying approach to unconventional models [2] likewise makes multisets an essential concept.

An algorithm with a proper initial state may evolve without multiple data
items. But any attempt to generally prevent actions to generate multiple instances of data items in data locations requires actions be given comprehensive
control over the location. Vice versa, interference free, concurrent access of actions to a data location demands that data locations potentially may carry multisets of data items. In fact, a number of unconventional models of computation
suggest actions on multisets.

In this paper we consider the most elementary model of (distributed) computing with multisets; i.e. high level Petri Nets. This model allows for useful
analysis techniques, based on the linear theory of multisets.

C.S. Calude et al. (Eds.): Multiset Processing, LNCS 2235, pp. 287–297, 2001.

2   The Linear Theory of Multisets

The Universe U

We employ the usual notation for sets, subsets, cartesian product, union, intersection, element, etc. $\mathbf{Z}$ denotes the integers; $\mathbf{N}$ the non-negative integers. The truth values are *true* or *false*.

For the sake of simplicity we assume a *universe* U, i.e. a collection of all relevant items. We construct cartesian products U^n and functions $f : U^n \to U$, but avoid constructing set theoretical anomalies.

Multisets

Formally, a multiset is a mapping

$$M : U \to \mathbf{N}.$$

For technical reasons we occasionally employ multisets with negative entries,

$$M : U \to \mathbf{Z}.$$

M is *finite* if $M(u) \neq 0$ for finitely many $u \in U$ only. A finite multiset M without negative entries may be written

$$M = [a_1, \ldots, a_n];$$

for each $u \in U$, there are $M(u)$ entries a_i in $[a_1, \ldots, a_n]$ identical to u. The empty multiset, $M(u) = 0$ for all $u \in U$, is written

$$[].$$

Each element $u \in U$ forms the multiset $[u]$. We often confuse u and $[u]$, context allowing. By

$$\mathcal{M}$$

we denote the collection of all multisets.

Operations on Multisets

Let M and N be multisets. M is smaller or equal to N

$$M \leq N,$$

if $M(u) \leq N(u)$ for all $u \in U$. The sum, $M + N$ is a multiset, defined by

$$(M + N)(u) := M(u) + N(u)$$

for all $u \in U$. As an analogy define $M - N$ by

$$(M - N)(u) := M(u) - N(u).$$

Hence, $M - N$ is a proper multiset (i.e. has no negative entries) iff $N \sqsubseteq M$. Finally, let

$$-M := [] - M,$$

i.e. $(-M)(u) = -(M(u))$.

$\mathcal{M}$-Valued Functions

We frequently consider functions f formed

$$f : U^n \to \mathcal{M}$$

called $\mathcal{M}$ - *valued*. As a particular function let $\mathbf{0}$ denote the empty function, i.e. $\mathbf{0}(u_1, \ldots, u_n) = []$. Each function $f : U^n \to U$ can canonically be extended to $\mathbf{f} : U^n \to \mathcal{M}$, by $\mathbf{f}(u_1, \ldots, u_n) := [f(u_1, \ldots, u_n)]$. We often write just f instead of $\mathbf{f}$.

$\mathcal{M}$-valued functions can be summed up. For $f, g : U \to \mathcal{M}$, let $f + g : U \to \mathcal{M}$ be defined as

$$(f + g)(u_1, \ldots, u_n) := f(u_1, \ldots, u_n) + g(u_1, \ldots, u_n).$$

Symbolic Representation of Elements and Functions

As usual, we frequently employ *terms* to denote data and functions. A term is composed from constant symbols, variables and operation symbols. A term without variables is called a *ground term* and can be *evaluated*, denoting an element of the universe. A term with one variable (multiple occurrences allowing) denotes a function, f. To compute $f(u)$, replace each occurrence of the variable by u. This results in a ground term, denoting an element. A term with n variables, together with an *order* on the variables, denotes an n-ary function f. To compute $f(u_1, \ldots, u_n)$, replace each occurence of the i-th variable by u_i $(i = 1, \ldots, n)$.

A multiset of terms with n variables denotes a multiset valued function. For example, the multiset $[x \cdot x, x + x]$ denotes a unary function. Applied to the integer 2, it returns the multiset $[4, 4]$.

3   Multiset Based Dynamic Systems

States

We start out with a set $P = \{p_1, \ldots, p_m\}$ of data stores, called *places*. A *state* S assigns to each place p its actual contents, a multiset S_p. Hence, a state S can be written as a vector

$$S = (S_p)_{p \in P}$$

of multisets.

Actions

An *action* transforms a state S into a new state, S'. An action

$$a = (a^-, a^+)$$

has two components, a^- and a^+. They describe the elements to be removed from places and to be augmented to places, respectively. Technically, actions are formed like states, i.e. as P-indexed vectors

$$a^- = (a^-_p)_{p \in P} \text{ and } a^+ = (a^+_p)_{p \in P}$$

of multisets.

There is a quite intuitive graphical representation of actions a, with a circle for each place and a box for a itself. An arc from the circle for p to the box and is inscribed by a^-_p. Likewise, an arc from the box to the circle for p is inscribed by a^+_p. Arcs inscribed by the empty multiset are skipped, as well as circles that eventually remain isolated.

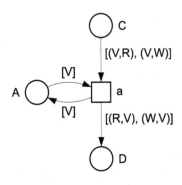

Fig. 1.

As part of the forthcoming case study, Fig. 1 shows an example. We assume places A, C and D, the constant V and two pairs of constants, (V, R) and (V, W), which also are elements of the universe. Then, let $a^-_A = a^+_A = [V]$, $a^-_C = [(V, R), (V, W)]$, $a^+_D = [(R, V), (W, V)]$, and $a^+_C = a^-_D = [\,]$.

Steps

A state S *enables an action* $a = (a^-, a^+)$ iff for each place $p \in P$

$$a^-_p \leq S_p,$$

i.e. iff for each place $p \in P$, the multiset a^-_p to be removed from p, is required to be available at p. In this case, S and a define a *step*

$$S \xrightarrow{a} S',$$

with S' defined for each place $p \in P$ by $S'_p = S_p - a^-_p + a^+_p$. The graphical representation of an action a, as e.g. in Fig. 1, mimics its role in steps $S \xrightarrow{a} S'$, i.e. its effect on S: Items flow along the arcs, as indicated in the arc inscriptions.

For example, assume three places A, C, D and a state S with $S_A = [V, W]$, $S_C = [(V, R), (V, W), (W, V)]$ and $S_D = []$. Then action a, as depicted in Fig. 1, is enabled and yields the step $S \xrightarrow{a} S'$, with $S'_A = S_A$, $S'_C = [(W, V)]$ and $S'_D = [(R, V), (W, V)]$.

Transitions

A system usually does with quite a large number of actions. Actions are usually *clustered* into *transitions*.

A transition is a *parameterized action*, with a set Q of *occurence modes* as parameters:

$$t = (t^-, t^+),$$

with $t^- = (t^-{}_p)_{p \in P}$ and $t^+ = (t^+{}_p)_{p \in P}$. Each $t^-{}_p$ and $t^+{}_p$ is a function

$$t^-{}_p : Q \to \mathcal{M} \quad \text{and} \quad t^+{}_p : Q \to \mathcal{M}.$$

A state S *enables t in a mode $q \in Q$* iff for each place $p \in P$, $t^-{}_p(q) \leq S$. This leads to the step

$$S \xrightarrow{t, q} S',$$

with S' defined for each $p \in P$ by

$$S'_p = S_p - t^-{}_p(q) + t^+{}_p(q).$$

A transition t can be conceived as a family

$$t = (t_q)_{q \in Q}$$

of actions $t_q = (t_q^-, t_q^+)$, defined for each $p \in P$ by

$$t_{q\,p}^- = t^-{}_p(q) \quad \text{and} \quad t_{q\,p}^- = t^-{}_p(q).$$

Fig. 2 depicts a transition. It employs the term representation as introduced at the end of Chapter 2, skiping multiset brackets. The action defined by occurence mode $q = V$ has been given in Fig 1.

Behaviors

As usual, a finite or infinite sequence

$$S_0 \xrightarrow{t_1, q_1} S_1 \xrightarrow{t_2, q_2} S_2 \longrightarrow \cdots$$

of steps $S_{i-1} \xrightarrow{t_i, q_i} S_i$ constitutes a *behavior*. Given a distinguished initial state *init*, a behavior is assumed to start with $S_0 = $ init

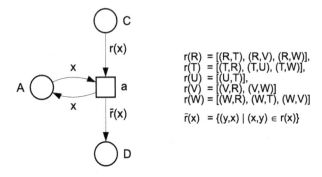

Fig. 2.

Dynamic Systems

We are now prepared to define the mathematical structure of multiset based dynamic systems. A system

$$N = (P, T, init)$$

consists of a set P of states, a set T of transitions as defined above, and an initial state, $init$. N characterizes a (finite or infinite) set of behaviors.

Graphical Representation

There is a quite intuitive graphical representation of dynamic systems: Each place p is represented us a circle (or ellipse), inscribed by $init(p)$. Each transition t is represented as a box, with arrows to and from circles. The arrow from p to t is inscribed by $t^-{}_p$; the arrow from t to p is inscribed by $t^+{}_p$. Arrows inscribed by the empty multiset function **0** are skipped.

4   A Case Study

Consensus in a Network of Agents

Let U be a set of *sites*. Each site u is capable of sending messages to a set of other sites (the "neighbors" of u). All sites together intend to reach some kind of contract or agreement. There is no mediator or broker; each site runs the same local algorithm. The graph in Fig. 3 depicts the set $\{R, T, U, V, W\}$ of sites. Neighborship is indicated by arcs. For example, R and W are neighbors, R and U are not neighbored. Each site is always in one of two states, *pending* or *agreed*. Initially, all sites are pending. In order to negotiate with its neighbors, a pending site may coincidently send *requests* to all its neighbors, or may receive a request from some neighbor. For example, pending site V coincidently sends a message to both R and W; or V receives a message from W. A site is to circulate at most one bunch of requests at a time. Each receiver of a request is obliged to answer it.

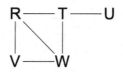

Fig. 3. A network of sites

In a situation where all requests sent by a site u have been answered, u may turn from *pending* to *agreed*. u remains *agreed* unless u obtains a message. Upon such a message, u turns pending again.

Call a state *stable* if *all* sites are agreed. We suggest an algorithm that does not guarantee a stable state will ever be reached. But it guarantees a stable state will retain: In a stable state, no site will turn *pending* again.

Fig. 4 shows the behavior of site V: The site is initially pending and its two requests (V, R) and (V, W) to its neighbors R and W are completed. Intuitively, (V, R) and (V, W) are "envelopes" that V may send to R and to W, respectively. Pending V has the choice of three actions, depending on its environment: With both requests completed, V may move to *agreed* by help of action d. Alternatively, V may initiate requests R and W, by action a, yielding (R, V) and (W, V) at *initiated requests*. And R may receive a message (V, y) from a neighbor y (with $y = R$ or $y = W$) and return the envelope (y, V) to this neighbor.

What has been done in Fig. 4 for the site V, can likewise be done for all othersites, too. The would be structurally identical; only inscriptions would change. Using a variable, x, for sites, Fig. 5 then squeezes all Figures in one: For a pending site x, transition a initiates the requests (y, x) to all neighbors y of x, provided all requests from x are completed.

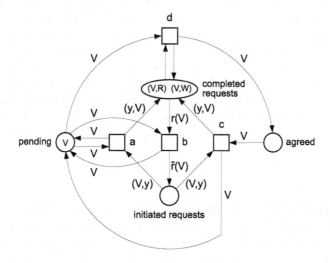

Fig. 4.

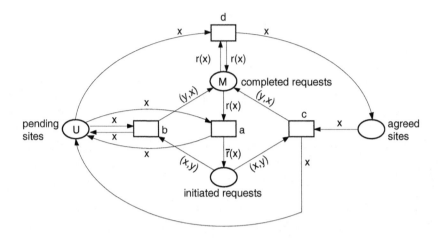

sort	site		fct	$r, \bar{r}$:site → set of messages
sort	message = site × site		var	x, y : site
const	U : set of sites			$r(x) = \{x\} \times M(x)$
const	M : set of messages			$\bar{r}(x) = M(x) \times \{x\}$

Fig. 5. Consensus Algorithm

A pending site x receives a request (x, y) along transition a, answers it by a message (y, x) and remains pending. An agreed site x receives a request (x, y) along transition c, answers it by a message (y, x), and turns pending. Finally, a pending site x may turn agreed along transition d, provided all requests $r(x)$ of x are completed. Notice the different roles of places in Fig. 5: To conceive the behavior of a single site u, replace x by u. Hence, conceive five copies of Fig 5, each replacing x by one of the sites R, T, U, V, W. Each site u has its local states pending(u) and agreed(u), and its local data store for *completed requests*. The *initiated requests* place links the five copies: It contains messages (u, v) sent by v, but not get received by u. The initial state enables transition a in each mode $x = R, \ldots, x = W$, because *pending site* contains $R, \ldots, W$ and *completed requests* contains $r(R) \cup \ldots \cup r(W)$. Likewise, transition d is enabled in all modes. Occurrence of d in all modes $x = R, \ldots, x = W$ leads to the stable state. Alternatively, $init \xrightarrow{a,R} S$ is a step with S (completed requests) $= r(S) \cup \ldots \cup r(W)$ and S *(initiated requests)* $= F(A)$.

The Consensus Algorithm

With $U = \{R, T, U, V, W\}$ and M as described above, Fig. 5 describes the consensus algorithm for the network of Fig. 3. For a different set U of agents and a different set M of neighbors, Fig. 5 likewise represents the consensus algorithm. Generally formulated, Fig. 5 represents the algorithm for *any* network: The specification of U, M, r and $\bar{r}$ allows for infinitely many interpretations.

5   Verification

The linear structure of steps allows for powerful analysis techniques; they even apply to the schematic level of system models, as discussed above.

Linear Functions

A function $f : \mathcal{M} \to \mathcal{M}$ is *linear* iff for all multisets M, N holds

$$f(M + N) = f(M) + f(N), \quad \text{and}$$
$$f(-M) = -f(M) \, .$$

To each unary $\mathcal{M}$-valued function $f : U \to \mathcal{M}$ there exists a unique linear extension $\mathbf{f} : \mathcal{M} \to \mathcal{M}$ with $\mathbf{f}([u]) = f(u)$.

Composition of Functions

A function $f : \mathcal{M} \to \mathcal{M}$ can be composed with a $\mathcal{M}$-valued function $g : U^n \to \mathcal{M}$, yielding $f \circ g : U^n \to \mathcal{M}$, defined as usual for $u_1, \ldots, u_n \in U$

$$f \circ g(u_1, \ldots, u_n) = f(g(u_1, \ldots, u_n)).$$

For the special case of $n = 0$, $g()$ is a multiset, written g. Then

$$f \bullet g = f(g)$$

which is application of f to a multiset.

Composition as Vector Product

For a vector $i = (i_p)_{p \in P}$ of functions $i_p : \mathcal{M} \to \mathcal{M}$ and a vector $t = (t_p)_{p \in P}$ of functions $t_p : U^n \to \mathcal{M}$ for some fixed n, we employ composition of their components as a product and define the usual vector product of i and t as

$$i \cdot t := \sum_{p \in P} i \circ t \, .$$

Obviously, $i \cdot t$ is a function $i \cdot t : U^n \to \mathcal{M}$. In case of 0-ary components t_p, i.e. multisets, we write

$$i(t) \quad \text{instead of} \quad i \circ t$$

Obviously, $i(t) \in \mathcal{M}$.

Place Invariants

A vector $(i_p)_{p\in P}$ of linear multiset functions $i_p : \mathcal{M} \to \mathcal{M}$ is a *place invariant of a dynamic system* $\sum$ iff for each transition t of $\sum$,

$$i \bullet t = \mathbf{0} .$$

A place invariant establishes a system invariant as follows:

Theorem 1. *Let* $\sum = (P, T, \text{init})$ *be a dynamic system and let* i *be a place invariant of* $\sum$. *For each reachable state* S *of* $\sum$ *holds:*

$$i(S) = i(\text{init}) .$$

Traps

A vector $i = (i_p)_{p\in P}$ of linear multiset function $i_p : \mathcal{M} \to \mathcal{M}$ is a *trap of* $\sum$ iff for all transitions t of $\sum$,

$$i \bullet t \geq \mathbf{0} .$$

Theorem 2. *Let* $N = (P, T, \text{init})$ *be a dynamic system and let* i *be a trap of* N *with* $i \bullet \text{init} \geq \mathbf{0}$ *For each reachable state* S *of* N *holds:*

$$i(S) \geq [] .$$

Verification of the Consensus Algorithm

We claimed that stable states of the consensus algorithm retain. Hence we have to show that at each reachable state of the algorithm holds: if all sites are agreed, then no request is initiated. Shortly, for all reachable S,

$$S \models (\textit{agreed sites} = U \quad \Rightarrow \quad \textit{initiated requests} = []). \tag{5.1}$$

Given a concrete set U and a concrete function r (and hence $\bar{r}$ and M), proposition (5.1) can be examined by visiting all reachable states S.

But we are not given concrete U and r. Instead, Fig. 5 is a *schema* for any choice of U and r. And (5.1) is claimed to hold for any such choice. As a matter of convenience, Fig. 6 replaces the long place inscriptions of Fig. 5 by A, B, C, D. What we have to show then reduces to

$$B = U \Rightarrow D = []. \tag{5.2}$$

This can now easily be done by help of two place invariants,

$$A + B = U \tag{5.3}$$

and

$$C + \overline{D} = M, \tag{5.4}$$

where $\overline{D} = \{(y, x)|\, (x, y) \in D\}$, as well as the trap

$$r(A) + C \geq r(U). \tag{5.5}$$

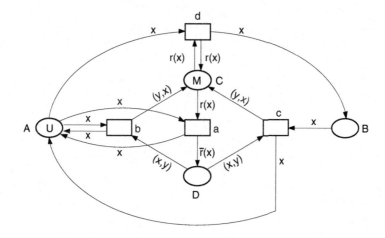

sort	site	fct	r, r̄ :site → set of messages
sort	message = site × site	var	x, y : site
const	U : set of sites		r(x) = {x} × M(x)
const	M : set of messages		r̄(x) = M(x) × {x}

Fig. 6.

$$\text{Obviously,} \quad r(U) = M. \tag{5.6}$$

Then (5.3) and (5.6) imply

$$r(A) + r(B) = M. \tag{5.7}$$

Subtracting (5.5) from the sum of (5.4) and (5.7)

$$\text{yields} \quad (5.4) + (5.7) - (5.5):$$
$$r(B) + \overline{D} \le M. \tag{5.8}$$

Now we conclude from (5.6)

$$B = U \not\to r(B) = r(U) = M$$
$$and \quad r(B) = M \to \overline{D} = [\,] , \text{ by } (5.8).$$

References

1. G. Berry and G. Boudol. The chemical abstract machine. *TCS*, 96:217–248, (1992).
2. R. Milner. Calculi for interaction. *Acta Informatica*, 33:707–737, (1996).
3. W. Reisig. *Elements of Distributed Algorithms*. Springer-Verlag, (1998).
4. W. Reisig and G. Rozenberg. *Lectures on Petri Nets I, II*, volume 1481 and 1492 of *LNCS*. Springer-Verlag, (1998).

Artificial Life Applications of a Class of P Systems: Abstract Rewriting Systems on Multisets

Yasuhiro Suzuki[1], Yoshi Fujiwara[2], Junji Takabayashi[3], and Hiroshi Tanaka[1]

[1] Bio-Informatics, Medical Research Institute, Tokyo Medical and Dental University, Yushima 1-5-45, Bunkyo, Tokyo 113 Japan
[2] Keihanna Center, Communications Research Laboratory, Kyoto 619-0289, Japan
[3] Center for Ecological Research, Kyoto University, Otsuka 509-3, Hirano, Kamitanakami, Otsu, 520-2113, Japan

Abstract. Artificial chemical system are well studied in Artificial Life and Complexity. Here we overview results about such a class of artificial chemical systems: Abstract Rewriting Systems on Multisets (ARMS), which are particular types of P Systems and exhibit complex behaviors such as non-linear oscillations. In short, an ARMS consists of a membrane which contains "chemical compounds" (denoted by symbols); these compounds evolve through chemical reactions in a cell-like device. To such systems we apply a genetic method and, by a simulation on computer, we find several results of Artificial Life interest.

1 Introduction

Artificial Life is the study of man-made systems that exhibit behaviors characteristic of natural living systems. It complements the traditional biological sciences concerned with the *analysis* of living organisms by attempting to *synthesize* life-like behaviors within computers and other artificial media [36].

C.Langton, Artificial Life.

It has been over a decade since Langton proposed "Artificial Life (ALife)" with the above definition. He also explained "by extending the empirical foundation upon which biology is based *beyond* the carbon-chain life that has evolved on Earth, Artificial Life can contribute to theoretical biology by locating *life-as-we-know-it* within the larger picture of *life-as-it-could-be*".

With the advent in biotechnology and increasing interests in genome information worldwide, genome databases progress so fast that the whole sequence database of human genome will be published soon. Hence now we are confronted with the problem of how to deal with the overwhelming genome information. As traditional biological analysis is only powerful in analyzing a single gene most of the case, ALife study turns out to be more and more important in carbon-chain life science.

C.S. Calude et al. (Eds.): Multiset Processing, LNCS 2235, pp. 299–346, 2001.
© Springer-Verlag Berlin Heidelberg 2001

One challenging and important theme in genome science is to discover gene-networks. Our knowledge of the genes' regulation has come from an accumulation of fundamental studies over the past. However recently with a new technology, the DNA chip provides us with genome wide information fast and informative. Combination with genomic database and use of ALife, now we are closer to find the correlation and regulation between genes. In order to understand the gene networks, we usually need to model a system by first synthesizing the empirical data, and in this work an Alife approach is necessary. One of the ALife models, Kauffuman's boolean network model, has been rather instructive in the analysis of gene networks [52].

Therefore, the subjects of ALife studies are changing. It is much easier these days to obtain numerous data from living systems, though we still need methods for observing or analyzing them systematically. Thus, it is necessary to construct theoretical/symbolic systems. This resemble the studies in cosmology. Although we cannot observe Big-bang, we can infer many features from empirical data through theoretical models.

Empirically, abstract chemical systems turn to be powerful in describing complex systems. In this paper first introduce the abstract chemical model we use, based on multiset rewriting, and then we present some experimental results. We will also briefly discuss an application in ecology. The used systems can be considered as P systems with only one membrane. After that, we enhance the adequacy of the system to ALife matters by introducing a membrane structure, similar to those used in P systems.

Abstract Chemical System. Life can be considered as a system in a specific class of chemical reaction systems, but real biochemical systems are so complex that it is difficult to reconstruct the precise dynamics of such a system. Thus, it is important to abstract the essential properties of biochemical systems in order to obtain insights into their dynamical properties.

There have been developed various Artificial Chemistries. In a broad sence, an artificial chemistry is a man-made system which is similar to a chemical system. It can be defined as a set of objects and a set of reaction rules which specify how the objects interact. It covers large field, such as:

- **Modeling:** living systems (e.g., origin of life), sociology and parallel processes,
- **Information processing:** control, automatic proving, chemical computing,
- **Optimization:** Combinatorial problems (e.g., TSP).

2   ARMS

Extending the concepts of abstract rewriting systems, we introduce the notion of an *abstract rewriting system on multisets* (ARMS). Intuitively, an *ARMS* is like a chemical solution in which floating *molecules* can interact with each other according to given reaction rules. Technically, a chemical solution is a finite multiset of elements denoted by symbols from a given alphabet, $A = \{a, b, \ldots, \}$;

these elements correspond to *molecules*. Reaction rules that act on the molecules are specified in *ARMS* by rewriting rules. As to the intuitive meaning of *ARMS*, we refer to the study of chemical abstract machines [5].

Let A be an *alphabet* (a finite set of abstract symbols). The set of all strings over A is denoted by A^*; the empty string is denoted by λ. (Thus, A^* is the free monoid generated by A under the operation of concatenation, with identity λ.) The length of a string $w \in A^*$ is denoted by $|w|$.

A *rewriting rule* over A is a pair of strings (u, v), $u, v \in A^*$. We write such a rule in the form $u \to v$. Note that u and v can also be empty. A *rewriting system* is a pair (A, R), where A is an alphabet and R is a finite set of rewriting rules over A.

With respect to a rewriting system $\gamma = (A, R)$ we define over A^* a relation $\Longrightarrow$ as follows: $x \Longrightarrow y$ iff $x = x_1 u x_2$ and $y = x_1 v x_2$, for some $x_1, x_2 \in A^*$ and $u \to v \in R$. The reflexive and transitive closure of this relation is denoted by $\Longrightarrow^*$.

¿From now on, we work with an alphabet A whose elements are called *objects*; the alphabet itself is called a *set of objects*.

A *multiset* over a set of objects A is a mapping $M : A \longrightarrow \mathbf{N}$, where $\mathbf{N}$ is the set of natural numbers, 0, 1, 2,.... The number $M(a)$, for $a \in A$, is the *multiplicity* of object a in the multiset M. Note that we do not accept here an infinite multiplicity. The set $\{a \in A \mid M(a) > 0\}$ is denoted by $supp(M)$ and is called the *support* of M. The number $\sum_{a \in A} M(a)$ is denoted by $weight(M)$ and is called the *weight* of M.

We denote by $A^{\#}$ the set of all multisets over A, including the empty multiset, $\emptyset$, defined by $\emptyset(a) = 0$ for all $a \in A$.

A multiset $M : A \longrightarrow \mathbf{N}$, for $A = \{a_1, \ldots, a_n\}$, can be naturally represented by the string $a_1^{M(a_1)} a_2^{M(a_2)} \ldots a_n^{M(a_n)}$ and by any other permutation of this string. Conversely, with any string w over A we can associate a multiset: denote by $|w|_{a_i}$ the number of occurrences of object a_i in w, $1 \leq i \leq n$; then, the multiset associated with w, denoted by M_w, is defined by $M_w(a_i) = |w|_{a_i}, 1 \leq i \leq n$.

The union of two multisets $M_1, M_2 : A \longrightarrow \mathbf{N}$ is the multiset $(M_1 \cup M_2) : A \longrightarrow \mathbf{N}$ defined by $(M_1 \cup M_2)(a) = M_1(a) + M_2(a)$, for all $a \in A$. If $M_1(a) \leq M_2(a)$ for all $a \in A$, then we say that multiset M_1 is included in multiset M_2 and we write $M_1 \subseteq M_2$. In such a case, we define the multiset difference $M_1 - M_2$ by $(M_2 - M_1)(a) = M_2(a) - M_1(a)$, for all $a \in A$. (Note that when M_1 is not included in M_2, the difference is not defined).

A rewriting rule

$$a \to a \ldots b,$$

is called a *heating rule* and denoted as $r_{\Delta > 0}$; it is intended to contribute to the stirring solution. It breaks up a complex *molecule* into smaller ones: *ions*. On the other hand, a rule such as

$$a \ldots c \to b,$$

is called a *cooling rule* and denoted as $r_{\Delta < 0}$; it rebuilds *molecules* from smaller ones. In this paper, reversible reactions, i.e., $S \rightleftharpoons T$, are not considered. We shall not formally introduce the refinement of *ions* and *molecules* though we use refinement informally to help intuition (for both types of rules we refer to [5]).

A *multiset rewriting rule* over a set A of objects is a pair (M_1, M_2), of elements in $A^\#$ (which can be represented as a rewriting rule $w_1 \to w_2$, for two strings $w_1, w_2 \in A^*$ such that $M_{w_1} = M_1$ and $M_{w_2} = M_2$). We use to represent such a rule in the form $M_1 \to M_2$.

An *abstract rewriting system on multisets* (in short, an *ARMS*) is a pair

$$\Gamma = (A, R),$$

where:

(1) A is a set of objects;
(2) R is a finite set of multiset evolution rules over A.

With respect to an *ARMS* Γ, we can define over $A^\#$ a relation: $(\Longrightarrow)$: for $M, M' \in A^\#$ we write $M \Longrightarrow M'$ iff

$$M' = (M - (M_1 \cup \ldots \cup M_k)) \cup (M'_1 \cup \ldots \cup M'_k,)$$

for some $M_i \to M'_i \in R, 1 \leq i \leq k, k \geq 1$, and there is no rule $M_s \to M'_s \in R$ such that $M_s \subseteq (M - (M_1 \cup \ldots \cup M_k))$; at most one of the multisets $M_i, 1 \leq i \leq k$, may be empty.

With respect to an *ARMS* $\Gamma = (A, R)$ we can define various types of multisets:

- A multiset $M \in A^\#$ is *halted* if there is no $M' \in A^\#$ such that $M \Longrightarrow M'$ (this is equivalent to the fact that there is no rule $M_1 \to M_2 \in R$ such that $M_1 \subseteq M$).
- A multiset $M \in A^\#$ is in *initial state* if there is no $M' \in A^\#$ such that $M' \Longrightarrow M$.

With respect to an ARMS Γ, we can define over $A^\#$ three *transition* relations:

(1) *sequential transition* $(\Longrightarrow_s)$: for $M, M' \in A^\#$ we write $M \Longrightarrow_s M'$ iff

$$M' = (M - M_1) \cup M'_1,$$

for some $M_1 \to M'_1 \in R$.

(2) *free transition* $(\Longrightarrow_f)$: for $M, M' \in A^\#$ we write $M \Longrightarrow_f M'$ iff

$$M' = (M - (M_1 \cup \ldots \cup M_k)) \cup (M'_1 \cup \ldots \cup M'_k,)$$

for some $M_i \to M'_i \in R, 1 \leq i \leq k, k \geq 1$; at most one of the multisets $M_i, 1 \leq i \leq k$, may be empty.

(3) *maximally parallel transition* $(\Longrightarrow_p)$: for $M, M' \in A^\#$ we write $M \Longrightarrow_p M'$ iff

$$M' = (M - (M_1 \cup \ldots \cup M_k)) \cup (M'_1 \cup \ldots \cup M'_k,)$$

for some $M_i \to M'_i \in R, 1 \leq i \leq k, k \geq 1$, and there is no rule $M_s \to M'_s \in R$ such that $M_s \subseteq (M - (M_1 \cup \ldots \cup M_k))$; at most one of the multisets $M_i, 1 \leq i \leq k$, may be empty.

(Of course, in all cases, the union and the difference operations are multiset operations.)

2.1   How *ARMS* Works

Example. Let us consider the *ARMS* defined as follows:

$$\Gamma = (A, R),$$
$$A = \{a, b, c, d, e, f\},$$
$$R = \{a, a, a \to c : r_1, b \to d : r_2, c \to e : r_3,$$
$$d \to f, f : r_4, a \to a, b, b, a : r_5, f \to h : r_6\}.$$

We assume the maximal size of a multiset is 4 and the initial state is given by $\{a, a, b, a\}$. In an *ARMS*, the rewriting rules are applied in parallel. When there are more than two applicable rules, then one rule is selected randomly. Figure 1 illustrates a a sequence rewriting steps, starting from the mentioned initial state.

$$\{a, b, a, a\} \subseteq a,\ a,\ a,\ a,\ b$$
$$\downarrow \quad \text{(the left hand side of } r_1, r_2, r_5)$$
$$\{c, d\} \quad \subseteq c,\ d \text{ (the left hand side of } r_3, r_4)$$
$$\downarrow$$
$$\{e, f, f\} \subseteq f \text{ (the left hand side of } r_6)$$
$$\downarrow$$
$$\{e, h, h\} \quad \text{There are no rule to apply, it reaches}$$
$$\text{the } halt \text{ state}$$

Fig. 1. Example of rewriting steps of *ARMS*

At the first step, the left hand side of rule of r_1, r_2 and r_5 are included in the initial state. In the next step, r_3 and r_4 are applied in parallel and $\{c, d\}$ is rewritten into $\{e, f, f\}$. In step 3, by using r_6, $\{e, f, f\}$ is transformed into $\{e, h, h\}$. There are no rules that can transform the multiset any further so, the multiset is in a *halt* state.

3   Typical Behaviors of ARMS

In this section we discuss the following two types of behaviors of ARMS: (i) ARMS in terms of sequential dynamical system and (ii) the simulation of Brusselator model which is a model of Belousov-Zyabotinsky reaction.

3.1   ARMS in Terms of Sequential Dynamical System

Sequential Dynamical Systems (SDs) are a new class of dynamical systems motivated by the formalization of simulation by composed local maps. Intuitively, SDs are dynamical systems produced by sequentially ordered compositions of

Table 1. Rule Ordering and Emergence of Cycles

Emergence of Cycles	Terminated
9.4%(68/720)	90.6%(652/720)

Table 2. Steps needed for Emergence of Cycles

Rewriting Steps	15	16	17	18	19	20	21	22
Number of Cycles	5	8	8	16	5	9	10	2

local maps. The dynamical properties of SDs delimit the behavioral repertoire of simulation. The concept of SDs can be found in some studies in Artificial Chemistries, [32] [17]; Barrett at el. have mathematically formalized this notion [2], [3]. They employ sequential cellular automata (sCA) over random graphs as a paradigmatic framework for elements of a theory of simulation and show that dynamical systems generated by simulations can be classified by equivalence classes on update sequences. On the other hand, we employ ARMS with sequential transition and show that update sequence affects the state transition.

ARMS with Sequential Transitions

Rule Order and Emergence of Cycles. We examine first the relationship between the ordering of rules and the emergence of cycles.

In the case of the ARMS from the previous example, the total number of possible orderings of rules is equal to 6! = 720. For each order, we assume that a will be introduced in the system after each rewriting step and we will check whether cycles will emerge or not.

Table 1 shows the results of this experiment. Surprisingly, cycles emerge in only 9.4 per cents (68/720) of 720 ordering, and in the remaining 90.6 per cents the rewriting calculus is halting. In 68 orders where cycles emerge, we need from 15 to 22 steps for generating a cycle and the average and the standard deviation is 19 and 3.3 steps (table 2), respectively. The average of the period length of cycles is 5.5 steps, whose standard deviation is 0.49 steps.

As to halting cases, the average of steps needed for termination is 39 steps, and the standard deviation is 6.48 steps. These results are summarized in Table 3.

Typical Examples. In this paragraph, we present two examples[1] which illustrates the importance of rule order. The first example is a typical case which generates a cycle, and it uses the following rule order:

$$O_{R1} = \{r_1 \to r_2 \to r_3 \to r_4 \to r_5 \to r_6\},$$

---

[1] It is notable that these two orderings provide illustrative examples on the formal property of non-termination of ARMS, which is discussed in Section 5.

Table 3. Summary of Statistical Characteristics

Steps Needed for Emerging Cycles	18.0 ± 3.3 Steps
Length of Cycles	5.5 ± 0.49 Steps
Steps Needed for Terminating	39.0 ± 6.48 Steps

whose state transition is shown below. After 18 steps, the system is stable in forming a cycle, with the period length of 5 steps.

$$1. \{\}$$
$$2. \{a\}$$
$$\downarrow \qquad\qquad 16 \text{ steps}$$
$$18. \{a, a, a, a, a, b, c, e\}$$
$$19. \{a, a, a, a, a, c, c, e\}$$
$$20. \{a, a, a, a, a, a, c, d, d, e\}$$
$$21. \{a, a, a, a, a, a, a, c, d\}$$
$$22. \{a, a, a, a, a, a, a, a, c, e\}$$
$$\overline{23. \{a, a, a, a, a, b, c, e\}}$$
$$24. \{a, a, a, a, a, c, c, e\}$$
$$25. \{a, a, a, a, a, a, c, d, d, e\}$$
$$26. \{a, a, a, a, a, a, a, c, d\}$$
$$27. \{a, a, a, a, a, a, a, a, c, e\}$$
$$\cdots$$

The next example is a typical case, which terminates in 42 steps. Although the ARMS applies the same rules, r_1 to r_5 for the state transition, the obtained result is completely different, as shown below. This example has the following rule order:

$$O_{R2} = \{r_3 \to r_1 \to r_2 \to r_4 \to r_5 \to r_6\}$$

Then, the state transition is as follows:

$$1. \{\}$$
$$2. \{a\}$$
$$\downarrow \qquad\qquad 16 \text{ steps}$$
$$18. \{a, a, a, a, a, a, a, a, c, c\}$$
$$19. \{a, a, a, a, a, b, c, c\}$$
$$20. \{a, a, a, a, a, c, c, c\}$$
$$21. \{a, a, a, a, a, a, c, c, c\}$$
$$\overline{22. \{a, a, a, a, a, a, a, a, c, c\}}$$
$$23. \{a, a, a, a, b, c, c, c\}$$
$$24. \{a, a, a, a, c, c, c, c\}$$
$$25. \{a, a, a, a, a, c, c, c, c\}$$
$$26. \{a, a, a, a, a, a, c, c, c, c\}$$
$$\cdots$$
$$\downarrow \qquad\qquad 16 \text{ steps}$$
$$42. \{a, a, c, c, c, c, c, c, c, c\}$$

In the next section, we further examine the effect of inputs and rule order. For this purpose, we adopt the above two kinds of order, O_{R1} and O_{R2}, because of the following two reasons: first, the behavior of O_{R1} is a typical cycle structure in 68 cases. Second, the behavior of O_{R2} is completely different from the above one, although the ARMS applies the same rules in both cases.

The Effect of Inputs and Rule Order. In the above section, we have assume that a is added after each rewriting step, with the ordering of rules fixed, and we have examined the effect of rule ordering. In this section, we only adopt the above two orderings: O_{R1} and O_{R2} and we examine the effect of random inputs. After that, we introduce randomness in rule ordering and examine its effect on the system behavior.

Characteristic Function. In order to incorporate randomness, we introduce characteristic function as follows.

Characteristic Function. Let x be a uniform random number selected in the interval $[0, 1)$. Then a characteristic function $h_\delta(x)$ is defined as:

$$h_\delta(x) = \begin{cases} 1, & x < \delta, \\ 0, & x \geq \delta, \end{cases}$$

where δ denotes a threshold from the following set, 0.1, 0.2, 0.3, 0.4, 0.5, 0.6, 0.7, 0.8, 0.9 .

For example, when δ is equal to 0.9, then the frequency, with $h_\delta(x)$ positive, is equal to 0.9. Thus, when this function is applied to inputs, a will be input for nine of ten rewriting steps, which also means that no input will be done in one of ten steps.

On the other hand, when the characteristic function is applied to rule order, the rule order will not be perturbed in nine of ten rewriting steps, which means that the rule order is perturbed in one of ten steps. For example, let us consider the case of O_{R1}. When the sixth step in ten rewriting steps is perturbed, the rules are applied as shown in the following sequence: r_1, r_2, r_3, r_4, r_5, $r_2(r_6)$, r_1, r_2, r_3, r_4, where r_6 is changed to r_2. Thus, both input randomness and rule randomness are equal to $1 - h_\delta(x)$.

Experimental Results. Using the above two kinds of ordering, we perform the following three kinds of experiments:

(1) Decrease the frequency of inputs, with the rules order fixed.
(2) Introduce randomness in the rules order, with inputs in every step.
(3) Decrease the frequency of inputs, and introduce randomness in the rules order.

In each case, we have performed 100 trials, each of which consists of 5005 rewriting steps, and we have measured the effect on emergence of cycles, extinction of cycles, and steps needed for termination.

Table 4. Relation between Frequency of Inputs and Termination of Calculus

Frequency of Input	0.9	0.8	0.7	0.6	0.5	0.4	0.3	0.2	0.1	
O_{R1}	0	0	0	0	0	0	0	0	0	(%)
O_{R2}	100	100	100	100	98	13	0	0	0	(%)

Rule Order: Fixed

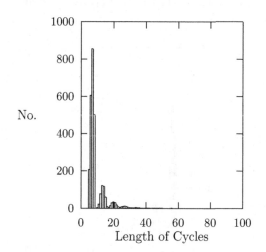

Fig. 2. An Illustrative Example in O_{R1}

Decrease Frequency of Input, with Rule Order Fixed. In the above three kinds of experiments, the most interesting one is the case when the frequency of inputs is decreased, with the rules order fixed. In the cases of the rules order O_{R1}, the rewriting steps are not terminated in all the trials, and cyclic structures also emerge. Interestingly, the period of cycle varies from 7 to 30 steps, and a period-doubling of cycles is observed. Figure 2 shows an illustrative example in the experiments, where the frequency of inputs is 0.2. The vertical axis indicates the number of cycles which emerge in 100 trials, and the horizontal axis shows the period length of cycles. In this figure, the peaks of the period length of cycles are located around 7, 13, 20, and 28 steps, which shows a period-doubling and for each period-doubling, a fusion of cycles is observed. On the other hand, in the case of the rules order O_{R2}, unstable cycles emerge and rewriting steps are terminated when the threshold is larger than 0.5. However, when the threshold is equal to 0.4, almost all rewriting trials are not terminated, and cycles become stable. Moreover, in this case, the variability of the period length of cycles is the largest among all trials. Figure 3 shows the distribution of the period length of cycles; the notation is the same as in Figure 3. The peaks of the period length of cycles is located around 11, 23, 34, 46, 57, and 67 steps, which indicates a period-doubling. If the threshold is less than 0.3, all the trials are stable, similar to the case of the rules order O_{R1}. These results are summarized in Table 4.

Table 5. Relation between Randomness of Rule Order $(1-\delta)$ and Number of Emerging Cycles

$1-\delta$	0.1	0.3	0.5	0.7	0.9
O_{R1}	1300(5)	325(5)	106(5)	106(5)	23(5)
O_{R2}	1269(11)	1118(5)	335(3)	223(3)	300(3)

A(B) : A and B denote the number of emerging cycles, and the median of the length of cycles, respectively.

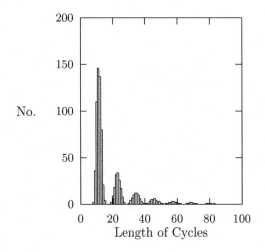

Fig. 3. An Illustrative Example in O_{R2}: Frequency of Input: 0.4, with Rule Order Fixed.

Randomness in Rule Order, with Inputs in Every Step. In these experiments, for the rules order O_{R1}, cycles become unstable and a period-doubling is observed when the randomness of the rules order is low. Figure 4 shows the distribution of the period length of cycles when randomness of rule order is equal to 0.2; the notation is the same as in Figure 2.

However, as randomness of the rules order grows, the number of emerging cycles decreases exponentially. Table 1 shows this tendency, the relation between randomness of rule order and the number of cycles emerging in 100 trials. When the randomness is larger than 0.5, almost the all trials generate only one cycle.

On the other hand, the behavior of cycles in O_{R2} is different. As shown in table 1, the number of emerging cycles decreases exponentially, as the randomness of the rules order grows. However, the behavior of cycles when randomness of rule is higher than 0.5 is different from that when the randomness is lower than 0.3: the median of the period length of cycles in the latter cases is half smaller than that in O_{R1}, while that in the former case is equal to or twice than that in O_{R1}. Interestingly, trials in the case of O_{R2} have two phase transitions: one is between 0.1 and 0.2, and the other is between 0.3 and 0.5.

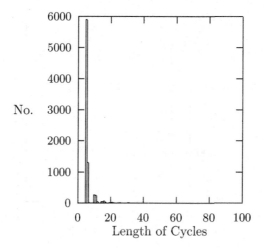

Fig. 4. An Illustrative Example in O_{R1}: Input: Fixed, Randomness of Rule Order: 0.2.

Figure 5 shows the distribution of the period length of cycles in O_{R2} when randomness of the rules order is equal to 0.1. It is notable that the cycles whose period length is 11.0 are dominant over those whose period length is 5.0.

Decrease Frequency of Input and Increase Randomness in the Rules Order. In the final cases when the frequency of inputs is decreased and randomness is introduced in the rules order, the obtained results are straightforward. In the cases when randomness is introduced in the rules order with an input in every step, cycles become unstable in O_{R1}, and only unstable cycles emerge in O_{R2}. However, in both cases, the number of emerging cycles decreases exponentially, as randomness of the rules order grows. Interestingly, the behavior of cycles in O_{R1} is very similar to that in O_{R2}, except when the frequency of inputs is very high and randomness of the rules order are very low.

Fusion of Cycles. A special case in the rules order O_{R2}, with the frequency of inputs 0.4, generates a complex behavior of cycles. A closer examination shows that a core cycle is generated first and that other cycles around this cycle fuse into a core cycle.
The generated core cycle c_8 is given as follows:

$$c_8 : 8 \rightarrow 9 \rightarrow 10 \rightarrow \cdots \rightarrow 18 \rightarrow 8 : (11 steps),$$

where each number denotes the number of rewriting steps. Around the c_c, the following cycle c_{54} is also observed:

$$c_{54} : 54 \rightarrow 55 \rightarrow 56 \rightarrow \cdots \rightarrow 66 \rightarrow 54 : (13 steps),$$

It is impossible to transit from State 8 to State 54 and State 66 to State 11, with the input fixed, but it will become possible when the randomness of the input is

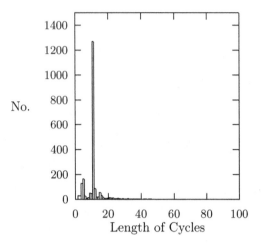

Fig. 5. An Illustrative Example in O_{R2}: Input: Fixed, Randomness of Rule Order: 0.1.

introduced. Then, c_8 fuses into c_{11}, and a new cycle will be generated as follows:

$$c_c : 8 \to 54 \to \cdots \to 66 \to 11 \to \cdots \to 8 : (21 steps)$$

Thus, the fusion of cycles depends on the randomness of the input: the randomness of the input influences the application of rules, and then the change of application of rules causes the change of the structure of a multiset, then it opens the hidden pathway to fusion of cycles.

Although the mechanism of fusion of cycles is not fully studied, it is notable that even such a simple structure produces a complex behavior of cycles. A further research on fusion, such as the classification of cycles, will be the object of a future work.

3.2   Brruselator Model on ARMS

We have performed an experiment about the Brusselator [43] model by using an ARMS with a free transition. The Brusselator is a model of a chemical oscillation related to the Belousov-Zabotinsky reaction (figure 6).

We interpret these abstract chemical reaction equations as the rewriting rules shown in figure 7.

$$
\begin{aligned}
A &\xrightarrow{k_1} X \\
B + X &\xrightarrow{k_2} Y + D \\
2X + Y &\xrightarrow{k_3} 3X \\
X &\xrightarrow{k_4} E
\end{aligned}
$$

Fig. 6. Abstract chemical model of Brusselator.

$$A \rightarrow X \qquad : r_1$$
$$B\,X \rightarrow Y\,D \qquad : r_2$$
$$X\,X\,Y \rightarrow X\,X\,X : r_3$$
$$X \rightarrow E \qquad : r_4$$

Fig. 7. Rewriting rules for Brusselator.

The ARMS for simulating Brusselator is defined as follows;

$$\Gamma = (A, R),$$
$$A = \{A, B, C, D, X, Y\},$$
$$R = \text{Rewriting rules for Brusselator.}$$

In this simulation, the reaction rate corresponds to the frequency of rule application. If r_1 has the highest reaction rate, then r_1 is applied at the highest frequency.

Simulation of the Brusselator Model. Let us examine the relationship between the frequency of rule application (reaction rate) and the concentration of X and Y in the multiset. The concentration of X and Y in the multiset is indicated by the number of X and Y present in the multiset.

As to the initial condition, we assume that the maximal multiset size is equal to 5000 and the initial state of the multiset is the empty multiset. We assume that the system gets inputs A and B continuously. Thus, this model can be regarded as a continuously-fed stirred tank reactor (CSTR).

In this simulation, we confirmed that oscillations between the number of X and Y in the multiset emerged. Furthermore, we discovered three types of oscillations, as follows:

(1) divergence and convergence (figure 11, figure 8),
(2) quasi-stable oscillations (figure 9) and
(3) unstable oscillations (figure 10).

The oscillation in the number of X and Y emerges as follows: first, the number of X increases; when the number of X decreases, the number of Y increases. We say that this type of oscillation is a *X-Y* **oscillation**, while oscillations in the number of X or Y are called *X*-**oscillations** and *Y*-**oscillations**, respectively.

Divergence. When the frequency of the application of rule r_2 is much larger than that of r_1, for example 0.1 for r_1 and 0.45 for r_2, the system becomes unstable and in many cases it diverges.

In this case, a system generates *X-Y* oscillations (figure 11). The system diverges easily by perturbation and it is difficult to predict whether we get oscillations or divergence.

Convergence. When the frequency of the application of r_1 is much larger than that of r_2, for instance, 0.1 versus 0.04, the system converges (figure 8). In this case almost all the oscillations are X or Y-oscillations, and only a few *X-Y* oscillations can be found.

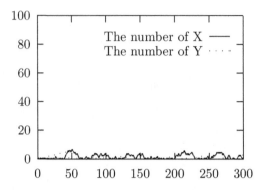

Fig. 8. Example of Convergence

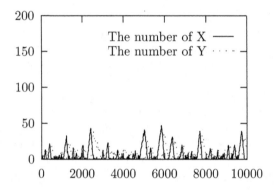

Fig. 9. Example of (quasi) Stable oscillation

(Quasi) Stable Oscillations. When the frequency of the applications of r_1 is slightly larger than that of r_2, for example, 0.1 versus 0.15, the system is able to generate X-Y oscillations. Even if we introduce perturbation to the system, it maintains this type of oscillation (figure 9).

In this case, X or Y-oscillations decrease and X-Y oscillations increase, and vice versa. X-Y oscillations can be classified into the following three categories;

(1) The highest peak of the number of Xs or Ys is between 35 and 50.
(2) The highest peak of the number of Xs or Ys is between 15 and 35.
(3) The highest peak of the number of Xs or Ys is between 1 and 15.

Each type (1), (2), (3) of oscillation appears irregularly. In other words, in this case the system generates a limit cycle, or an "attractor".

Unstable Oscillation. When the frequency of r_2 is slightly larger than that of r_1, for instance, 0.1 versus 0.35, the system exhibits an unstable oscillation. In this case, there are a few X or Y-oscillations and almost all the oscillations become X-Y oscillations. A great variety of X-Y oscillation can be found (figure 10).

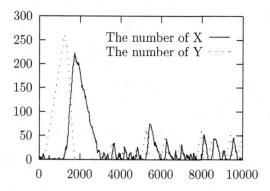

Fig. 10. Example of Unstable oscillation

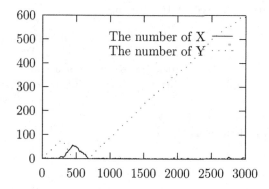

Fig. 11. Example of Divergence

4 Theoretical Analysis of ARMS

In this section we address some theoretical aspects of ARMS. First, the order
parameter that describes the degree of emergence of cycles is introduced. This
parameter is related to the Langton's λ parameter and Edge of Chaos. After
that, we consider rewriting systems from the viewpoint of stochastic transitions
of states.

4.1   Order Parameter of an ARMS

From an examination of the effectiveness of the termination property, we obtained the λ_e parameter as an order parameter for the qualitative behavior of
ARMS.

We define "order" here as being given by the *diversity* of cycles. Thus, in
this paper, "ordered state" refers to a case where the system yields simple cycles
(such as the limit cycle), while "disordered state" refers to the case where the
system yields chaotic or complex cycles.

The λ_e Parameter. Let us define the λ_e parameter as follows:

$$\lambda_e = \frac{\Sigma r_{\Delta S > 0}}{1 + (\Sigma r_{\Delta S < 0} - 1)} \tag{1}$$

where $\Sigma r_{\Delta S > 0}$ corresponds to the number of *heating rules* used, and $\Sigma r_{\Delta S < 0}$ to the number of *cooling rules* used. This parameter is defined when the number of rules used is greater than 1.

When the ARMS only uses rules of the type $r_{\Delta S < 0}$, λ_e is equal to 0.0. On the contrary, if the ARMS uses rules of the type $r_{\Delta S > 0}$ and $r_{\Delta S < 0}$ with the same frequency, λ_e is equal to 1.0. Finally, when the ARMS only uses rules of the type $r_{\Delta S > 0}$, λ_e is greater than 1.0.

Simulation. We confirmed the appropriateness of the λ_e parameter through a simulation of the ARMS, and verified that the parameter reflects the diversity of cycles that are generated by the system.

Setup. A simulation was carried out under the following environment: In this simulation, an ARMS with a free transition is used. We intend to focus on qualitative features of the rewriting rule, while not assuming any inputs. A rule was selected randomly according to the following protocol: the probability of selecting a $r_{\Delta S > 0}$ rule is given by the probability $0 \leq p \leq 1$, while a $r_{\Delta S < 0}$ rule is selected with probability $(1 - p)$.

The ARMS is defined as follows;

$$\Gamma = (A, R),$$
$$A = \{a, b\},$$

Rules. The length of the left- or right-hand-side of a rule was between one and five. Both sides of the rules were obtained by sampling with replacement of two symbols a and b. A set of rewriting rules was constructed as the overall permutation of both sides of the rules. The number of rules is given by Equation (2), where n corresponds to the kinds of symbols that we take and m is the range of lengths of strings.

$$\left\{ \sum_{k=1}^{m} \binom{n + k - 1}{k} \right\}^2 \tag{2}$$

As Equation (2) illustrates, if many kinds of symbols are used, the number of rules increases rapidly. That is why we use two symbols in this simulation. We assume that the length of strings can range between 1 and 5. Then, the number of rules for this simulation is equal to 30976.

Method of Simulation. We assume that the maximal multiset size is 10. At the beginning of a simulation, the value of p is set to 0 and it is increased by steps of 0.01. At each value of p, 100 new initial states with base number between 1 and 10 are generated by selecting the symbol a or b randomly. The base number of the initial state of a multiset is decided randomly. For each initial state the simulation is performed for 1000 steps.

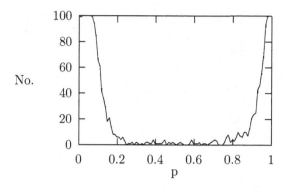

Fig. 12. The correlation between the number of terminating calculations and p

Experimental Results. Let us present the experimental results, focusing on the following two points:

(1) the correlation between the system's terminating property and the value of p,
(2) the correlation between the diversity of periods of the generated cycles and the value of p.

Termination. Figure 12 illustrates the correlation between p and the number of terminating calculations. In Figure 12, the vertical axis corresponds to the number of terminating calculations and the horizontal axis corresponds to p.

In this simulation, before p exceeded 0.1, most calculations terminated. When p exceeded 0.1, the number of terminating calculations decreased rapidly, while when p was greater than 0.2, this decrease leveled off. Then, for p between 0.3 and 0.85, only a few calculations terminated, while with p greater than 0.85, the number of terminating calculations increased rapidly again. In other words, for p near 0 or 1, the system strongly terminated, while with when p away from 0 or 1, the termination property became weak.

The Number of Generated Cycles. Figure 13 illustrates the relationship between the number of generated cycles and p.

As p increases, the number of generated cycles also increases rapidly. This rise levels off when p exceeds 0.3. For p between 0.3 and 0.8, the number of generated cycles remains at the same level, however, when p exceeds 0.75, this number rapidly decreased again.

This result indicates that the termination property indeed is related to the number of generated cycles. As we mentioned in the previous section, when p was close to 1 or 0, the system strongly terminated while for p away from 0 or 1, this property became weak. Also, for p near 1 or 0, only a few cycles were generated while for p away from 1 and 0, the system generated many cycles. We may thus conclude that the degree of termination influences the number of cycles generated by the system.

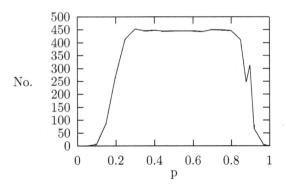

Fig. 13. Correlation between the number of generated cycles (vertical axis) and p (horizontal axis).

It is interesting that once the number of generated cycles reaches around 450, it remains at the same level, even while p was changed from 0.3 continuously up to 0.8. We believe that for p between 0.3 and 0.8, the system is in an equilibrium state.

To investigate the system's behavior in this equilibrium state, let us examine the relationship between the kinds of periods generated by the cycles and the value of p.

The Kinds of Periods Generated by the Cycles. The experimental result indicates that even if the system is in an equilibrium state, the kinds of periods are different for each value of p. Figure 14 displays the average number of different kinds of periods. As we can see in this figure, when p reaches about 0.5, the number of different kinds of periods is maximal. In other words, when *cooling rules* and *heating rules* are used at the same frequency, many kinds of periods are generated.

Discussion. We shall show that the λ_e parameter is related to the termination property, which implies that Langton's λ parameter is also related to termination.

Before discussing this issue, let us describe cellular automata, Langton's λ parameter [36], and Wolfram classes [66] in more detail. Then, we display the relation among these parameters and Wolfram classes.

Cellular Automata. A Cellular Automaton (CA) is a discrete mathematical model [65] whose behavior is caused by the interactions among neighboring sites. This model has been applied to various fields, for instance, statistical mechanics, mathematical biology, public hygiene (as a model of infectious disease, for example), in medicine and so on.

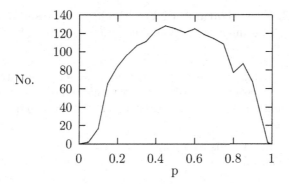

Fig. 14. Correlation between the kinds of periods of generated cycles (vertical axis) and p (horizontal axis).

Formally, a cellular automaton is a D-dimensional lattice automaton. Each lattice site is updated by a deterministic rule involving a local neighborhood of sites, in discrete time steps.

The possible state of a site is chosen from the alphabet $K = \{0, 1, \ldots, k-1\}$, and the value of site i at step t is denoted by $a_i^{(t)}$. At each time step, each site value is updated according to the particular neighborhood of N sites around it by a local rule

$$\phi : K^N \to K, \tag{3}$$

of the form

$$pa_i^{(t)} = \phi\{a_{1-r}^{(t-1)}, a_{1-r+1}^{(t-1)}, \ldots, a_{1+r}^{(t-1)}\}. \tag{4}$$

This local rule leads to a global mapping

$$\Phi : K^N \to K^N \tag{5}$$

on all cellular automaton configurations. Then, in general

$$\Omega^{(t+1)} = \Phi\Omega^{(t)} \subseteq \Omega^{(t)}, \tag{6}$$

where

$$\Omega^{(t+1)} = \Phi^t N^K \tag{7}$$

is the set of configurations generated after t iterated applications of Φ (Wolfram [65], Langton [36]).

For example, a local rule for the one dimensional cellular automaton with neighborhood two is:

$$000\ 001\ 010\ 011\ 100\ 101\ 110\ 111$$
$$\ \ 0\ \ \ \ 1\ \ \ \ 0\ \ \ \ 0\ \ \ \ 1\ \ \ \ 0\ \ \ \ 0\ \ \ \ 0.$$

By using this rule, a step of global mapping is;

$$000101011011100$$
$$0100000000001.$$

Qualitative Characterizations of CA. Wolfram's classification and Langton's λ parameter as qualitative characterizations of the behavior of cellular automata are well known.

Wolfram Classes. Wolfram ([66]) proposed four classes of qualitative behavior patterns of cellular automata based on his investigation of a large sample of CA rule tables. He maintained that any cellular automaton would fall into one of the four basic classes:

- **Class 1**: Evolution leads to a homogeneous state.
- **Class 2**: Evolution leads to a set of separated simple stable or periodic structures.
- **Class 3**: Evolution leads to a chaotic pattern.
- **Class 4**: Evolution leads to chaotic, localized structures, sometimes long-lived.

These four classes lead to an analogy between the CA's behavior and the classification of dynamical systems:

- **Class 1**: Limit points.
- **Class 2**: Limit cycles.
- **Class 3**: Chaotic behavior of the kind associated with strange attractors.
- **Class 4**: No direct analogue.

With respect to Class 4, Wolfram suspected that CA in this class are capable of universal computation, so that properties of its infinite-time behavior would be undecidable [66].

Langton's λ Parameter. Langton criticized Wolfram's classification by saying: "it is obvious that such a classification can only serve as rough approximations to the more subtle, underlying structure" (Langton [36], p.46), and proposed his λ parameter to describe CA rules, to obtain "a deeper understanding of the structure of cellular automata rule spaces, one that provides an explanation for their relationships to one another"(Langton [36], p. 46). The λ parameter is defined as follows:

$$\lambda = \frac{K^N - n_q}{K^N},\qquad\qquad (8)$$

where K corresponds to the number of symbols used, N to the size of the neighborhood, and n_q to the number of local rules which transform cellular automata to the *quiescent* state, respectively. The quiescent state is picked arbitrarily, and is usually associated with a "special" state, such as the "zero" state. If all rules transform to the quiescent state, then $K^N = n_q$, and the the λ parameter is equal to 0. When no rules transform to the quiescent state on the other hand, λ is equal to 1.

The values $\lambda = 0$ and $\lambda = 1 - \frac{1}{K}$ represent the most homogeneous and the most heterogeneous rule tables. As the value of the λ parameter increases, the dynamical activity of cellular automata becomes chaotic. The correspondence between the value of the λ parameter and the dynamical activity of cellular automata is as follows:

$$\text{fixed point} \rightarrow \text{periodic} \rightarrow \text{``complex''} \rightarrow \text{chaotic.}$$

In terms of the Wolfram classes, the sequence is:

$$\text{Class 1} \Rightarrow \text{Class 2} \Rightarrow \text{Class 4} \Rightarrow \text{Class 3.} \tag{9}$$

Langton ([36]) demonstrated that the complex rules are located in between the periodic and the chaotic rules, and that there is a clear phase-transition between periodic and chaotic behavior.

Computational Algebraic Characteristics of λ Parameter. We demonstrate here that the λ parameter indicates a degree of *termination* and that when cellular automata yield complex or chaotic behavior, this termination property of the system becomes weak. When λ is close to 0, the dynamical activity of the cellular automaton quickly dies out or reaches a uniform fixed point. In this regime, thus, the termination property of cellular automata is strong. As the value of the λ is increasing, cellular automata evolve to periodic structures or chaotic aperiodic patterns. As the value of the λ increases even more, the termination property becomes weak. The larger the parameter becomes, the later the calculation terminates. Since the transitional space of cellular automata is finite, if a calculation does not terminate or is difficult to stop, cyclic structures must emerge in the process of the computation.

When λ is equal to 0.125, for example, we could find the following rule:

$$000\ 001\ 010\ 011\ 100\ 101\ 110\ 111$$
$$0\quad 0\quad 1\quad 0\quad 0\quad 0\quad 0\quad 0.$$

As this parameter increases, the rule might change to that given below:

$$000\ 001\ 010\ 011\ 100\ 101\ 110\ 111$$
$$1\quad 0\quad 1\quad 0\quad 0\quad 0\quad 0\quad 1.$$

The λ parameter of this rule is equal to 0.375. As the value of λ parameter increases, the number of rules such as,

$$* * * (* = 0 \text{ or } 1)$$
$$0 ,$$

decreases while, on the other hand, the number of rules such as,

$$* * * (* = 0 \text{ or } 1)$$
$$1 ,$$

increases. The CA rule yields a complex or chaotic pattern when λ is near $1 - \frac{1}{K}$, and rules that lead to the quiescent state are used at the same frequency as rules that lead *away* from that state.

Relation between λ_e, λ, Wolfram Classes, and the Termination Property. The qualitative dynamics of CA and ARMS that we have described is summarized in Table 6. It suggests that each parameter and class indicate a degree of terminating.

Table 6. Relation among parameters, complexity classes

terminating	λ_e	λ	Wolfram classes
Strong	$\lambda_e \cong 0.0,$ $\lambda_e \gg 1.0$	$\lambda \cong 0.0$	I or II
Fair strong	$1.0 > \lambda_e > 0.0,$ $\lambda_e > 1.0$	$0.0 < \lambda < 1 - \frac{1}{K}$	II or IV
Weak	$\lambda_e \cong 1.0$	$\lambda \cong 1 - \frac{1}{K}$	III

Comments on the 'Edge of Chaos'-Regime. We have demonstrated that an equal frequency of $r_{\Delta S > 0}$ rules to $r_{\Delta S < 0}$ rules yields dynamical patterns for both cellular automata and ARMS. We can now see that the principle of "edge of chaos" (Langton [36], Kauffman [32]), according to the dynamics just described, results from a biased ratio of $r_{\Delta S > 0}$ rules to $r_{\Delta S < 0}$ rules.

If these two types of rules are used at the same frequency, then the system yields a chaotic pattern. However, when the ratio changes slightly, the "edge of chaos" regime emerges. Thus, a lack of symmetry in the application of rules generates the diversity of cycles in these systems.

4.2   Markov Analysis and Beyond

In this section we consider rewriting systems from the viewpoint of stochastic transitions of states. Specifically, it is shown that the dynamics of a rewriting system, in some situations, can be analyzed as a Markov process. In the limit of large system size, one obtains a set of deterministic differential equations, which often provides a useful tool, complementary to the computational analysis. We also describe what are not assumed and are difficult to examine in the Markov analysis, and shall shed light on some unique aspects of ARMS computation.

As we also said above, a multiset over a set of n objects can be identified with a vector in the n-dimensional space. Consider the alphabet $A = (a_1, \cdots, a_n)$; a state can be expressed as

$$a_1{}^{x_1} a_2{}^{x_2} \cdots a_n{}^{x_n} \longleftrightarrow \boldsymbol{x} = (x_1, x_2, \cdots, x_n), \tag{10}$$

where all components x_i are integers.

Each rewriting step in an ARMS is a state transition, which can be described as a displacement of the state vector:

$$\boldsymbol{x} \longrightarrow \boldsymbol{x} + \boldsymbol{r}. \tag{11}$$

The displacement vector $\boldsymbol{r} = (r_1, r_2, \cdots, r_n)$ simply represents how the number of each object a_i changes in a single rewriting step.

Consider the Brusselator system from Figure 6, which we shall use as a working example throughout this section. Since we are interested in the behavior of the numbers of X and Y, a state is a two-dimensional vector $\boldsymbol{x} = (x, y)$ where x and y are the numbers of X and Y respectively. The system has four rules, namely $A \to X$, $B + X \to Y + D$, $2X + Y \to 3X$, and $X \to E$, which correspond to the four transitions:

$$\boldsymbol{r}_1 = (1, 0), \ \boldsymbol{r}_2 = (-1, 1), \ \boldsymbol{r}_3 = (1, -1), \ \boldsymbol{r}_4 = (-1, 0), \tag{12}$$

respectively.

Now it would be natural to assume that in a single step of rewriting rules are applied *randomly* with probabilities depending on the present state and each rule. For the above example, a possible model of the transition probabilities might be as follows:

$$\begin{aligned}
\mathrm{Prob}(\boldsymbol{x} \to \boldsymbol{x} + \boldsymbol{r_1}) &= k_1 \, \Delta t, \\
\mathrm{Prob}(\boldsymbol{x} \to \boldsymbol{x} + \boldsymbol{r_2}) &= k_2 \, x \, \Delta t, \\
\mathrm{Prob}(\boldsymbol{x} \to \boldsymbol{x} + \boldsymbol{r_3}) &= k_3 \, x^2 y \, \Delta t, \\
\mathrm{Prob}(\boldsymbol{x} \to \boldsymbol{x} + \boldsymbol{r_4}) &= k_4 \, x \, \Delta t, \\
\mathrm{Prob}(\boldsymbol{x} \to \boldsymbol{x}) &= 1 - (k_1 + k_2 x + k_3 x^2 y + k_4 x) \Delta t,
\end{aligned} \tag{13}$$

where Δt is the small interval of time in which a single step of computation takes place. For example, the probability of the third rule application is proportional to the combinatorial number for encounters of the three *chemicals*, X, X and Y.

We should emphasize the fact that it is our *assumption* that one can treat a rewriting system in such a stochastic manner. Especially, note that the order of applications of a sequence of rewriting rules is supposed to be not important. Under this assumption, the transition probabilities depend not on the remote past of the history of states but on the state of one-step before. That is, the stochastic process is *Markovian,* and for studying such processes we have abundant tools. Let us briefly recapitulate a few results of the Markov analysis in what follows (see [31], [27], for example). We shall see how a set of differential equations such as Equation (26) shown below, which is often written down rather intuitively and immediately from Figure 6 and Equation (13), is derived.

Suppose that at time t_0 the state is $\boldsymbol{x}_0$. One can consider the conditional probability distribution for later time $t > t_0$, $P(\boldsymbol{x}, t | \boldsymbol{x}_0, t_0)$. Given the transition probabilities, one can write the so-called *master equation* for the conditional probability distribution by using the Markov property mentioned above. Using the abbreviation, $P(\boldsymbol{x}, t) = P(\boldsymbol{x}, t | \boldsymbol{x}_0, t_0)$, this is

$$\frac{\partial P(\boldsymbol{x}, t)}{\partial t} = {\sum_{\boldsymbol{z}}}' W(\boldsymbol{x} | \boldsymbol{z}) P(\boldsymbol{z}, t) - {\sum_{\boldsymbol{z}}}' W(\boldsymbol{z} | \boldsymbol{x}) P(\boldsymbol{x}, t), \tag{14}$$

where $W(\boldsymbol{x}' | \boldsymbol{x})$ is the transition probability from $\boldsymbol{x}$ to $\boldsymbol{x}'$ per unit time. ${\sum_{\boldsymbol{z}}}'$ in Equation (14) represents the summation over all possible transitions except $\boldsymbol{z} = \boldsymbol{x}$. The master equation should be accompanied with appropriate boundary conditions such as vanishing probability for a negative number of some x_i.

Usually an ARMS is working for a relatively large number of objects, say dozens and hundreds. For such a large system, one has an approximate treatment of the master equation, namely the *system size expansion* [31].

First, let us have a formal expansion of Equation (14). We use the notation

$$t(\boldsymbol{\Delta x}, \boldsymbol{x}) \equiv W(\boldsymbol{x} + \boldsymbol{\Delta x} | \boldsymbol{x}), \tag{15}$$

then rewrite Equation (14) as

$$\frac{\partial P(\boldsymbol{x}, t)}{\partial t} = \sum_{\boldsymbol{r}} [t(\boldsymbol{r}, \boldsymbol{x} - \boldsymbol{r}) P(\boldsymbol{x} - \boldsymbol{r}, t) - t(\boldsymbol{r}, \boldsymbol{x}) P(\boldsymbol{x}, t)]$$

$$= \sum_{\boldsymbol{r}} \sum_{n=1}^{\infty} \frac{1}{n!} (-\boldsymbol{r} \cdot \nabla_{\dot{\boldsymbol{x}}})^n [t(\boldsymbol{r}, \boldsymbol{x}) P(\boldsymbol{x}, t)], \tag{16}$$

where Taylor expansion was done in the second equation, and note that $\sum_{\boldsymbol{r}}$ means the summation over all possible rules. We also define

$$\boldsymbol{r} \cdot \nabla_{\boldsymbol{x}} \equiv \sum_{i=1}^{n} r_i \frac{\partial}{\partial x_i} \equiv r_i \frac{\partial}{\partial x_i}. \tag{17}$$

Henceforth, as shown in the second definition of Equation (17), summation over repeated indexes is assumed unless otherwise stated. Up to the second order, Equation (16) reads

$$\frac{\partial P(\boldsymbol{x}, t)}{\partial t} = -\frac{\partial}{\partial x_i} [A_i(\boldsymbol{x}) P(\boldsymbol{x}, t)] + \frac{1}{2} \frac{\partial^2}{\partial x_i x_j} [B_{ij}(\boldsymbol{x}) P(\boldsymbol{x}, t)], \tag{18}$$

where

$$A_i(\boldsymbol{x}) \equiv \sum_{\boldsymbol{r}} r_i \, t(\boldsymbol{r}, \boldsymbol{x}), \tag{19}$$

$$B_{ij}(\boldsymbol{x}) \equiv \sum_{\boldsymbol{r}} r_i r_j \, t(\boldsymbol{r}, \boldsymbol{x}). \tag{20}$$

We introduce the system size parameter Ω, which represents the typical number of objects, or chemicals involved in the system. Note that the typical size of the jump Δx_i is some small integers, say ± 1, compared with Ω. In the transition probability expressed in Equation (15), it would be natural to assume that the functional form is such that

$$t(\boldsymbol{\Delta x}, \boldsymbol{x}) = \Omega \, \psi(\boldsymbol{\Delta x}, \boldsymbol{x}/\Omega). \tag{21}$$

That is, the transition probability is proportional to Ω and is a function of "concentration" $\boldsymbol{x}/\Omega$ and displacement $\boldsymbol{\Delta x}$. For example, in the Brusselator system described by Equation (13), Equation (21) requires an appropriate dependence of each k_i on Ω.

It turns out to be useful to express the state variable $\boldsymbol{x}$ as

$$\boldsymbol{x} = \Omega\,\boldsymbol{\phi} + \sqrt{\Omega}\,\boldsymbol{z}, \tag{22}$$

in terms of "deterministic" variable $\boldsymbol{\phi}$ and "fluctuation" variable $\boldsymbol{z}$. In fact, as easily verified, under the change of variable by Equation (22), the lowest-order ($\sqrt{\Omega}$) of Equation (16) or Equation (18) is

$$\frac{d\phi_i(t)}{dt} = \tilde{A}_i(\boldsymbol{\phi}(t)), \tag{23}$$

and the second-order (Ω^0) reads

$$\frac{\partial \tilde{P}(\boldsymbol{z},t)}{\partial t} = -\partial_j \tilde{A}_i(\boldsymbol{\phi}(t))\frac{\partial}{\partial z_i}[z_j \tilde{P}(\boldsymbol{z},t)] + \frac{1}{2}\tilde{B}_{ij}(\boldsymbol{\phi}(t))\frac{\partial^2 \tilde{P}(\boldsymbol{z},t)}{\partial z_i z_j}, \tag{24}$$

where $P(\boldsymbol{x},t) = P(\boldsymbol{x}(\boldsymbol{z},t),t) \equiv \tilde{P}(\boldsymbol{z},t)$. Also defined are $A_i(\boldsymbol{x}) = \Omega\tilde{A}_i(\boldsymbol{x}/\Omega)$ and $B_{ij}(\boldsymbol{x}) = \Omega\tilde{B}_{ij}(\boldsymbol{x}/\Omega)$ in accordance with Equation (21). $\partial_j\tilde{A}_i(\cdot)$ is the partial derivative with respect to the j-th component of the argument of $A_i(\cdot)$.

Equation (23) gives the differential equation for the deterministic part, and Equation (24) describes the stochastic fluctuation around it. The latter would simply yield a Gaussian distribution around the deterministic trajectory, whose variance changes with time in most cases. A more interesting case is when the fluctuation grows rapidly in time, or the so-called anomalous fluctuation, which needs alternative treatment rather than that derived here.

For the Brusselator system, the deterministic part is given by the differential equation for the state $\boldsymbol{x} = (x,y)$:

$$\frac{dx(t)}{dt} = k_1 - k_2\,x + k_3\,x^2 y - k_4\,x \tag{25}$$

$$\frac{dy(t)}{dt} = k_2\,x - k_3\,x^2 y \tag{26}$$

as easily found from Equation (12) and Equation (23). This result is not surprising. One could write such a differential equation immediately from Equation (13), but rather intuitively. We have seen that it is actually the deterministic part of the stochastic process in the limit of large system-size.

Depending on the paramters k_i, Equation (26) has interesting behaviors [43]. Let us focus on the limit cycle corresponding to chemical oscillation for illustration. Figure 15 shows a set of solutions for the parameters, $k_1 = 100$, $k_2 = 3$, $k_3 = 10^{-4}$, $k_4 = 1$. On the other hand, in the ARMS formulation, one has the result shown in Figure 16 for the same set of parameters with the same initial condition (cf. [57]). The simulation was done according to the set of rules in Equation (13) ($\Delta t = 10^{-5}$ for system sizes of hundreds). Comparison in the phase space of $(x(t), y(t))$ explicitly shows a similar behavior of limit cycle as depicted in Figure 17.

Although the result should be accepted as natural, we think that the Markov approach can provide a useful tool and a non-trivial bridge between the computational approach and the viewpoint of dynamical systems. First of all, the

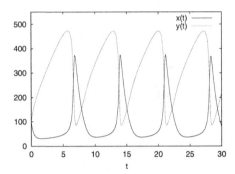

Fig. 15. Limit cycle behavior of Brusselator differential equation. Parameters: $k_1 = 100$, $k_2 = 3$, $k_3 = 10^{-4}$, $k_4 = 1$.

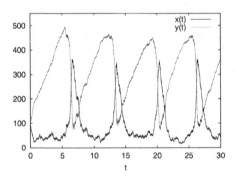

Fig. 16. Limit cycle behavior of ARMS for the Brusselator system. Parameters: $k_1 = 100$, $k_2 = 3$, $k_3 = 10^{-4}$, $k_4 = 1$.

comparison of ARMS with dynamical differential equations, such as illustrated as above, is practically useful. Several points can be immediately considered including

- phase flow,
- qualitative change of phase flow depending on parameters, or bifurcation,
- stability analysis,
- asymptotic behavior,
- analysis of small fluctuation around deterministic trajectories,

and so on. These points might not be easy to study in the formulation of ARMS, and could also be examined from a computational viewpoint. Another interesting problem would be to consider the "computational ability" of ARMS from the dynamical viewpoint.

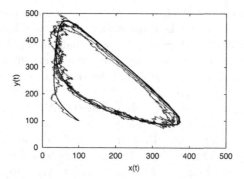

Fig. 17. Trajectories in the phase space $(x(t), y(t))$. The bold line and thin line correspond to the differential equation and to the ARMS simulation respectively. The initial values and the parameters are the same in both cases.

We would, however, like to emphasize that such a Markov analysis as presented here is only a complementary approach to understand certain aspects of ARMS. In particular, we should point out a few things.

- The order of successive application of rewriting rules can be important, while it is assumed to be completely random in this section.
- If the size of the system is not very large, say a dozen, one needs to analyze the Markov process without a useful approximation, though possible. For a not-so-large system, novel behaviors of ARMS are expected rather than what are shown in the deterministic approach (for example, anomalous fluctuations).
- One could consider the case in which the transition probability does not obey Equation (21). How to apply a rewriting rule in ARMS is a non-trivial problem, and may cause different behaviors.

A further study including these points is necessary and will shed light on unique aspects of ARMS computation.

5 Application of ARMS for Biological Systems

In this section, we will address following two applications of ARMS for investigating biological issues:

- Origin of Life in terms of a computation.
- Modeling an Ecological system.

5.1 Origin of Life

In order to develop a computational model of the origin of life, we use the concept of Chemical Autopoiesis proposed by L.Luisi [37].

Since the seminal work of Oparin [45] the question of the origin of life has usually been discussed within the framework of molecular Darwinism. The main ingredient of the molecular Darwinism is a mechanism which leads to selection in a large variety of structures: starting from small molecules, compounds with increasing molecular complexity and with emergent novel properties (binding, recognition, catalysis, informations) would have evolved, until the most extraordinary of emergent properties, life itself, originated. A key step is the emergence of self-replication, since accidentally built structures would decay and disappear under prebiotic conditions. Based on our present knowledge, we ascribe self-replication of DNA and most of all the other functions (catalysis, recognition and transport, energy strange, and transduction) to proteins, although this last notion has hanged somewhat with the discovery of the catalytic power of RNA. Coevolution of nucleic acids and proteins in a system of mutually dependent macromolecules, able to replicate, is thus considered as the genesis of life [37].

Chemical Autpoiesis. Life, as we know it, is cellular and although in principle other life forms are possible, we will ignore them. A cell is characterized, first of all, by a boundary which discriminates the "self" from the environment. Within this boundary, life is a metabolic network. Based on nutrients coming from the outside world, a cell sustains itself by a network of reactions that take place inside the boundary and which produce the cell components, including those which are assembled as the boundary itself. Thus, the cell is self-generating (i.e., cellular components are synthesized by the cell) and self perpetuating (i.e., these components are in turn transformed into new compounds, which are part of the reaction network that continuously replaces them).

Maturana and Varela tried to translate these general observations into scientific definition of "minimal life" [38]. The starting proposition is to define life as a unitary operation rather than as a structure. This unitary operation is what they have called an *autopoetic unit* [38]. An autopoetic unit is a system [38] that determines its own making, due to a network of reactions which take place within its own well-defined boundary. To use Varela's words: "an autopoetic machine continuously generates and specifies its own organization through its operation as a system of production of its own components" [38].

Let us now turn to consider an autopoetic unit in chemical terms, qualifying for example the notion of boundary in terms of chemical structure and then qualifying the process of self-generation and self-perpetuation in terms of relative rate constants. As already mentioned, the notion of the operational closure is central: it refers to the containment of system operations within the system boundary. A boundary means a physical berried, which imposes a diffusion step on the entering/outgoing metabolites.

To apply the concepts of autopoesis to concrete chemical systems, it is necessary to adhere to the original definition of boundary. Accordingly, the term boundary will be restricted to a three-dimensionally closed structure. This definition eliminates other possible kinds of boundaries, sugh as an interface between two phase (liquid/liquid or liquid/solod), and the simple surface of a molecule:

thus intermolecular binding and recognition of molecules is not *per se* an autopoetic process. Neither are autopoetic boundaries the surfaces of a growing crystal or the contours of cyclic molecules. Rather, vesicles, or micelles, are bounded in the autopoetic sense, in that they form tree-dimensionally closed structures with a membrane which is, in principle, able selectively to discriminate among entering/outgoing molecules.

Having defined in more chemical terms the notion of boundary, let us now consider the simplest possible autopoetic system. A minimal autopoetic system would require a boundary composed by at least one component, C; it will be characterized by only one entering metabolite, A, and by two unimolecular reactions, a self-generation reaction leading to C (at the expense of A); and a decomposition reaction which transforms C into a product P, which then goes out of the boundary. These chemical reactions are determined by the bounded unit, i.e., the transformation of A into C and of C into P take place only inside the boundary; we will assume that such spontaneous reactions are chemically irreversible (i.e., that we can neglect the back-reactions).

Let us also assume, for the sake of simplicity, that all diffusion processes throughout the boundary, as well as the rate of self-assembling of C, are very fast with respect to the chemical transformations $A \rightarrow C$ and $C \rightarrow P$. The system then produces its own component C, thus regenerating its own structure [37].

Luisi's team experimentally implemented this principle in the self-replicating micellar system which is based on a reaction between a regent, which is strictly localized in the micellar interior, and a co-surfactant. The co-surfactant is present in large excess in the bulk, from where it is continuously delivered to the interface of the micelle. In this experiment, the reaction takes place at the interface of the micelles; the chemical reactivity is a direct result of the presence of a boundary. Then they realized an autopoietic situation, in which the product of the reaction is the consequence of the boundary constraints of the original structure, and the surfactant product then assembles spontaneously in the boundary itself [37].

5.2 Artificial Cell Systems

We can develop a computational model related to Luisi's chemical autopoesis system in terms of P systems, as introduced by Gh. Păun in [49]. Further details can be found in [8], [47], [10], [46], [48], etc. In this model, a membrane contains "chemical compounds" (denoted by symbols) which are processed through chemical reactions specific to the region/compartment defined by the membrane.

P systems were introduced with computability goals. Based on the principles outlined above, we develop here an *Artificial Cell System* (*ACS*). It consists of a multiset of symbols, a set of rewriting rules (reaction rules) and membranes.

A membrane is an important structure for living systems. It distinguishes "self" from its environment and hierarchical structures inside the system (like cells, organs and so on) are composed by membranes. Membranes change their structure dynamically and constitute a system. We are interested in their dynamical structure in terms of computations.

Although not many researches have tackled this topic previously [39], [40], the focus of these researches is on the formation of a membrane. The aim of this study is to investigate the role of membrane in terms of computations, thus we do not treat its formation.

By simulations, we find that a chemical evolution like a behavior emerges and cells evolve to a structure consisting of several cell-like membranes. We investigate the correlation among the type of reaction rules (rewriting rules), characteristic to a membrane and the evolution of cells, and then find the characteristics of a membrane effect on its evolution and the λ_e parameter to describe the correlation.

The Membrane Structure (MS). To describe the membrane and its structure, we proceed as in [49] and we first define the language MS over the alphabet $\{[,]\}$ whose strings are recurrently defined as follows:

(1) $[,] \in MS$
(2) if $\mu_1, ..., \mu_n \in MS$, n ≥ 1, then $[\mu_1...\mu_n] \in MS$;
(3) there is nothing else in MS.

Consider now the following relation on MS: for $x, y \in MS$ we write $x \sim y$ if and only if we can write the two strings in the form $x = [_1...[_2...]_2[_3...]_3...]_1, y = [_1...[_3...]_3[_2...]_2...]_1$, i.e., if and only if two pairs of parentheses which are not contained in one other can be interchanged, together with their contents.

This is clearly an equivalence relation. We denote by $\overline{MS}$ the set of equivalence classes of MS with respect to this relation. The elements of $\overline{MS}$ are called *membrane structures*.

It is easy to see that the parentheses $[,]$ appearing in a membrane structure are matching correctly in the usual sense. Conversely, any string of correctly matching pairs of parentheses $[,]$, with a matching pair at the ends, corresponds to a membrane structure.

Each matching pair of parentheses $[,]$ appearing in a membrane structure is called a *membrane*. The number of membranes in a membrane structure μ is called the *degree* of μ and is denoted by $\deg(\mu)$. The external membrane of a membrane structure μ is called the *vessel;* membrane of μ. When a membrane which appears in $\mu \in \overline{MS}$ has the form $[\ ]$ and no other membranes appear inside the two parentheses then it is called an *elementary* membrane.

ACS and ACSE. We will now define two types of ACS;

(1) ACS, and
(2) ACS with an Elementary membrane (ACSE).

Descriptions of an ACS. A transition ACS is a construct

$$\Gamma = (A, \mu, M_1, ..., M_n, R, MC, \delta, \sigma),$$

where:

(1) A is a set of objects;
(2) μ is a membrane structure (it can be changed during a computation);
(3) $M_1, ..., M_n$, are multisets associated with the regions 1,2, ... n of μ;
(4) R is a finite set of multiset evolution rules over A.
(5) MC is a set of membrane compounds;
(6) δ is the threshold value of dissolving a membrane;
(7) σ is the threshold value of dividing a membrane;

μ is a membrane structure of degree n, n $\geq$ 1, with the membranes labeled in a one-to-one manner, for instance, with the numbers from 1 to n. In this way, also the regions of μ are identified by the numbers from 1 to n.

Rewriting rules are applied in following manner:

(1) The same rules are applied to every membrane. There are no rules specific to a membrane (this is a difference from P systems, where the rules are specific to membranes, and correspond to the so-called *P systems with global rules*).
(2) All the rules are applied in parallel. In every step, all the rules are applied in all membranes to all objects to which they can be applied. If there are more than two rules that can be applied to an object, then one rule is selected randomly.
(3) If a membrane dissolves, then all the objects in its region are left free in the region immediately above it.
(4) All objects and membranes not specified in a rule and which do not evolve are passed unchanged to the next step.

The set R of rewriting rules is a finite set of multiset rewriting rules over A. Both the left and the right side of a rule are obtained by sampling with replacement of symbols. A set of reaction rules is constructed as the overall permutation of both sides of the rules.

Input and Output. Chemical compounds are supported from outside of the system to and some compounds are exhausted from the system. All chemical compounds are transferred among cells, a randomly selected chemical compound is sent into the membrane just above or below it. Although a membrane does not allow specificity of transport across the membrane, a cell can control its chemical environment by chemical reactions.

Dissolving and Dividing a Membrane of an ACS. A membrane has a contexts specified as a multiset of symbols. To maintain a membrane, it needs to have a certain minimal volume. A membrane disappears if the volume of its contents decreases below the needed volume to maintain it. Dissolving the membrane is defined as follows:

$$[_h a, ...[_i b, ...]_i]_h \rightarrow [_h a, b, ...]_h,$$

where the ellipsis {...} illustrates chemical compounds inside the membrane. Dissolving takes place when

$$\frac{|w_i|_{MC}}{|M_i|} < \delta$$

where δ is a threshold value for dissolution. All chemical compounds in its region are then set free and they are merged into the region immediately above it.

On the other hand, when the volume of the membrane contents increases to a certain extent, then a membrane is divided. Dividing a membrane is realized by dividing it in multisets of random sizes. The frequency at which a membrane is divided is decided depending on its size. As the size of a multiset becomes larger, the cell is divided more frequently. Technically, this is defined as follows:

$$[_h a, b, ...]_h \rightarrow [_h a, ... [_i b, ...]_i]_h$$

Dividing takes place when

$$\frac{|w_h|_{MC}}{|M_h|} > \sigma$$

where σ is a threshold for dividing the membrane. All chemical compounds in its region are separated randomly into new membranes.

Description of an ACSE. An ACSE is different from an ACS only in the way of dividing and dissolving cells. Dissolving the membrane is defined as follows:

$$[_h a, b, ...]_h \rightarrow [_0 a, b, ...,]_0$$

Dissolving takes place when

$$\frac{|w_h|_{MC}}{|M_h} < \delta$$

where δ is a threshold value for dissolving the membrane. All chemical compounds in its region are then set free and they are merged into the region of M_0.

Dividing is defined as follows;

$$[_h a, b, ...]_h \rightarrow [_h a, ...]_h [_i b, ...]_i.$$

Dividing takes place when

$$\frac{|w_h|_{MC}}{|M_h|} > \sigma,$$

where σ is a threshold for dividing the membrane. All chemical compounds in its region are separated randomly in the old and new membranes. Hence, in an ACSE, a structured cell such as $[a, b[c, [d, e]]]$ does not appear.

Evolution of Cells. When a cell grows and it exceeds the threshold value for dividing, it divides into parts of random sizes. This can be seen as a kind of *mutation*. If a divided cell does not have any membrane compounds, it must disappear.

Furthermore, to maintain the membrane through chemical reactions inside the cell can be seen as a *natural selection*. If a cell cannot maintain the membrane compounds, it must disappear.

Thus, both dividing membranes and dissolving membranes produce evolutionary dynamics. These correspondences are summarized as follows:

Natural Selection	Dissolving a membrane,
Mutation	Dividing a cell into parts of random size.

Behavior of *ACSE* and *ACS*. In this section, we will show some experimental results about ACSE and ACS.

ACSE. The evolution of elementary cells can be regarded as an approximate model of the chemical evolution in the origin of life.

To check this intuition, the following *ACSE* was simulated:

$\Gamma = (A = \{a, b, c\}, \mu = \{ [,]_0, \dots [,]_{100}\} M_0 = \{[a^{10}, b^{10}, c^{10}]^{100}\}, R, MC = \{b\}, \delta = 0.4, \sigma = 0.2)$,

where:

(1) R, the length of the left- or right-hand-side of a rule is between one and three. Both sides of the rules are obtained by sampling with replacement of the three symbols a, b and c;

(2) The membrane structure are assumed to be $(\mu = \{[_1]_1 \dots [_{100}]_{100}\})$.

Through the simulation we discovered that the strength of a membrane affects the behavior of cells. The strength of a membrane is defined as the frequency of decreasing membrane compounds.

When a Membrane is Strong. When a membrane is strong, the most stable cell consists of only one membrane, cells of this type become "mother" cells and they produce "daughter" cells.

In order to display a state of a cell we represent it by a number by using the following transformation function: $f(M(a), M(b), M(c)) = 10^2 \times M(a) + 10^1 \times M(b) + 10^0 \times M(c)$. For example, the state $\{a, a, b, c, c\}$ is transformed into $10^2 \times 2 + 10^1 \times 1 + 10^0 \times 2 = 212$.

Figure 18 illustrates the evolution of cells when a membrane is strong.

The cells that are close to the horizontal axis are mother cells. Some daughter cells depart from the group and evolve different types of cells, even though almost all cells are in the group. In this case, dissolving a membrane compound takes place per 100 steps.

Figure 19 is focused to the mother cells.

At the beginning there are about ten groups, and some of them become extinct: after 200 steps there remain about four groups.

When a Membrane is Weak. Figure 20 illustrates the case when a membrane is weak, and it dissolves every 3 steps.

In this case, the system cannot form a group of mother cells such as in the previous case. Since the group of cells drifts to more stable cells, the cells grow larger. Even if a large cell divides into parts of random sizes, the probability of including enough membrane compounds to maintain its membrane is larger than for a small cell.

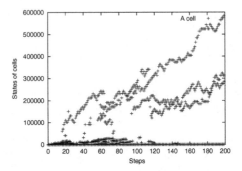

Fig. 18. When a membrane is strong. The lines illustrate the regions where cells exist and points correspond to the state of cells.

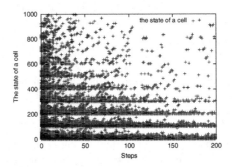

Fig. 19. Evolution of mother cells. The points correspond to cells.

We believe that this behaviors of evolution is similar to the evolution of viruses [59]. The settings of this simulation are so rough, however, that the possibility remains open that chemical evolution in origin of life is similar to virus evolution. This issue will be addressed in a future research.

The Correlation between the Behavior of an ACS and the Characteristics of Rewriting Rules

Description of a Simulation. Next, the following *ACS* was simulated:

$\Gamma = (A = \{a, b, c\}, \mu = \{\emptyset\}, M_0 = \{a^{10}, b^{10}, c^{10}\}, R, MC = \{b\}, \delta = 0.4, \sigma = 0.2)$,

where R, the length of the left- or right-hand-side of a rule is between one and three. Both sides of the rules are obtained by sampling with replacement of the three symbols a, b and c and membrane structures are not assumed.

λ_e *Parameter.* In order to investigate the correlation between the rewriting rules and the behavior of the model, we will introduce the λ_e parameter [54]. λ_e

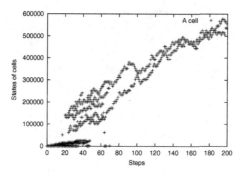

Fig. 20. When a membrane is weak

indicates the degree of reproduction in a cell. When λ_e close to 0.0, the degree is quite low and as λ_e is getting larger than 1.0, the degree becomes higher.

Let recall the λ_e parameter here. It was defined as follows;

$$\lambda_e = \frac{\Sigma r_{\Delta S > 0}}{1 + (\Sigma r_{\Delta S < 0} - 1)} \qquad (27)$$

where $\Sigma r_{\Delta S > 0}$ corresponds to the number of *heating rules*, and $\Sigma r_{\Delta S < 0}$ to the number of *cooling rules*. This parameter is well-defined when the number of rules is greater than 1.

When the *ARMS* only uses rules of the type $r_{\Delta S < 0}$, λ_e is equal to 0.0. On the contrary, if the *ARMS* uses rules of the type $r_{\Delta S > 0}$ and $r_{\Delta S < 0}$ with the same frequency, λ_e is equal to 1.0. Finally, when the *ARMS* only uses rules of the type $r_{\Delta S > 0}$, λ_e is greater than 1.0.

λ_e Parameter and System's Behavior. The behavior of *ACS* is classified into four classes by this parameter as follows;

- Type I: A cell does not evolve and disappears.
- Type II: The period of dividing membranes and dissolving membranes appears cyclically.
- Type III: A cell evolves to a complex, hierarchically structured cell.
- Type IV: All chemical compounds inside a cell grow rapidly. However, cells hardly divide.

Each λ_e value is not very important. Although they change in the different environments, these four classes are unchanged.

Type I (λ_e Close to 0.0). When λ_e is close to 0.0 the *cooling rule* is mainly used, and cells will hardly grow up. Thus membranes will hardly be divided (figure 21).

step state
0. $[a^{10}, b^{10}, c^{10}]$
1. $[a^9, b^5, c^{10}]$
2. $[a^{10}, b^2, c^7]$
3. $[a^{11}, b^3, c^7]$
4. $[a^6, b^4, c^5]$

....

10. $[a^1, b^4, c^2]$

....

16. $[a^1, b^4]$

Fig. 21. An example of state transition of ACS: Type I (λ_e close to 0.0)

step state
0. $[a^{10}, b^{10}, c^{10}]$
1. $[a^6, b^9, c^9]$

....

91. $[a^4, b^4, c^1]$
92. $[[b^2] \; [b^3, c^1]]$
93. $[[b^2] \; [b^1, c^1]]$
94. $[a^6, b^1, c^1]$

....

114. $[a^2, b^5]$
115. $[[a^1, b^2][a^1, b^3]]$
116. $[[\; a^1, b^2][b^2][a^1, b^1]]$

....

140. $[[[\; a^1, b^1 \;][b^1, c^1]][[b^2 \;][a^1, b^1]][b^1, c^1]]$
141. $[[a^3, c^3][a^3, c^2][a^1, c^2]$
142. $[\; a^6, c^5 \;]$

Fig. 22. An example of state transition of ACS: Type II (λ_e in between 0.5 and $\sim$ 1.0)

Type II (λ_e in between 0.5 and $\sim$ 1.0). When λ_e is in between 0.5 $\sim$ 1.0, membranes are more likely to be divided than when λ_e is close to 0.0. But, since *cooling rules* are likely to be used, the membrane compounds do not increase very much. Thus, when a cell evolves to a certain size the membrane compounds of each cell decrease and they are dissolved (figure 22). Thus there emerges a *cell cycle* behavior.

Type III (λ_e in between 1.05 $\sim$ 2.33). When λ_e exceeds 1.0 and the *heating rule* is likely to be used, a cell grows up easier and the frequency of cell division becomes high. Thus a cell evolves into a large complex cell (figure 23).

Type IV (λ_e More than 2.5). When the λ_e parameter becomes much larger than 1.0, membranes will hardly be divided, because the number of compounds of all kinds in the cell increase and it is difficult to specifically increase the number of only membrane compounds. Consequently, a cell will hardly divide (figure 24).

step state
0. $[\,a^{10},b^{10},c^{10}]$
1. $[\,a^{10},b^{10},c^{9}]$
....
41. $[a^2,b^7,c^5]$
42. $[[b^2]\ [b^6,c^1]]$
43. $[[b^4]\ [b^3,c^1]]$
44. $[[b^4]\ [b^2\ ][b^2,c^1]]$
45. $[[b^4,c^2][b^2,c^2][b^1,c^3]]$
46. $[[[b^3,c^1][b^1,c^1]][b^2,c^2][b^1,c^3]]$
47. $[[[[b^2][a^1,b^1]][b^1,c^1]][a^1,b^3,c^2]][a^1,b^2,c^3\ ]]$
140. $[[[a^1,b^1][b^1,c^1]][b^2][a^1,b^1]][b^1,c^1]]$
214. $[[[[a^{16},b^3,c^2][a^5,b^5,c^1][a^3,b^2]]\ [a^4,b^3]][[[b^2][b^2][a^1,b^1]][a^3,b^1]]\ [a^3,b^1][b^4][a^3,b^1]]]$

Fig. 23. An example of state transition of ACS: Type III (λ_e in between $1.05 \sim 2.33$)

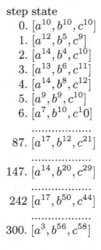

step state
0. $[a^{10},b^{10},c^{10}]$
1. $[a^{12},b^5,c^9]$
2. $[a^{14},b^4,c^{10}]$
3. $[a^{13},b^6,c^{11}]$
4. $[a^{14},b^8,c^{12}]$
5. $[a^9,b^9,c^{10}]$
6. $[a^7,b^{10},c^10]$
....
87. $[a^{17},b^{12},c^{21}]$
....
147. $[a^{14},b^{20},c^{29}]$
....
242 $[a^{17},b^{50},c^{44}]$
....
300. $[a^3,b^{56},c^{58}]$

Fig. 24. An example of state transition of ACS: Type IV (λ_e more than 2.5)

We believe that the division in four classes describes the basic behavior of the system. However, when δ or σ are changed the λ_e value that corresponds to each type is also changed. Thus a deeper investigation is needed with respect to the correlation between δ, σ and λ_e.

5.3   Genetic ACS (GACS)

Since a rewriting rule promotes a reaction, it can be regarded as an enzyme. Here we extend ACS with evolutionary mechanism, and we call the system Genetic ACS (GACS).

Descriptions of GACS. A transition $GACS$ is a construct

$$\Gamma = (A, \mu, M_1, ..., M_n, R, \delta, \sigma),$$

where:

(1) A is a set of objects;
(2) μ is a membrane structure (it can be changed during a computation);
(3) $M_1, ..., M_n$, are multisets associated with the regions 1,2, ... n of μ;
(4) R is a finite set of multiset evolution rules over A;
(5) δ is the threshold of dissolving membrane;
(6) σ is the threshold of dividing membrane.

The way of applying rewriting rules, the way of dissolving and dividing, as well as the input and output are the same as for ACS and ACSE.

An Enzyme. We denote a set of reaction rules as follows:

	a	b	c
a	x_{aa}	x_{ab}	x_{ac}
b	x_{ba}	x_{bb}	x_{bc}
c	x_{ca}	x_{cb}	x_{cc},

where x_{ij} means the number of compounds i which are transformed from j. For example, 2_{ab} means a rewriting rule, $b \rightarrow a, a$. We call the table a transformation table.

Transmission of an Enzyme. When a membrane is divided, the enzyme which is inside the membrane is copied and passed down to a new divided cell. At that time, a point mutation occurs only in the copied enzyme and it is passed down to the new cell. The enzyme remains in the old membrane as well as the new one. Point mutations occur every time a membrane divides. When a membrane is dissolved the enzyme which is inside the membrane loses its activity. A point mutation is a rewriting of the number of x_{ij}. Thus, it changes the number of transforming compounds to i. We assumed $x_{ij} \in \{0, 1, 2\}$. In ACS and ACSE the system has only one set of rewriting rules, however, in GACS, each membrane has its set of rules (like in P systems).

An Experimental Result of a GACS. We will discuss some experimental results about GACS. Specifically, the following *GACS* was simulated:
$\Gamma = (A = \{a, b, c\}, \mu = \{\emptyset\}, M_0 = \{a^{10}, b^{10}, c^{10}\}, R, MC = \{c\}, \delta = 0.4, \sigma = 0.2)$,
where the transformation table (R) is as follows:

	a	b	c
a	0_{aa}	0_{ab}	1_{ac}
b	1_{ba}	0_{bb}	0_{bc}
c	0_{ca}	1_{cb}	0_{cc}.

The productivity of membrane compounds p is defined as the ratio of the total number of non-membrane compounds to be produced to the total number of membrane compounds to be produced:

$$p = \frac{\sum_{j=a}^{j=b} x_{cj}}{\sum_{j=a}^{i=b} \sum_{j=a}^{j=c} x_{i,j}},$$

When $p = 0$ the enzyme does not produce any membrane compounds, when $p = 1$, it produces the same number of membrane compounds as the non-membrane compounds, and when $p > 1$, it produces more membrane compounds than non-membrane compounds.

Figure 25 illustrates the time series of productivity, where the vertical axis illustrates the productivity, the horizontal axis illustrates the steps and each dot is an enzyme. It shows that at first almost all enzymes evolve to $p > 1$. However, after 100 steps, the productivity of the enzymes decreases. After 100 steps, both the number of cells and the sizes of cells increase exponentially. Furthermore, the structure of cells becomes complicated.

Figure 26 illustrates the correlation between the number of cells, the size of cells and the number of steps, where each dot corresponds to a cell. Figure 27 illustrates the internal nodes of the whole system. If we regard the outermost membrane as the root and other cells as internal nodes and leaves, we can regard the whole system as a tree. In order to indicate the complexity of the tree, we use the number of internal nodes in the tree. Figure 27 illustrates that the number of internal nodes increases exponentially, after 150 steps.

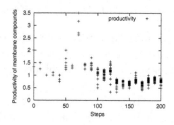

Fig. 25. Productivity of enzymes.

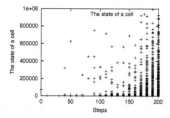

Fig. 26. State transition of the system

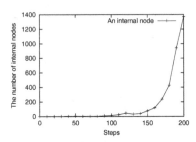

Fig. 27. Internal nodes of the cell

It is interesting that when a cell grows into a hierarchical cell, the enzyme evolves to a low productivity one.

The reason of this behavior can be considered as follows; the enzyme whose productivity is high always suffers from mutations, because it promotes membrane division, it generates more mutations than low productivity ones. If every cell is an elementary cell, an enzyme have to keep producing membrane compounds at a high rate. However, when the cell forms a structure, an enzyme with a high productivity is no longer necessary, because if an internal cell dissolves, the cell that includes the dissolved cell obtains its membrane compounds.

Therefore, during the evolution of a cell into a structured cell, the cell needs a highly productive enzyme. However, once it evolves a structured cell, highly productive enzymes are weeded out. This is the role of membranes in terms of computation.

5.4   Biological Inspired Computational Models

In this section, we will address computer science aspects of ACS and GACS. We are interested in living systems in terms of computation but not interested in solving optimization problems, solving NP-problems, and so on.

The aim of this study is to create a biological inspired computing model, *Computational living systems*. In such a system, in order to obtain the result, we observe the behavior and change the environment conditions, without stopping the "computations". In other words, we steer the system to the right direction by changing the environment and lead the system to our settled goal. Although such a system may not be suitable to obtain an optimizer, it may be suitable to create a system like living systems. This contribution is a preliminary research toward this goal.

Genetic Programming by Using GACS

How We Will Compose It? By using GACS, we attempted to generate a program by using a GACS. In the GACS, a program corresponds to an enzyme, thus we breed an enzyme which can solve a particular problem. We apply this approach to the simple problem of *doubling*: calculate the double value of the number of a and b then show the result as the number of c $(c = 2(a + b))$.

Description of GACS. The used *GACS* is defined as follows:
$\Gamma = (A = \{a, b, c\}, \mu = \{[_1]_1 \ldots [_{100}]_{100}\}, M_0 = \{[a^2, b^2, c^0]^{100}\}, R)$. In the initial state, all transformation tables R are

	a	b	c
a	0_{aa}	0_{ab}	1_{ac}
b	1_{ba}	0_{bb}	0_{bc}
c	0_{ca}	1_{cb}	0_{cc},

and one hundred of elementary cells are assumed inside M_0. No compounds are transformed among cells and no input and output are assumed. Although we performed simulation by using different types of cells in the initial state, the experimental results are the same, thus we will address only the case when in the initial state each cell is $[a^2, b^2, c^0]$.

Dissolving and Dividing a Membrane. The way of dissolving and dividing are the same as in ACSE. After n rewriting steps, if the number of c is smaller than 7, the membrane is dissolved and the enzyme inside it loses its activity. For example,

$$[_h a^3, b^5, c^8]_h [_i a^4, b^5, c^1]_i \rightarrow [_h a^3, b^5, c^8]_h.$$

In the above example, the number of c inside the membrane i is smaller than 7, so the membrane is dissolved.

After n rewriting steps, if the number of c is larger than 9, the membrane is divided and a point mutation takes place in its enzyme. Furthermore, a new enzyme is passed down to a new cell. When a cell divided, the inside multiset of the divided cell and its parent cell are set to $\{a^2, b^2, c^0\}$ again, so they try to solve the problem again. The results is:

$$[_h a^3, b^5, c^{10}]_h \rightarrow [_h a^2, b^2, c^0]_h [_i a^2, b^2, c^0]_i$$

In the example, because the number of c inside the membrane h is larger than 9, the membrane h is divided and a new membrane i emerges. Then compounds which are inside both membranes i and h are set to $\{a^2, b^2, c^0\}$.

The Fitness of the GACS. The fitness of an enzyme is defined as the number of steps to reach the solution. By using this fitness, good enzymes that can solve the problem within a smaller number of steps than the others are selected.

Experimental Result. At first, all enzymes are set to:

$$a \rightarrow b,$$
$$b \rightarrow c,$$
$$c \rightarrow a.$$

After 5000 rewriting steps, the enzymes that reach the solution within 8 steps are selected:

$$\begin{array}{lll}
a \to & a,c, & a \to a,b,c,c, \quad a \to a,c,c, \\
b \to & a,b,c,c, & b \to a,b,c,c, \quad b \to \quad b, \\
c \to & c,c; & c \to \quad c,c; \quad\quad c \to \quad c,c.
\end{array}$$

For each rule, one hundred of elementary cells that are $[a^2, b^2, c^0]$ are set again and calculations are performed again. Next, the enzymes that can solve the problem within 5 steps are selected. Then, there remains only one enzyme:

$$\begin{array}{l}
a \to a,c,c, \\
b \to \quad c,c, \\
c \to \quad c,c.
\end{array}$$

This is a solution of this simulation. In fact, a simpler solution of this problem is

$$\begin{array}{l}
a \to c,c, \\
b \to c,c, \\
c \to \quad c;
\end{array}$$

Thus the survived enzyme evolved to a similar enzyme in the solution.

This model is a candidate of the computational living system. In the future we plan to create a GACS as an artificial living system of computation that can solve more complicated problems.

5.5   Modeling and Analyzing an Ecological System by Using ARMS

ARMS can be a tool to model a complex system such as ecological systems, social systems and so on. We will consider here an example of application of an ARMS for modeling an ecological system.

A phenomenon in which plants may respond to herbivore feeding activities by producing volatiles that in turn attract carnivorous enemies of the herbivores has been reported recently [11] [44] [12]. These volatiles are not the mere result of mechanical damage, but are produced by the plant as a specific response to herbivore damage.

In theoretical biological studies on such systems, one notable study was done by Sabelis and de Jong [51] who showed that there was a wide range of conditions for signaling plants to coexist with plants that spend their enemy in alternative ways, the kinds of volatiles become polymorphic within species. Through the use of game theory, Sabelis and de Jong showed that ESS corresponds to the case when each species of plant produces the volatiles in a polymorphic way. In order to investigate the population dynamics of tritrophic systems, we employ the abstract rewriting system on multisets, (ARMS) [54][53].

Implementing an Ecological System by Using the ARMS. In this section, we model the tritrophic system by using *ARMS*. Let the symbol "a" be a leaf, "b" be a herbivore, "d" be a carnivore and "c" be the density of herbivore-induced volatiles. Furthermore, we define "e" be an empty state in order to introduce the death state. A plant is defined implicitly as a number of leaves. Evolution rule R_1 is defined as follows:

$$a \xrightarrow{k_1} a, a \quad r_1 \quad (\text{increase in the number of leaves}),$$

$$a, b \xrightarrow{k_2} b, b, c \quad r_2 \quad (\text{herbivore eats a leaf}),$$

$$d, b, c \xrightarrow{k_3} d, d \quad r_3 \quad (\text{carnivore catches herbivore}),$$

$$d \xrightarrow{k_4} e \quad r_4 \quad (\text{death of a carnivore}),$$

$$b \xrightarrow{k_5} e \quad r_5 \quad (\text{death of a herbivore}).$$

Here, $k_1, ..., k_5$ denotes the reaction rate, corresponding to the frequency of rule application. For example, when $k_4 = 0.1$ and $k_5 = 0.2$, r_5 will be applied twice as often as r_4.

In this case, k_2 is defined according to the state of multiset, i.e.,

$$k_2 = \frac{M(b)}{M(a) + M(b) + M(d)}.$$

Rule r_1 corresponds to the sprouting and growth of a plant, r_2 corresponds to the case when a herbivore eats a leaf and the leaf generates volatiles, r_3 corresponds to the herbivore being preyed on by a carnivore, r_4 corresponds to the death of a carnivore and r_5 corresponds to the death of a herbivore. More precisely, r_2 denotes the case when a leaf exists (a), and a herbivore eats the leaf and breeds there (b). The leaf produces volatiles compounds (c) as a result of being eaten, so as to attract carnivores. Rule r_3 denotes the case when there is a herbivore present (b) with the volatiles (c), and a carnivore (d) is attracted by the volatiles of r_2, catches the herbivore and breeds there (dd). The breeding rate of the carnivore is expressed as the changing the right-hand side of ds, e.g., $d, b, c \xrightarrow{k_3} d, d, d, d$.

Using this model, we compared the case when leaves generate volatiles with the case where they do not. The evolution rules of the system without volatiles R_2 are defined as follows:

$$a \xrightarrow{k_6} a, a \quad k_6 \quad (\text{increase in the number of leaves}),$$

$$a, b \xrightarrow{k_7} b, b \quad k_7 \quad (\text{herbivore eats a leaf}),$$

$$d, b \xrightarrow{k_8} d, d \quad k_8 \quad (\text{carnivore catches herbivore}),$$

$$d \xrightarrow{k_9} e \quad k_9 \quad (\text{death of a carnivore}),$$

$$b \xrightarrow{k_{10}} e \quad k_{10} \quad (\text{death of a herbivore}).$$

We set the reaction rates of k_1, k_4, k_5, k_6, k_9 and k_{10} as 0.5, 0.1, 0.1, 0.5, 0.1 and 0.1, respectively.

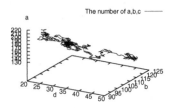

Fig. 28. Time course of population size of plants (a), herbivores (b) and carnivores (c) $(R_1, r_3 = 0.55)$

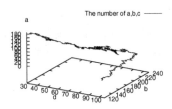

Fig. 29. Time course of population size of plants (a), herbivores (b) and carnivores (c) (until plants extinction, $R_2, r_8 = 0.55$)

Simulation and Results. In both R_1 and R_2, the symbiotic relationship consisting of plants, herbivores and carnivores can be identified. It is interesting that R_2 (the relationship under system without volatiles) is likely to collapse under the same conditions as for R_1 depending on the number of plants, herbivores and carnivores in the initial state, and the reaction rate of r_3 and r_8.

The time course of population size of plants, herbivores and carnivores of R_1 (figure 28) is very different from R_2 (figure 29).

In R_1 there are three attractors (a, b, c); (200, 100, 22)(1), (180, 105, 27)(2) and (150, 120, 46)(3). Tritrophic population is first attracted by (1), then (2) and finally (3). The progression for R_2 is attracted to (120, 200, 82), but departs from there as the system quickly degenerates.

These results show that the system with volatile generation by plants is more robust than the system without volatile generation under otherwise identical conditions. Thus, the role of volatiles warrants further investigation.

6   Final Remarks

A. Turing has abstracted the essence of a *computation* in terms of (human) calculation by *pen* and *paper*; in this way the notion of a Turing Machine has emerged, the main computing model investigated in computer science. However, the model is considered restrictive and many people have tried to consider non-Turing computing models.

It is obvious that living systems never do calculation by using *pen* and *paper*. Thus, if we could abstract the idea of a computation in terms of living systems, then it is possible to reach a the computational principle of emerging hierarchical structure "out of nothing".

We have not found the principle yet, but we have obtained some hints about its basic features. They are:

(1) Emergence of the hierarchical structure of interactions
(2) Parallelism.

Emergence of the Hierarchical Structure of Interactions. We saw that the membrane structure produces a hierarchical structure. However, in the present model, the produced membrane structures do not interact with each other. We must consider the framework that allows further interactions among these emergent structures.

Parallelism. In a living system, every cell and every organ is acting in parallel. Moreover, if some organs or cells lose their functionality, another organs or cells try to increase their activity for recovering/compensate and the system keeps its activity.

Acknowlodgements

This research is supported by RCAST (at Doshisha University) from Ministry of Education, Science and Culture in Japan and was partly supported by the Program for Promotion of Basic Research Activities for Innovative Biosciences (Bio-oriented Technology Research Advancement Institution).

References

1. W. Banzaf, Self-replicating sequences of numbers – foundation I and II: General and strings of length n = 4, Biological Cybernetics, (69), 269-281, 1993.
2. C.L. Barrett and C.M. Reidys, Elements of a theory of computer simulation I: Sequential CA over random graphs, Appl. Math. and Comp., (to appear).
3. C.L. Barrett, H.S. Mortiveit H.S. and C.M. Reidys, Elements of a theory of computer simulation II: Sequential dynamical systems, Appl. Math. and Comp., (to appear).
4. R. Begley and D. Farmer, Spontaneous emergence of a metabolism, Artificial Life II, pp141-158, Addison-Wesley, 1991
5. G. Berry and G. Boudol. 1992. The chemical abstract machine. *Theoretical Computer Science* 96: 217–248.
6. H. Bersini, Reaction mechanisms in the oo chemistry, Artificial Life VII, pp39-48, MIT press, 2000
7. B. Bollobas, Random Graphs, Academic Press, 1985

8. C. Calude, Gh. Păun, *Computing with Cells and Atoms*, Taylor and Francis, London, 2000 (Chapter 3: "Computing with Membranes").
9. J. Dassow, Gh. Păun, *Regulated Rewriting in Formal Language Theory*, Springer-Verlag, Berlin, 1989.
10. J. Dassow, Gh. Păun, On the power of membrane computing, *J. of Universal Computer Sci.*, 5, 2 (1999), 33–49 (www.iicm.edu/jucs).
11. J. Takabayashi and Dicke, M. Plant-carnivore mutualism through herbivore-induced carnivore attractants, Trends in Plant Science 1: 109-113, 1996.
12. M. Dicke, J. Takabayashi, C. Schütte,O.R. Krips, Behavioral ecology of plant-carnivore interactions: variation in response of phytoseiid mites to herbivore-induced plant volatiles. *Experimental and Applied Acarology 22*: 595-601, 1997.
13. P. Dittrich, J. Ziegler and W. Banzaf, Artificial Chemistries – A Review, http://ls11-www.cs.uni-dortmund.de
14. P. Dittrichi and Banzaf W., Self-evolution in constructive binary string system, Artificial Life 4(2), pp203-220, MIT press, 1998.
15. M. Eigen and P. Schuster, The Hypercycle, Springer-Verlag, 1979.
16. D.S. Fenizio, A less abstract artificial chemistry, Artificial Life VII, pp49-53, MIT press, 2000
17. W. Fontana, Algorithmic Chemistry, Artificial Life II, 160-209, Addison Wesley, 1994.
18. W. Fontana, L. W. Buss, The arrival of the fittest: Toward a theory of biological organization, *Bulletin of Mathematical Biology*, 56 (1994), 1–64.
19. D. Frijtyers and A. Lindenmayer, L systems, Lecture Notes In Computer Science, vol. 15, Springer Verlag, 1994.
20. J. H. Gallier, Logic for Computer Science, p89, John Wiley & Sons, 1987.
21. M. R. Garey, D. J. Johnson, *Computers and Intractability. A Guide to the Theory of NP-Completeness*, W. H. Freeman and Comp., San Francisco, 1979.
22. M. J. O'Donnell, Computing in system described by equations, Lecture Note in Computer Science, Vol.58, Springer Verlag, 1977.
23. C. Epstain, and R.Butow, Microarray technology -enhanced versatility, persistent challenge, Curr. Opin. Biotech, 11, 36-41, 2000
24. P. Erdös and A. Reny, On random graphs, Publicationes Mathematicae, 6, 290-297, 1959.
25. W. Feller, An Introduction to Probability Theory and Its Applications, I, 1957.
26. R.J. Field and M. Burger. 1985. *Oscillations and Traveling Waves in Chemical Systems*. John Wiley and Sons.
27. C. W. Gardiner, *Handbook of Stochastic Methods*, 2nd edition, (Springer-Verlag, 1985).
28. J. Guare, Six Degrees of Separation, A Play, Vintage, 1990.
29. J. E. Hopcroft and J. D. Ullman, Introduction to Automata theory, Languages and Computation, Addison-Wesley, 1979.
30. G. Huet and D. S. Lankford, On the Uniform halting problem for Term Rewriting Systems, Rapport 359, IN-RIA, 1978.
31. N. G. van Kampen, *Stochastic Processes in Physics and Chemistry*, (North-Holland, 1981).
32. S. A. Kauffman, The Origins of Order, Oxford University Press, 1993.
33. J. W. Klop, Term Rewriting System, in S. Abramsky, Don M. Gabby and T. S. E. Maibaum, edit, Handbook of Logic in Computer Science, 3-62, Clarendon Press, 1992.
34. D. E. Knuth and P. B. Bendix, Simple word problems in universal algebras, North-Holland, 1985.

35. S. N. Krishna, R. Rama, A variant of P systems with active membranes: Solving NP-complete problems, *Romanian J. of Information Science and Technology*, 2, 4 (1999), 357–367.
36. C. G. Langton, Life at the edge of chaos. In *Artificial Life II*, edited by C. G. Langton, C. Taylor, J. D. Farmer, and S. Rasmussen. Redwood City, CA: Addison Wesley, 1991.
37. P.L. Luisi, The chemical implementation of autopoiesis, Self-production of Supramolecular structures, pp 179-197, Kluwer Academic publ. 1994
38. H. R. Maturana and F. J. Varela, Autopoiesis and Cognition, D. Reidel Publishing Company, 1980.
39. B. McMullin and F. Varela, Rediscovering Computational Autopoiesis, *ECAL'97*. 1997.
40. K. Mirazo, A. Moreno, F. Moran, et. al., Designing a Simulation Model of a Self-Maintaining Cellular System *ECAL'97*. 1997.
41. R. B. Nachbar, Molecular evolution: Automed manipulation of hierarchical chemical topology and its application to average molecular structures, Genetic Programming and Evolvable Machines, 1(1/2):54-94, 2000
42. M.E.J. Newman, Small Worlds, Santa Fe Institute Working Paper, 1999.
43. G. Nicolis and I. Prigogine. 1989. *Exploring Complexity, An Introduction*. San Francisco: Freeman and Company.
44. T. Maeda, J. Takabayashi, J.,Yano, A. Takafuji, Factors affecting the resident time of the predatory mite Phytoseiulus persimilis (Acari: Phytoseiidae) in a prey patch, *Applied Entomology and Zoology 33*: 573-576, 1998
45. AI. Oparin, KB. Serebrovskaya, SN. Pantskhava and NV. Vesil'eva. Enzymatic synthesis of polyadenylic acid in coacervate drops. Biokhimiya 28, 4, 671-643, 1963
46. Gh. Păun, Y. Suzuki, H. Tanaka, P Systems with energy accounting, Intern. J. Computer Math., 79, 3/4 (in print)
47. Gh. Păun, P systems with active membranes: Attacking NP complete problems, *J. Automata, Languages and Combinatorics*, 6, 1 (2001), 75–93.
48. Gh. Păun, Y. Sakakibara, T. Yokomori, P systems on graphs of restricted forms, submitted, 1999.
49. Gh. Păun, Computing with membranes, *Journal of Computer and System Sciences*, 61, 1 (2000), 108–143, and *Turku Center for Computer Science-TUCS Report* No 208, 1998 (www.tucs.fi).
50. G. Rozenberg and A. Salomaa, *The Mathematical Theory of L Systems*, Academic Press, New York, 1980.
51. Sabelis, M. W. amd De jong, M.C.M. shyould all plants recruit bodygurads? Conditions for a polymorphic ESS of synomone production in plants. Oikos, 53: 247-252, 1988.
52. R. Smogyl, et.al., Cluster Analysis and data visualization of largescale gene expression data, Pacific Symposium on Biocomputing,3,42-53,1999.
53. Y. Suzuki and H. Tanaka, Symbolic chemical system based on abstract rewriting system and its behavior pattern, *Journal of Artificial Life and Robotics*, 1 (1997), 211–219.
54. Y. Suzuki and H. Tanaka, Order parameter for a symbolic chemical system, *Artificial Life VI*, MIT Press, 1998, 130–139.
55. Y. Suzuki and H. Tanaka, Artificial proto-cell based on symbolic chemical systems, submitted, 1999.
56. Y. Suzuki, S. Tsumoto, H. Tanaka, Analysis of cycles in symbolic chemical systems based on abstract rewriting systems on multisets, *Artificial Life V*, MIT Press, 1996, 522-528.

57. Y. Suzuki and H. Tanaka. 1997. Chemical oscillation on symbolic chemical systems and its behavioral pattern. In *Proceedings of the International Conference on Complex Systems*, Nashua, NH, 21-26 Sept 1997.

58. Y. Suzuki and H. Tanaka, On a LISP implementation of a class of P systems, *Romanian J. of Information Science and Technology*, 3, 2 (2000).

59. H. Tanaka, F. Ren, S. Ogishima, Evolutionary Analysis of Virus Based on Inhomogeneous Markov Model, ISMB'99, p 148, 1999.

60. http://www.cs.virginia.edu/oracle/

61. P. Walde, A. Goto, A. Monnard, et.al., Oparin's reactions revised: enzymatic synthesis of poly i micelles and self-reproducing vesicles, J. Am. Chem. Soc., 116, 7541-7574, 1994

62. D. J. Watts, Small Worlds, Princeton Univ. Press, 1999

63. J. D. Watson, N. H. Hopkins at el, Molecular biology of the gene, The Benjamin/Cummings publishing Company, Inc, 1992.

64. S. Wolfram, Cellular Automata and Complexity, 1994.

65. S. Wolfram, 1984a. Computation theory of cellular automaton. *Commun. Math. Phys.* 96: 15–57.

66. S. Wolfram, 1984b Universality and complexity in cellular automata. *Physica D* 10: 1–35.

Mathematics of Multisets

Apostolos Syropoulos*

Department of Civil Engineering
Democritus University of Thrace
GR-671 00 Xanthi, Greece
apostolo@obelix.ee.duth.gr

Abstract. This paper is an attempt to summarize the basic elements of the multiset theory. We begin by describing multisets and the operations between them, then we present hybrid sets and their operations. We continue with a categorical approach to multisets, and then we present fuzzy multisets and their operations. Finally, we present partially ordered multisets.

1 Introduction

Many fields of modern mathematics have been emerged by *violating* a basic principle of a given theory only because useful structures could be defined this way. For example, modern non-Euclidean geometries have been emerged by assuming that the *Parallel Axiom*[1] does not hold. Similarly, *multisets* [8,13,16] have been defined by assuming that for a given set A an element x occurs a finite number of times. Multisets are also known as "bags" (but many consider this term too vulgar...), "heap", "bunch", "sample", "occurrence set", "weighted set" and "fireset"—finitely repeated element set. An argument against the position that the term "bag" is too vulgar is that this term is a plain English word—something in which we put things to carry them around. Besides, in English language mathematical literature it is a tradition to use plain words –group, set, ring, ... – unlike other sciences, where people invent new long ones by sticking Greek and Latin words together. Also, we must note that the term "multiset" has been coined by N.G. de Bruijn [8]. The first person who actually used multisets is Richard Dedekind in his well-known paper "Was sind und was sollen die Zahlen?" ("The nature and meaning of numbers") [6]. This paper was published in 1888. The reader interested to read a rather complete account of the development of multiset theory should read Blizard's excellent survey [5].

From a practical point of view multisets are very useful structures arising in many areas of mathematics and computer science. The prime factorization of an integer $n > 0$ is a multiset $\mathcal{N}$ whose elements are primes. Every monic polynomial $f(x)$ over the complex numbers corresponds in a natural way to the

---

* Dedicated to the fond memory of my brother Mikhail Syropoulos.
[1] Which can be stated as follows: Given a point P not incident with line m, there is exactly one line incident with P and parallel to m.

C.S. Calude et al. (Eds.): Multiset Processing, LNCS 2235, pp. 347–358, 2001.

multiset $\mathcal{F}$ of its "roots". Other examples of multisets include the zeros and poles of meromorphic functions, the invariants of matrices in a canonical form, the invariants of finite Abelian groups, etc. The terminal strings of a context-free grammar form a multiset which is a set iff the grammar is unambiguous. Processes in an operating system can be thought of as multisets. The mathematical treatment of concurrency involves the use of multisets. In social sciences, multisets can be used to model social structures, etc.

There are three methods to define a set and we are recalling them now, since they will be heavily used in the rest of the text:

1. A set is defined by naming all its members (*the list method*). This method can be used only for finite sets. Set A, whose members are $a_1, a_2, \ldots, a_n$, is usually written as
$$A = \{a_1, a_2, \ldots, a_n\}.$$

2. A set is defined by a property satisfied by its members (*the rule method*). A common notation expressing this method is
$$A = \{x \mid P(x)\},$$
where the symbol $\mid$ denotes the phrase "such that", and $P(x)$ designates a proposition of the form "x has the property P."

3. A set is defined by a function, usually called the *characteristic function*, that declares which elements of a universal set X are members of set A and which are not. Set A is defined by its characteristic function, χ_A, as follows:
$$\chi_A(x) = \begin{cases} 1, \text{ if } x \in A \\ 0, \text{ if } x \notin A \end{cases}$$

In what follows we present the definition of multisets and the basic operations between multisets. Moreover, we briefly present hybrid sets, i.e., multisets which may have negative integers as multiplicities as well as nonnegative integers. Then, we proceed with a categorical approach to multisets by defining categories of multisets. Next, we present fuzzy multisets and their operations. We finish by presenting pomsets and their basic operations.

2   Multisets and Their Operations

Ordinary sets are composed of pairwise different elements, i.e., no two elements are the same. If we relax this condition, i.e., if we allow multiple but finite occurrences of any element, we get a generalization of the notion of a set which is called a *multiset*.

There are two different kinds of sets with a finite number of repeated elements: sets with distinguishable repeated elements, e.g., people sharing a common property, and sets with indistinguishable repeated elements, e.g., a "soup" of elementary particles. Monro [12] calls the first kind of sets *multisets* and the second *multinumbers*. However, in order to avoid confusion, we will use the term multiset for Monro's multinumbers and the term *real* multisets for Monro's multisets.

Real multisets and multisets are associated with a (ordinary) set and an equivalence relation or a function, respectively. Here are the formal definitions:

Definition 1. *A real multiset $\mathcal{X}$ is a pair (X, ρ), where X is a set and ρ an equivalence relation on X. The set X is called the field of the real multiset. Elements of X in the same equivalence class will be said to be of the same sort; elements in different equivalence classes will be said to be of different sorts.*

Given two real multisets $\mathcal{X} = (X, \rho)$ and $\mathcal{Y} = (Y, \sigma)$, a morphism of real multisets is a function $f : X \to Y$ which respects sorts; that is, if $x, x' \in X$ and $x \rho x'$, then $f(x) \sigma f(x')$.

Definition 2. *Let D be a set. A multiset over D is just a pair $\langle D, f \rangle$, where D is a set and $f : D \to \mathbb{N}$ is a function.*

The previous definition is the characteristic function definition method for multisets.

Remark 1. Any ordinary set A is actually a multiset $\langle A, \chi_A \rangle$, where χ_A is its characteristic function.

Since multisets are sets with multiple but finite occurrences of any element, one can define a multiset by employing the list method. However, in order to avoid confusion we will use square brackets for multisets and braces for sets. In what follows we will employ the most suitable definition method for each case we encounter.

An important notion in set theory is the notion of a subset. Moreover, for ordinary sets there are certain operations one can perform between sets, such as set intersection, union, etc. We proceed with the definitions of the notion of the subset of a multiset and the operations between multisets.

Definition 3. *Suppose that $\mathcal{A} = \langle A, f \rangle$ is a multiset; the subset B of A is called the support of $\mathcal{A}$ if for every x such that $f(x) > 0$ this implies that $x \in B$, and for every x such that $f(x) = 0$ this implies that $x \notin B$.*

It is clear that the characteristic function of B can be specified as:

$$\chi_B(x) = \min(f(x), 1).$$

Example 1. Given the multiset $\mathcal{A} = [a, a, a, b, c, c]$, then its support is the set $\{a, b, c\}$.

Definition 4. *Assume that $\mathcal{A} = \langle A, f \rangle$ and $\mathcal{B} = \langle A, g \rangle$ are two multisets. We say that $\mathcal{A}$ is a sub-multiset of $\mathcal{B}$, denoted $\mathcal{A} \subseteq \mathcal{B}$ if for all $a \in A$ we have*

$$f(a) \leq g(a).$$

A is called a *proper sub-multiset* of B, denoted $A \subset B$, if in addition for some $a \in A$ we have

$$f(a) < g(a).$$

Obviously, it follows that for any two multisets, $A = B$ iff $A \subseteq B$ and $B \subseteq A$.

Definition 5. *Let* $A = \langle A, f \rangle$ *be a multiset;* A *is the empty multiset if for all* $a \in A$, $f(x) = 0$.

Definition 6. *Suppose that* $A = \langle A, f \rangle$ *is a multiset; its cardinality, denoted* $\mathrm{card}(A)$, *is defined as*

$$\mathrm{card}(A) = \sum_{a \in A} f(a).$$

If A is a set, then $\mathcal{P}^A$ is the set of all multisets which have A as their support set. Moreover, A is the smallest non-empty multiset in $\mathcal{P}^A$ in the sense that if $B \in \mathcal{P}^A$, then

$$\mathrm{card}(B) \geq \mathrm{card}(A).$$

We are now turning our attention to the operations between multisets. We define, in this order, the sum, the removal, the union and the intersection of two multisets.

Definition 7. *Suppose that* $A = \langle A, f \rangle$ *and* $B = \langle A, g \rangle$ *are two multisets. Their sum, denoted* $A \uplus B$, *is the multiset* $C = \langle A, h \rangle$, *where for all* $a \in A$:

$$h(a) = f(a) + g(a).$$

It can be easily shown that the multiset sum operation has the following properties:

1. Commutative: $A \uplus B = B \uplus B$;
2. Associative: $(A \uplus B) \uplus C = A \uplus (B \uplus C)$;
3. There exists a multiset, the null multiset $\emptyset$, such that $A \uplus \emptyset = A$.

It is important to note that there exists no inverse and multiset sum is not idepotent.

Definition 8. *Suppose that* $A = \langle A, f \rangle$ *and* $B = \langle A, g \rangle$ *are two multisets. The removal of multiset* B *from* A, *denoted* $A \ominus B$, *is the multiset* $C = \langle A, h \rangle$, *where for all* $a \in A$:

$$h(a) = \max\Big(f(a) - g(a), 0 \Big).$$

Definition 9. *Suppose that* $A = \langle A, f \rangle$ *and* $B = \langle A, g \rangle$ *are two multisets. Their union, denoted* $A \cup B$, *is the multiset* $C = \langle A, h \rangle$, *where for all* $a \in A$:

$$h(a) = \max\Big(f(a), g(a) \Big).$$

Definition 10. *Suppose that* $\mathcal{A} = \langle A, f \rangle$ *and* $\mathcal{B} = \langle A, g \rangle$ *are two multisets. Their intersection, denoted* $\mathcal{A} \cap \mathcal{B}$, *is the multiset* $\mathcal{C} = \langle A, h \rangle$, *where for all* $a \in A$:

$$h(a) = \min\Big(f(a), g(a)\Big).$$

The following properties can be easily established for union, intersection, and sum of multisets:

1. Commutativity:

$$\mathcal{A} \cup \mathcal{B} = \mathcal{B} \cup \mathcal{A}$$
$$\mathcal{A} \cap \mathcal{B} = \mathcal{B} \cap \mathcal{A};$$

2. Associativity:

$$\mathcal{A} \cup (\mathcal{B} \cup \mathcal{C}) = (\mathcal{A} \cup \mathcal{B}) \cup \mathcal{C}$$
$$\mathcal{A} \cap (\mathcal{B} \cap \mathcal{C}) = (\mathcal{A} \cap \mathcal{B}) \cap \mathcal{C};$$

3. Idempotency:

$$\mathcal{A} \cup \mathcal{A} = \mathcal{A}$$
$$\mathcal{A} \cap \mathcal{A} = \mathcal{A};$$

4. Distributivity:

$$\mathcal{A} \cup (\mathcal{B} \cap \mathcal{C}) = (\mathcal{A} \cup \mathcal{B}) \cap (\mathcal{A} \cup \mathcal{C})$$
$$\mathcal{A} \cap (\mathcal{B} \cup \mathcal{C}) = (\mathcal{A} \cap \mathcal{B}) \cup (\mathcal{A} \cap \mathcal{C});$$

5.

$$\mathcal{A} \uplus (\mathcal{B} \cup \mathcal{C}) = (\mathcal{A} \uplus \mathcal{B}) \cup (\mathcal{A} \uplus \mathcal{C})$$
$$\mathcal{A} \uplus (\mathcal{B} \cap \mathcal{C}) = (\mathcal{A} \uplus \mathcal{B}) \cap (\mathcal{A} \uplus \mathcal{C});$$

6.

$$\mathcal{A} \cap (\mathcal{A} \uplus \mathcal{B}) = \mathcal{A}$$
$$\mathcal{A} \cup (\mathcal{A} \uplus \mathcal{B}) = \mathcal{A} \uplus \mathcal{B};$$

7.

$$\mathcal{A} \uplus \mathcal{B} = (\mathcal{A} \cup \mathcal{B}) \uplus (\mathcal{A} \cap \mathcal{B}).$$

Let $\mathcal{A} = \langle X, f \rangle$ be a multiset and $B \subseteq X$. We are interested in forming the multiset $\mathcal{C} = \langle X, g \rangle$, where

$$g(x) = \begin{cases} f(x), & \text{if } x \in B, \\ 0, & \text{if } x \notin B, \end{cases}$$

or, in other words, $g(x) = f(x) \cdot \chi_B(x)$.

A closely related problem is that of forming a multiset by removing all the elements from $\mathcal{A}$ which are in the set B. That is, we are interested in forming the multiset $\mathcal{D} = \langle X, h \rangle$, where

$$h(x) = \begin{cases} 0, & \text{if } x \in B, \\ f(x), & \text{if } x \notin B, \end{cases}$$

which can be expressed compactly as follows:

$$h(x) = f(x) \cdot (1 - \chi_B(x)), \forall x \in B.$$

However, one may note that $1 - \chi_B(x)$ is the characteristic function of the *complement* set of B, denoted $\bar{B}$. So, the previous equation becomes

$$h(x) = f(x) \cdot \chi_{\bar{B}}(x), \forall x \in B.$$

Thus,

$$\mathcal{D} = \mathcal{A} \circledast \bar{B}.$$

We shall call the operation $\mathcal{A} \circledast B$ *multi-intersection*. In general, it holds that if $\mathcal{A} = \langle X, \chi_A \rangle$, $A \subseteq X$, and B is a set, then $\mathcal{A} \circledast B = A \cap B$. Moreover, the following properties hold:

$$A \circledast X = A$$
$$A \circledast \emptyset = \emptyset$$
$$(\mathcal{A}_1 \cap \mathcal{A}_2) \circledast B = (\mathcal{A}_1 \circledast B) \cap (\mathcal{A}_2 \circledast B)$$
$$(\mathcal{A}_1 \cup \mathcal{A}_2) \circledast B = (\mathcal{A}_1 \circledast B) \cup (\mathcal{A}_2 \circledast B)$$

3   Hybrid Sets

Hybrid sets and new sets are generalization of multisets and sets respectively. In a hybrid set the multiplicity of an element can be either a negative number, zero, or a positive number. A new set is to hybrid sets what is a set to a multiset, i.e., a special case. We give now the definition of the hybrid set due to Loeb [9]:

Definition 11. *Given a universe U, any function $f : U \to \mathbb{Z}$, where $\mathbb{Z}$ is the set of all integers, is called a hybrid set. The value of $f(u)$ is said to be the multiplicity of the element u. If $f(u) \neq 0$ we say that u is a member of f and we write $u \in f$; otherwise, we write $u \notin f$. We define the number of elements, $\#f$, to be the sum $\sum_{u \in U} f(u)$. f is said to be an $\#f$ (element) hybrid set.*

Hybrid sets are denoted by employing the list method and by inserting a bar to separate elements with negative multiplicity from those with a non-negative multiplicity. Elements occurring with a positive multiplicity are written on the left of the bar, and elements occurring with a negative multiplicity are written on the right of the bar.

Example 2. If $f = \{a, b, b \mid d, e, e\}$ is a hybrid set, then $f(a) = 1$, $f(b) = 2$, $f(d) = -1$, and $f(e) = -2$.

The empty hybrid set, denoted $\emptyset$, is the unique hybrid set for which all elements have multiplicity equal to zero. We can specify the empty hybrid set by $\{|\}$. The definition of subset-hood in the case of hybrid sets is also due to Loeb:

Definition 12. *Let f and g be hybrid sets. We say that f is a subset of g and that g contains f and we write $f \subseteq g$ if either $f(u) \ll g(u)$ for all $u \in U$, or $g(u) - f(u) \ll g(u)$ for all $u \in U$, where $\ll$ is a partial ordering of the integers defined as follows: $i \ll j$ iff $i \leq j$ and either $i < 0$ or $j \geq 0$.*

We proceed now with the definition of the various operations between hybrid sets.

Definition 13. *Assume that f and g are two hybrid sets over the same universe U. Then, their intersection, $f \cap g$, is the hybrid set h, such that $h(u) = \max(f(u), g(u))$, their union, $f \cup g$, is the hybrid set h, such that $h(u) = \min(f(u), g(u))$, and their sum, $f \uplus g$, is the hybrid set h, such that $h(u) = f(u) + g(u)$.*

We easily verify the correctness of these definitions by using the definition of subset-hood.

4 Categorical Models of Multisets

Let $\mathcal{C}$ be a category. A functor $E : \mathcal{C}^{\mathrm{op}} \to \mathbf{Set}$ is called a *presheaf* on $\mathcal{C}$. Thus, a presheaf on $\mathcal{C}$ is a contravariant functor. The presheaves on $\mathcal{C}$ with natural transformations as arrows form a category denoted $\mathbf{Psh}(\mathcal{C})$. Suppose that C is a set, i.e., a discrete category; then, the presheaf $F : C \to \mathbf{Set}$ denotes a multiset, since $F(c)$ is a set whose cardinality is equal to the number of times c occurs in the multiset. So, for any set C, the category $\mathbf{Psh}(C)$ denotes the category of all multisets of C. These remarks lead us to the definition of a category of all possible multisets:

Definition 14. *Category* $\mathbf{MSet}$ *is a category of all possible multisets.*

1. *The objects of the category consist of pairs (A, P), where A is a set and $P : A \to \mathbf{Set}$ a presheaf on A.*
2. *If (A, P) and (B, Q) are two objects of the category, an arrow between these objects is a pair (f, λ), where $f : A \to B$ is a function and $\lambda : P \to Q \circ f$ is a natural transformation, i.e., a family of functions.*
3. *Arrows compose as follows: suppose that $(A, P) \xrightarrow{(f,\lambda)} (B, Q)$ and $(B, Q) \xrightarrow{(g,\mu)} (C, R)$ are arrows of the category, then $(f, \lambda) \circ (g, \mu) = (g \circ f, \mu \times \lambda)$, where $g \circ f$ is the usual function composition and $\mu \times \lambda : P \to R \circ (g \circ f)$.*
4. *Given an object (A, P), the identity arrow is $(\mathrm{id}_A, \mathrm{id}_P)$.*

The last part of the definition is a kind of wreath product (see [4]). However, it is not clear at the moment how this definition fits into the general theory of wreath products.

This is not the only way one can categorically define multisets. Suppose that $F : A \to \mathbf{Set}$ is a presheaf and that A is a set. Then if we form the set $X = \bigcup_{i \in A} X_i$, where $X_i = F(i)$, we can define the function $p : X \to A$. This function is equivalent to the presheaf F. Moreover, $p^{-1}(a)$, i.e., the preimage of p, is the set of copies of a in the multiset. Now, we can define another category of all possible multisets:

Definition 15. *Category* $\mathbf{Bags}$ *is a category of all possible multisets.*

1. *The objects of the category consist of pairs (A, p), where $p : \bigcup_{i \in A} X_i \to A$.*

2. *An arrow between two objects (A,p) and (B,q) is a pair (f,g), where $f :$ $A \to B$ and $g : X \to Y$, such that the following diagram commutes:*

3. *Suppose that $(A,p) \xrightarrow{(f,g)} (B,q)$ and $(B,q) \xrightarrow{(f',g')} (C,r)$ are two arrows. Then $(f,g) \circ (f',g') = (f' \circ f, g' \circ g)$ such that in the following diagram*

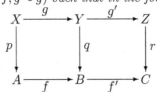

the outer rectangle commutes iff the inner squares commute.

4. *Given an object (A,p) the identity arrow is $(\mathrm{id}_A, \mathrm{id}_X)$.*

It is obvious that categories **MSet** and **Bags** are *equivalent*. Moreover, one can study the properties of the categories **MSet** and **Bags**, but we feel this is not the appropriate place for such a presentation. We now investigate the way one can embed the category **Bags** into a Chu category [15].

Given an arbitrary object $\perp$ in a category **A**, we construct the category $\mathbf{Chu}(\mathbf{A}, \perp)$ as follows:

1. The objects of $\mathbf{Chu}(\mathbf{A}, \perp)$ consist of triplets (A_1, r, A_2), where A_1, A_2 are objects in A and $r : A_1 \otimes A_2 \to \perp$ is an arrow in **A**.
2. An arrow from (A_1, r, A_2) to (B_1, s, B_2) is a pair $(f, \hat{f})$, where $f : A_1 \to B_1$ and $\hat{f} : B_2 \to A_2$ are arrows in **A** such that the square

$$
\begin{array}{ccc}
A_1 \otimes B_2 & \xrightarrow{f \otimes \mathrm{id}_{B_2}} & B_1 \otimes B_2 \\
\downarrow{\scriptstyle \mathrm{id}_{A_1} \otimes \hat{f}} & & \downarrow{\scriptstyle s} \\
A_1 \otimes A_2 & \xrightarrow{r} & \perp
\end{array}
$$

commutes.
3. Arrow composition is defined pairwise.

If **A** is any $*$-autonomous category (see [1,2]), then $\mathbf{Chu}(\mathbf{A}, \perp)$ is $\mathbf{A}^2$, where $\perp$ is a dualizing object [3]. It is now easy to define a full embedding of **Bags** into $\mathbf{Chu}(\mathbf{Rel}, 1)$, where **Rel** is the category of sets and binary relations between them and 1 is any singleton set.

We start by defining the object part of the functor:

Definition 16. *(**Object part**) Functor $\mathcal{M}$ maps each object (A, p) of **Bags** into the Chu space $(A, \tilde{p}, X)$, where $X = \text{dom}\,p$, and $\tilde{p}$ is the relation obtained from function p and a p X_i iff the multiplicity of a is equal to the cardinality of X_i.*

We now proceed with the arrow part of the functor:

Definition 17. *(**Arrow part**) Let (A, p) and (B, q) be two objects of the category **Bags**. Moreover, suppose that $(A, p) \xrightarrow{(f,g)} (B, q)$ is an arrow between these objects; then $\mathcal{M}(f, g) = (\tilde{f}, \tilde{g}^{-1})$, where $\tilde{f}$ is the relation obtained from function f and $\tilde{g}^{-1}$ the inverse of the relation obtained from function g.*

The possible consequences of this embedding are explored in [15].

5   Fuzzy Multisets

Fuzzy set theory has been introduced as a means to deal with vagueness in mathematics. The theory is well-established and we will not get into the trouble of presenting it. We just note that fuzzy set theory was an attempt to develop a formal apparatus to involve a partial membership in a set, mainly to arm people in the modeling of empirical objects and facts. In other words, fuzzy set theory is, sort to say, a generalization of the notion of set membership.

Definition 18. *Suppose that X is a set. Any function $A : X \to I$, where $I = [0, 1]$, is called a fuzzy subset of X. Function A is usually called the membership function of the fuzzy subset A.*

Fuzzy multisets have been introduced by Yager [16] and have been studied by Miyamoto [10,11] and others. A fuzzy multiset of some set X is just a multiset of $X \times I$. We are now defining summation of fuzzy multisets:

Definition 19. *If $\mathcal{A} = \langle X \times I, f \rangle$ and $\mathcal{B} = \langle X \times I, g \rangle$ are two fuzzy multisets, then their sum, denoted $\mathcal{A} \uplus \mathcal{B}$, is the fuzzy multiset $\mathcal{C} = \langle X \times I, h \rangle$, where for all $(x, \mu_x) \in X \times I$:*

$$h(x, \mu_x) = f(x, \mu_x) + g(x, \mu_x).$$

As in the case of crisp[2] multisets, there is more that one way to define a fuzzy multiset. In order to define the basic operations between fuzzy multisets, we define fuzzy multisets by the list method. If $\mathcal{A} = \{(x_i, \mu_i)\}_{i=1,\ldots,p}$ be a fuzzy multiset, then we can write the same set as $\mathcal{A} = \{\{\mu_{11}, \ldots, \mu_{1\ell_1}\}/x_1, \ldots, \{\mu_n, \ldots, \mu_{n\ell_n}\}/x_n\}$. Note that $\{\mu_{11}, \ldots, \mu_{1\ell_1}\}$ is actually a multiset of I. Next, we rearrange the multisets $\{\mu_{11}, \ldots, \mu_{1\ell_1}\}$ so that the elements appear in decreasing order. Finally, we need to add zeroes so that the length of all multisets $\{\mu_{11}, \ldots, \mu_{1\ell_1}\}$ is the same. This representation is called the *graded sequence*. To make things clear we give an example:

---

[2] In fuzzy set theory the term crisp is used to characterize anything that is non-fuzzy.

Example 3. Let

$$\mathcal{A} = [(a, 0.2), (b, 0.5), (b, 0.1), (a, 0.2), (a, 0.3), (d, 0.7)]$$

be a fuzzy multiset. Then its graded sequence is

$$\mathcal{A} = [\{0.3, 0.2, 0.2\}/a, \{0.5, 0.1, 0\}/b, \{0.7, 0, 0\}/d]$$

In case we want to perform certain operations on two or more fuzzy multisets, all multisets $\{\mu_{11}, \ldots, \mu_{1\ell_1}\}$ must have the same length. Moreover, even if one fuzzy multiset does not contain an element c we must add an entry of the form $\underbrace{\{0, 0, \ldots, 0}_{p \text{ times}}/c\}$, where p is the length of all other multisets. We are now giving the definitions of the various operations between fuzzy multisets:

Definition 20. *Assume that* $\mathcal{A} = [[\mu_{1p}, \ldots, \mu_{11}]/x_1, \ldots, [\mu_{np}, \ldots, \mu_{n1}]/x_n]$ *and* $\mathcal{B} = [[\mu'_{1p}, \ldots, \mu'_{11}]/x_1, \ldots, [\mu'_{np}, \ldots, \mu'_{n1}]/x_n]$ *are two fuzzy multisets; then*

1. $\mathcal{A} \subseteq \mathcal{B}$ *iff for every* x_i, $\mu_{ij} \leq \mu'_{ij}$, $j = 1, \ldots, p$.
2. $\mathcal{A} = \mathcal{B}$ *iff for every* x_i, $\mu_{ij} = \mu'_{ij}$, $j = 1, \ldots, p$.
3. $\mathcal{C} = \mathcal{A} \cup \mathcal{B}$, *where* $\mathcal{C} = [[\mu''_{1p}, \ldots, \mu''_{11}]/x_1, \ldots, [\mu''_{np}, \ldots, \mu''_{n1}]/x_n]$ *iff for every* x_i, $\mu''_{ij} = \max(\mu'_{ij}, \mu_{ij})$, $j = 1, \ldots, p$.
4. $\mathcal{C} = \mathcal{A} \cap \mathcal{B}$, *where* $\mathcal{C} = [[\mu''_{1p}, \ldots, \mu''_{11}]/x_1, \ldots, [\mu''_{np}, \ldots, \mu''_{n1}]/x_n]$ *iff for every* x_i, $\mu''_{ij} = \min(\mu'_{ij}, \mu_{ij})$, $j = 1, \ldots, p$.

When the functions max and min are replaced by a t-norm t and a t-conorm s respectively, we obtain the definitions for $\cap_t$ and $\cup_s$, respectively. The union and intersection of arbitrary fuzzy multisets $\mathcal{A}$, $\mathcal{B}$, and $\mathcal{C}$ satisfy the following laws:

1. Commutativity.
$$\mathcal{A} \cup \mathcal{B} = \mathcal{B} \cup \mathcal{A}$$
$$\mathcal{A} \cap \mathcal{B} = \mathcal{B} \cap \mathcal{A}$$

2. Associativity.
$$(\mathcal{A} \cup \mathcal{B}) \cup \mathcal{C} = \mathcal{A} \cup (\mathcal{B} \cup \mathcal{C})$$
$$(\mathcal{A} \cap \mathcal{B}) \cap \mathcal{C} = \mathcal{A} \cap (\mathcal{B} \cap \mathcal{C})$$

3. Distributivity.
$$(\mathcal{A} \cup \mathcal{B}) \cap \mathcal{C} = (\mathcal{A} \cap \mathcal{B}) \cup (\mathcal{A} \cap \mathcal{C})$$
$$(\mathcal{A} \cap \mathcal{B}) \cup \mathcal{C} = (\mathcal{A} \cup \mathcal{B}) \cap (\mathcal{A} \cup \mathcal{C})$$

Next, we define the α-cut for fuzzy multisets. We first recall the notion of the α-cut for a fuzzy set:

Definition 21. *Let* U *be a set, let* C *be a partially ordered set and let* $A : U \to C$. *For* $\alpha \in C$, *the* α-cut *of* A, *is* $A^{-1}(\uparrow \alpha) = \{u \in U \mid A(u) \geq \alpha\}$. *The subset of* U *will be denoted by* A_α.

Definition 22. *Assume that* $\mathcal{A} = \langle X \times I, f \rangle$ *is a fuzzy multiset, and that* $\alpha \in (0, 1]$. *Then* $\mathcal{A}_\alpha = \langle X, f' \rangle$, *i.e., the* α-cut *of* $\mathcal{A}$, *is a multiset such that*

$$f'(x) = \sum_{\mu_x \geq \alpha} f(x, \mu_x).$$

Consequently, the α-cut of a fuzzy multiset is just a multiset.

Given a fuzzy multiset $\mathcal{A} = \langle X \times \mathrm{I}, h \rangle$ and function $f : X \to Y$ we can define two images:

$$f[\mathcal{A}] = \uplus_{x \in \mathcal{A}}\{f(x)\}$$
$$f(\mathcal{A}) = \bigcup_{x \in \mathcal{A}}\{f(x)\}$$

Note that in case the fuzzy multiset is just a fuzzy subset, the second image corresponds to the *extension principle* of fuzzy set theory.

6 Partially Ordered Multisets

Partially ordered multisets (or just pomsets) have been used by Pratt [14] as a means to model concurrency. In this model a process is a set of pomsets. Here we will only present the definition of a pomset and the basic operations between pomsets. The reader interested in learning more on their use on modeling concurrency is refereed to Pratt's paper. The following definition of pomset is due to Gischer [7] and is copied verbatim from Pratt's paper:

Definition 23. *A labeled partial order (lpo) is a 4-tuple $(V, \Sigma, \leq, \mu)$ consisting of*

1. *a vertex set V, typically modeling events;*
2. *an alphabet Σ (for symbol set), typically modeling actions such as the arrival of integer 3 at port Q;*
3. *a partial order $\leq$ on V, with $e \leq f$ typically being interpreted as event e necessarily preceding event f in time; and*
4. *a labeling function $\mu : V \to \Sigma$ assigning symbols to vertices, each labeled event representing an occurrence of the action labeling it, with the same action possibly having multiply occurrences, that is, μ need not be injective.*

A pomset is then the isomorphism class of an lpo, denoted $[V, \Sigma, \leq, \mu]$.

Now we are ready to define the basic operations between pomsets:

Definition 24. *Assume that $p = [V, \Sigma, \leq, \mu]$ and $p' = [V', \Sigma', \leq', \mu']$ are two pomsets. Then:*

1. *their concurrence $p\|p'$ is the pomset $[V \cup V', \Sigma \cup \Sigma', \leq \cup \leq', \mu \cup \mu']$, where V and V' are assumed to be disjoint;*
2. *their concatenation $p; p'$ is as for concurrence except that instead of $\leq \cup \leq'$ the partial order is taken to be $\leq \cup \leq' \cup(V \times V')$; and*
3. *their orthocurrence $p \times p'$ is the pomset $[V \times V', \Sigma \times \Sigma', \leq \times \leq', \mu \times \mu']$.*

Acknowledgments

I thank Charles Wells, Paul Taylor, and Steve Vickers for answering all of my questions regarding categorical models of multisets.

References

1. M. Barr, *-Autonomous Categories, Lecture Notes in Mathematics, Springer-Verlag, Berlin, 752, 1979.
2. M. Barr, The Chu construction, Theory and Applications of Categories, 2 (1996), 17–35.
3. M. Barr, Personal communication, 1999.
4. M. Barr, Ch. Wells, Category Theory for Computer Science, Les Publ. CRM, Nontréal, third ed., 1999.
5. W.D. Blizard, The development of multiset theory, Modern Logic, 1 (1991), 319–352.
6. W.D. Blizard, Dedekind multisets and function shells, Theoretical Computer Sci., 110 (1993), 79–98.
7. J. Gischer, Partial Orders and the Axiomatic Theory of Shuffle, PhD Thesis, Computer Science Dept., Stanford Univ., 1984.
8. D.E. Knuth, The Art of Computer Programming, vol. 2: Seminumerical Algorithms, Addison-Wesley, 1981.
9. D. Loeb, Sets with a negative number of elements, Advances in Mathematics, 91 (1992), 64–74.
10. S. Miyamoto, Fuzzy multisets and applications to rough approximation of fuzzy sets, in Proc. Fourth Intern. Workshop on Rough Sets, Fuzzy Sets, and Machine Discovery, RSFD'96, 1996, 255–260.
11. S. Miyamoto, Two images and two cuts in fuzzy multisets, in Proc. Eight Intern. Fuzzy Systems Ass. World Congress, IFSA'99, 1999, 1047–1051.
12. G.P. Monro, The concept of multiset, Zeitcshr. f. math. Logic und Grundlagen d. Math., 33 (1987), 171–178.
13. Z. Manna, R. Waldinger, The Logical Basis for Computer Programming, vol. 1: Deductive Reasoning, Addison-Wesley, 1985.
14. V. Pratt, Modelling concurrency with partial orders, Intern. J. Parallel Programming, 15, 1 (1986), 33–71.
15. A. Syropoulos, Multisets as Chu spaces, manuscript, 2001.
16. R. Yager, On the theory of bags, Intern. J. General Systems, 13 (1986), 23–37.

Author Index

Lecture Notes in Computer Science

For information about Vols. 1–2175
please contact your bookseller or Springer-Verlag

Vol. 2213: M.J. van Sinderen, L.J.M. Nieuwenhuis (Eds.), Protocols for Multimedia Systems. Proceedings, 2001. XII, 239 pages. 2001.

Vol. 2214: O. Boldt, H. Jürgensen (Eds.), Automata Implementation. Proceedings, 1999. VIII, 183 pages. 2001.

Vol. 2215: N. Kobayashi, B.C. Pierce (Eds.), Theoretical Aspects of Computer Software. Proceedings, 2001. XV, 561 pages. 2001.

Vol. 2216: E.S. Al-Shaer, G. Pacifici (Eds.), Management of Multimedia on the Internet. Proceedings, 2001. XIV, 373 pages. 2001.

Vol. 2217: T. Gomi (Ed.), Evolutionary Robotics. Proceedings, 2001. XI, 139 pages. 2001.

Vol. 2218: R. Guerraoui (Ed.), Middleware 2001. Proceedings, 2001. XIII, 395 pages. 2001.

Vol. 2219: S.T. Taft, R.A. Duff, R.L. Brukardt, E. Ploedereder (Eds.), Consolidated Ada Reference Manual. XXV, 560 pages. 2001.

Vol. 2220: C. Johnson (Ed.), Interactive Systems. Proceedings, 2001. XII, 219 pages. 2001.

Vol. 2221: D.G. Feitelson, L. Rudolph (Eds.), Job Scheduling Strategies for Parallel Processing. Proceedings, 2001. VII, 207 pages. 2001.

Vol. 2223: P. Eades, T. Takaoka (Eds.), Algorithms and Computation. Proceedings, 2001. XIV, 780 pages. 2001.

Vol. 2224: H.S. Kunii, S. Jajodia, A. Sølvberg (Eds.), Conceptual Modeling – ER 2001. Proceedings, 2001. XIX, 614 pages. 2001.

Vol. 2225: N. Abe, R. Khardon, T. Zeugmann (Eds.), Algorithmic Learning Theory. Proceedings, 2001. XI, 379 pages. 2001. (Subseries LNAI).

Vol. 2226: K.P. Jantke, A. Shinohara (Eds.), Discovery Science. Proceedings, 2001. XII, 494 pages. 2001. (Subseries LNAI).

Vol. 2227: S. Boztaş, I.E. Shparlinski (Eds.), Applied Algebra, Algebraic Algorithms and Error-Correcting Codes. Proceedings, 2001. XII, 398 pages. 2001.

Vol. 2228: B. Monien, V.K. Prasanna, S. Vajapeyam (Eds.), High Performance Computing – HiPC 2001. Proceedings, 2001. XVIII, 438 pages. 2001.

Vol. 2229: S. Qing, T. Okamoto, J. Zhou (Eds.), Information and Communications Security. Proceedings, 2001. XIV, 504 pages. 2001.

Vol. 2230: T. Katila, I.E. Magnin, P. Clarysse, J. Montagnat, J. Nenonen (Eds.), Functional Imaging and Modeling of the Heart. Proceedings, 2001. XI, 158 pages. 2001.

Vol. 2232: L. Fiege, G. Mühl, U. Wilhelm (Eds.), Electronic Commerce. Proceedings, 2001. X, 233 pages. 2001.

Vol. 2233: J. Crowcroft, M. Hofmann (Eds.), Networked Group Communication. Proceedings, 2001. X, 205 pages. 2001.

Vol. 2234: L. Pacholski, P. Ružička (Eds.), SOFSEM 2001: Theory and Practice of Informatics. Proceedings, 2001. XI, 347 pages. 2001.

Vol. 2235: C.S. Calude, G. Păun, G. Rozenberg, A. Salomaa (Eds.), Multiset Processing. VIII, 359 pages. 2001.

Vol. 2237: P. Codognet (Ed.), Logic Programming. Proceedings, 2001. XI, 365 pages. 2001.

Vol. 2239: T. Walsh (Ed.), Principles and Practice of Constraint Programming – CP 2001. Proceedings, 2001. XIV, 788 pages. 2001.

Vol. 2240: G.P. Picco (Ed.), Mobile Agents. Proceedings, 2001. XIII, 277 pages. 2001.

Vol. 2241: M. Jünger, D. Naddef (Eds.), Computational Combinatorial Optimization. IX, 305 pages. 2001.

Vol. 2242: C.A. Lee (Ed.), Grid Computing – GRID 2001. Proceedings, 2001. XII, 185 pages. 2001.

Vol. 2244: D. Bjørner, M. Broy, A.V. Zamulin (Eds.), Perspectives of System Informatics. Proceedings, 2001. XIII, 548 pages. 2001.

Vol. 2245: R. Hariharan, M. Mukund, V. Vinay (Eds.), FST TCS 2001: Foundations of Software Technology and Theoretical Computer Science. Proceedings, 2001. XI, 347 pages. 2001.

Vol. 2246: R. Falcone, M. Singh, Y.-H. Tan (Eds.), Trust in Cyber-societies. VIII, 195 pages. 2001. (Subseries LNAI).

Vol. 2247: C. P. Rangan, C. Ding (Eds.), Progress in Cryptology – INDOCRYPT 2001. Proceedings, 2001. XIII, 351 pages. 2001.

Vol. 2248: C. Boyd (Ed.), Advances in Cryptology – ASIACRYPT 2001. Proceedings, 2001. XI, 603 pages. 2001.

Vol. 2249: K. Nagi, Transactional Agents. XVI, 205 pages. 2001.

Vol. 2250: R. Nieuwenhuis, A. Voronkov (Eds.), Logic for Programming, Artificial Intelligence, and Reasoning. Proceedings, 2001. XV, 738 pages. 2001. (Subseries LNAI).

Vol. 2251: Y.Y. Tang, V. Wickerhauser, P.C. Yuen, C.Li (Eds.), Wavelet Analysis and Its Applications. Proceedings, 2001. XIII, 450 pages. 2001.

Vol. 2252: J. Liu, P.C. Yuen, C. Li, J. Ng, T. Ishida (Eds.), Active Media Technology. Proceedings, 2001. XII, 402 pages. 2001.

Vol. 2253: T. Terano, T. Nishida, A. Namatame, S. Tsumoto, Y. Ohsawa, T. Washio (Eds.), New Frontiers in Artificial Intelligence. Proceedings, 2001. XXVII, 553 pages. 2001. (Subseries LNAI).

Vol. 2254: M.R. Little, L. Nigay (Eds.), Engineering for Human-Computer Interaction. Proceedings, 2001. XI, 359 pages. 2001.

Vol. 2256: M. Stumptner, D. Corbett, M. Brooks (Eds.), AI 2001: Advances in Artificial Intelligence. Proceedings, 2001. XII, 666 pages. 2001. (Subseries LNAI).

Vol. 2258: P. Brazdil, A. Jorge (Eds.), Progress in Artificial Intelligence. Proceedings, 2001. XII, 418 pages. 2001. (Subseries LNAI).

Vol. 2259: S. Vaudenay, A.M. Youssef (Eds.), Selected Areas in Cryptography. Proceedings, 2001. XI, 359 pages. 2001.

Vol. 2260: B. Honary (Ed.), Cryptography and Coding. Proceedings, 2001. IX, 416 pages. 2001.

Vol. 2264: K. Steinhöfel (Ed.), Stochastic Algorithms: Foundations and Applications. Proceedings, 2001. VIII, 203 pages. 2001.